Nanoscience

Friction and Rheology
on the Nanometer Scale

Nanoscience
Friction and Rheology
on the Nanometer Scale

E Meyer

University of Basel

RM Overney

University of Washington

K Dransfeld

University of Konstanz

T Gyalog

University of Basel

World Scientific
Singapore • New Jersey • London • Hong Kong

Published by

World Scientific Publishing Co. Pte. Ltd.

5 Toh Tuck Link, Singapore 596224

USA office: 27 Warren Street, Suite 401-402, Hackensack, NJ 07601

UK office: 57 Shelton Street, Covent Garden, London WC2H 9HE

British Library Cataloguing-in-Publication Data
A catalogue record for this book is available from the British Library.

NANOSCIENCE: FRICTION AND RHEOLOGY ON THE NANOMETER SCALE

ISBN-13 978-981-02-2562-9
ISBN-10 981-02-2562-8
ISBN-13 978-981-238-062-3 (pbk)
ISBN-10 981-238-062-0 (pbk)

Foreword

"Nanoscience: Friction and Rheology on the Nanometer Scale" is intended to give an introduction for students and to give an overview of previous work for beginners in the field. It may also help to distribute knowledge amomg experienced researchers. Friction and rheology are multidisciplinary research topics. The journals, where these subjects are covered, are related to physics, chemistry, mechanical and electrical engineering and others. An introduction to the topics will be given in Chapter I and subsequent chapters will treat subjects, such as the description of instruments (Chapter II), the understanding of normal (Chapter III) and lateral forces (Chapter IV) and various application examples and instrumental aspects of friction force microscopy (Chapter VIII and Appendix). Chapter VI gives an overview of rheology on the nanometer-scale. Chapter VII gives insights into the relationship between ultrasonics and friction. Chapters I, IV and V were written by Tibor Gyalog and E.M. Chapter VI was written by René Overney and Chapter VII by Klaus Dransfeld. Chapters II, III, VIII and the Appendix by E.M. The book was written during the years 1995–1998.

R.O. wishes to thank Mingyan He and Lily Quan for their help with artwork and editing and the Exxon Research Foundation and the Royal Research Foundation of the University of Washington for financial support.

Special thanks to Roland Lüthi, Jane Frommer, Masamichi Fujihira, Gregor Overney, Loris Scandella, Lukas Howald, Winfried Gutmannsbauer, Thomas Bonner, Dominique Brodbeck, Harry Heinzelmann and Henry Haefke, who contributed a lot to this book. Alexis Baratoff, Hans-Joachim Güntherodt, Harry Thomas are gratefully acknowledged for their discussions and support and Jacqueline Vetter for all the secretary work and many nice discussions. Also, special thanks to Martin Bammerlin, Martin Guggisberg, Christian Loppacher, Thomas Lehmann, Felice Battiston, Stefan Messmer, Lukas Eng, Martina Kubon, Jing Lü, Simon Rast, Peter Streckeisen, Christian Wattinger, Urs Gysin, Hans Hug, Roland Bennewitz. One of the first contacts of E.M. and R.O. to the world of tribology was the NATO-ASI on "Fundamentals of Friction" in Braunlage in 1992. The stimulating atmosphere was extraordinary. There, they also met Gary McClelland and James Belak, who contributed so much to the field of nanotribology. The contacts with Mathew Mate were always stimulating. Other important persons were Gerd Binnig, Heinrich Rohrer, Christoph Gerber, Gerhard Meyer, John Pethica, Mark Welland, Nic Spencer and Ricardo Garcia. Stimulating contacts with M. Salmeron, G.A. Somorjai, B.J. Briscoe, K.L. Johnson, D. Tabor, W.T. Tysoe, B. Bhushan, P. Fleischauer, N.M. Gardos, S. Granick, J.N. Israelachvili, J. Krim, J.L. Lauer, I. Singer, O. Marti, C. Quate, K. Zum Gahr, J.-M. Georges, E.H. Freitag, and N. Wüthrich and the members of their groups are gratefully acknowledged.

Ernst Meyer
Basel, May 1998

Bibliography of the authors

Klaus Dransfeld, *University of Konstanz, Institute of Physics*. Klaus Dransfeld is a Professor of physics. He received his Ph.D. at the University of Köln in 1954. Became Postdoctoral Fellow at the Clarendon Laboratory, Oxford, England. He was scientific staff member at Bell Telephone Laboratories, Murray Hill from 1957-1960. 1960-1965 Associate Professor of physics at the University of California, Berkeley, USA. 1965-73 Professor of physics at the Technical University Munich. 1973-1977 Director of the High-Field-Magnetic-Laboratory of the Max-Planck-Society, Grenoble. 1977-1981 Director of the Max-Planck-Institute of Condensed Matter, Stuttgart. 1982-1993 Professor of physics at the University of Konstanz. Since 1994, Professor em. at the University of Konstanz. His main interests are: Experimental investigations of superfluid helium, glasses, polymers and bio-molecules in high magnetic fields, using ultrasonic and acoustic surface waves at high frequencies, at low temperatures. Also applications of tunneling, force and friction microscopy. General interests in high-frequency dynamics of organic and anorganic material on the nanometer scale. In 1989, he received the German-French Gentner-Kastler price. He is a member of the Heidelberger Akad. der Wissenschaften, German Akad. der Naturforscher Leopoldina, Halle. Dr. h.c. University of Grenoble and Augsburg, Honorarprofessor at the University of Nanjing and Tongji University in Shanghai, China.

Klaus Dransfeld

Tibor Gyalog, *University of Basel, Institute of Physics* Tibor Gyalog is a Ph.D. student in Ernst Meyer's condensed matter physics group at the University of Basel. He works together with Harry Thomas, Professor of the condensed matter division at the University of Basel. Gyalog's main subject is the theory of atomic friction mechanisms, i.e., topology, dissipation, commensurability and computer simulations in order to understand the basic mechanisms which are important in friction theory, such as instabilities and hysteresis. Gyalog will finish his Ph.D. in summer 1998.

Tibor Gyalog

René M. Overney, *University of Washington, Department of Chemical Engineering* Dr. René M. Overney, Assistant Professor in Chemical Engineering, focuses in his research on issues related to rheology and tribology on the nanoscale. After having received his Ph.D. in physics from the University of Basel in 1992, he spent a postdoctoral year at the Tokyo Institute of Technology in Yokohama, Japan, in the Department of Bioengineering. In 1994, he joint Exxon's Corporate Research, Annandale, NJ, where he applied for the first time scanning probe microscopy to measure liquid properties on the nanoscale. He started his present position at the University of Washington in the Department of Chemical Engineering in 1996. Dr. Overney's on-going research is concerned with the applicability and restriction of laws in rheology and contact mechanics, acquired from bulk material measurements, in the nanometer vicinity of interactive and material confining interfaces. His multidisciplinary nanoscale research extends from biophysics, polymer physics, surface chemistry to bioengineering and chemical engineering. His work on nanorheology and polymer physics was awarded by Exxon Chemicals with the Best-Paper-Award during the annual CRC meeting in 1995. Dr. René Overney is a member of the American Physical Society, the American Chemical Society and the Material Research Society.

René Overney

Ernst Meyer, *University of Basel, Institute of Physics* Dr. Ernst Meyer is a Professor of physics. He received his Ph.D. at the University of Basel in 1990. The topic of force microscopy on ionic crystals and layered materials was treated in his thesis. He worked at the IBM Research Center Zurich (1992-1994). In 1997, he started his present position at the University of Basel. His present research interests are the development of surface science techniques, such as friction force microscopy and dynamic force microscopy with true atomic resolution. He is also active in the field of sensors based upon micromechanics, magnetic spin resonance detection with force microscopy. He received an award from the Swiss Physical Society for the development of friction force microscopy. He is member of the Swiss and American Physical Society, member of the Editorial Board of Tribology Letters and co-editor of the book "Forces in Scanning Probe Methods", Kluwer Academic Publishers, 1995.

Ernst Meyer

Contents

Chapter 1

Introduction and motivation

1.1 Introduction

Friction is one of the oldest phenomena in the history of mankind and in particular of natural science. Already, in the stone ages, frictional heat was used to create fire. Experiences with the difficult frictional properties of ice and snow, led to technological inventions, such as skis or sleds. From old Egypt, the first tribologist is known, who poured water in front of a collosus, who was pulled by hundreds of Egyptians. Today, we know that wood on wet sand (friction coefficient $\mu \approx 0.2$) gives lower friction compared to wood on dry sand (friction coefficient $\mu \approx 0.22\text{-}0.5$), which made it possible that the collosus could be moved by only 172 persons. Therefore, this early tribologist did a very good job and facilitated the life of his colleagues. We also can learn that the central questions of friction arise in public and goods transportation. In the middle ages, the use of pork fats was quite common for the lubrication of axes of wagons or chariots. Leonardo da Vinci introduced the first modern concepts of friction. He found the dependence of friction on load and the independence of geometrical contact area. Later, Amontons rediscovered and extended da Vincis observations. Thus, they are called da Vinci-Amontons laws. The third friction law is named after Coulomb who found that dry friction is independent of velocity. Already, Coulomb investigated the origins of friction. He suggested that roughness on the micrometer scale is responsible for the occurrence of friction. However, there was experimental evidence against this hypothesis of Coulomb: Highly polished surfaces did not exhibit low, but high friction. An alternative explanation was given by Desaguliers who suggested that molecular adhesion might be the relevant phenomenon. However, molecular adhesion was known to be proportional to contact area, whereas friction was found to be independent of contact area.

It took about 200 years until this controversy was solved. Around 1950, Bowden and Tabor performed systematic, tribological experiments which showed that the contact of a macroscopic body is formed by a number of small asperities. Thus, another contact area, the real area of contact had to

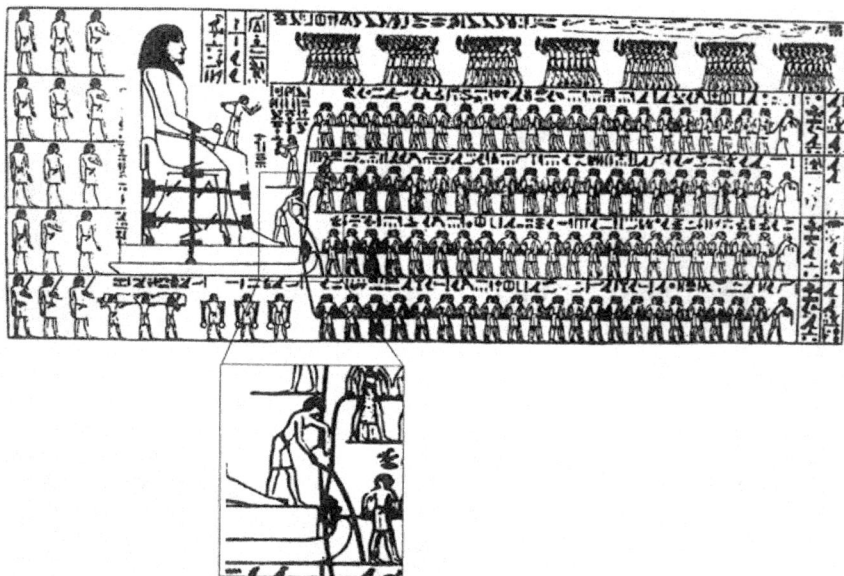

Figure 1.1: 172 Egyptians pull a collosus. One man pours a liquid on the ground in order to reduce friction. (From[1])

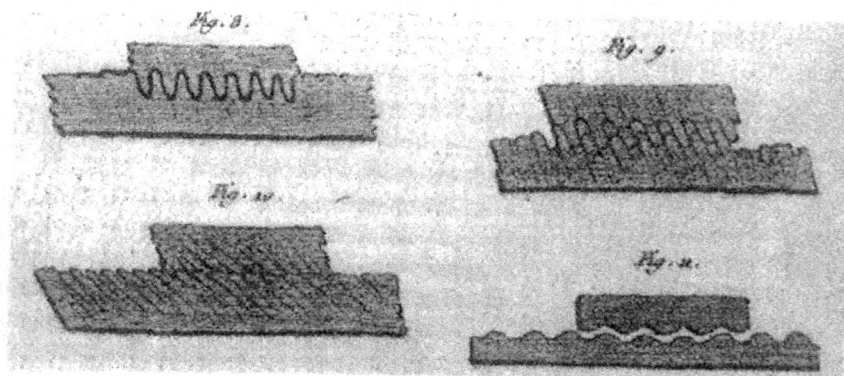

Figure 1.2: Roughness model: Coulomb suggested that roughness is determining friction. (From[1])

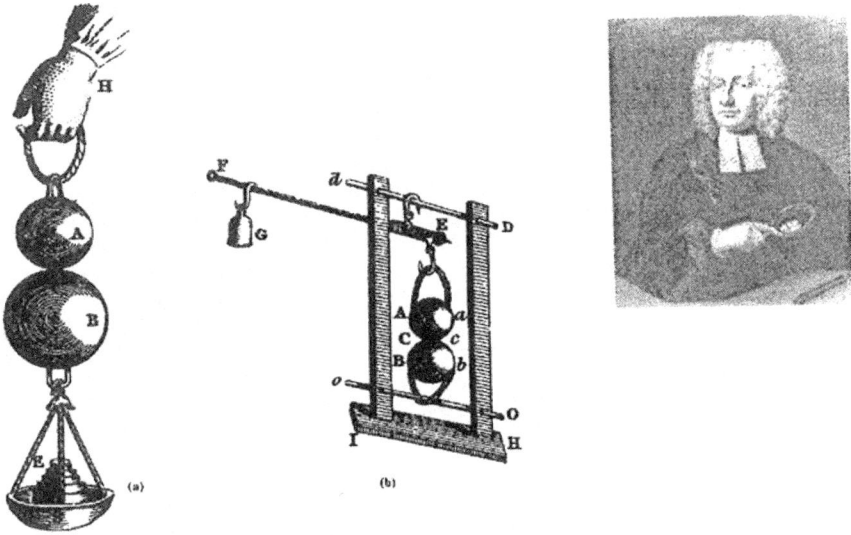

Figure 1.3: In 1725, J.T. Desaguliers demonstrated the cohesion of lead. He also suggested that adhesion might be relevant for friction. (From[1])

be introduced. This new concept was extremely successful and is the basics of most present tribological studies. Essentially, the Bowden-Tabor model states that friction is proportional to the real area of contact

$$F_R = \tau \cdot A_R \tag{1.1}$$

The proportionality constant τ is called shear strength and is related to some intrinsic, more fundamental properties of the interface. From this point of view, Desaguliers was right to assume that adhesion, which is also proportional to the contact area, is more related to friction than roughness. Therefore, the model is also called Bowden-Tabor adhesion model. In first approximation, the real area of contact does not depend on the apparent contact area. By increasing the load, the number of contacting asperities increases with load. The Bowden-Tabor adhesion model explains the da Vinci-Amontons laws of the macroscopic world. However, a basic understanding is still lacking. On which properties does the shear strength τ depend? What are the microscopic mechanisms of friction? How is energy dissipated? How do lubricants affect the shear properties? Can we calculate friction from molecular interaction potentials in a quantitative way? During the last 10 years, the field of tribology on the atomic scale became of interest to a bigger scientific community. Instruments, such as the surface force apparatus, quartz microbalance and the friction force microscope were built for this specific question. Some new phenomena, such as stick-slip on the atomic scale or stick-slip in relation to phase transitions were discovered. Quantitative measurements under well-defined conditions were achieved and compared to theoretical models. Actually, it turned out that the computer simulations, especially the molecular dynamics calculations, were extremely useful for an understanding and visualization of the complex processes. The aim of this book is to provide an overview of tribology. Chapter I gives a brief overview of the history of tribology. Chapter II is an overview of instruments in tribology, where tribometers, surface force apparatus, quartz crystal microbalance and friction force microscopy experiments are described. Then, chapter III and IV will give an overview about the normal and lateral forces which are relevant for tribology. Chapter V will discuss the energy dissipation mechanisms. Chapter VI will give an overview of Nano-Rheology. Chapter VII gives some insights into the close relationship of friction and ultrasonics. The appendix gives some more details on the calibration procedure of friction force microscopy.

1.2   Short outline of the history of tribology

Friction is an every-day experience and almost everybody is aware of its existence. Thus, it is natural that, since a couple of centuries, many researchers tried to get a fundamental understanding. Already, the great pioneers of tribology found, that friction plays a special role in the field of physics and they found phenomenological friction laws, which seemed to be against intuition, e.g., the independence of friction of the contact area. Today, we still learn these three macroscopic laws of friction in school, which were established by Leonardo da Vinci, Guillaume Amontons and Charles Augustin Coulomb:

1. **Independence of the area of contact**
   Friction is independent of the apparent area of contact.

2. **Amonton's Law**
   Friction is proportional to the applied load. The ratio $\mu = F_L/F_N$ is called coefficient of friction. It is larger for static friction than for kinetic friction.

3. **Coulomb's Law**
   Kinetic friction is independent of the velocity.

These three fundamental laws of friction, which are based upon macroscopic experiments, are still not fully understood in terms of more fundamental microscopic processes. In the following, a brief historical review will give a short insight into the work of tribological pioneers. For detailed informations see in the references[1].

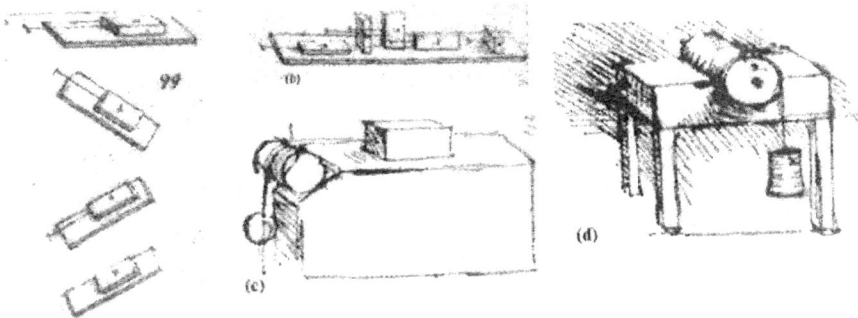

Figure 1.4: Schematics of the friction experiments by Leonardo. He measured the inclination angle of the plane when the block starts sliding. (From[1])

1.3 Leonardo da Vinci (1452-1519)

Leonardo da Vinci has the credit to be the first who made quantitative studies of the problem of friction. Leonardo's experimental setup for friction measurements was rather simple. He measured the angle α of an inclined plane, where a body, put on the plane, started sliding and the weight needed to make a block on a table moving (see Fig. 1.4). With his methods he was only able to measure static friction and most probably he wasn't aware of the difference between static and kinetic friction. Leonardo found the following two laws of friction, in which we essentially recover friction laws 1 and 2.

Leonardo da Vinci

1. The friction made by the same weight will be of equal resistance at the beginning of its movement although the contact may be of different breadths and lengths.

2. Friction produces double the amount of effort if the weight be doubled.

Leonardo defined a friction coefficient as the ratio of the friction divided by the mass of the slider. Experimentally, he found an universal friction coefficient of 0.25 independent of the material. This universal friction coefficient of 0.25 is called Bilfinger value. Many other friction scientists after Leonardo believed in the existence of an universal material independent friction coefficient. However, most of them found another value but all in the range 0.1 - 0.6. Amontons' sketch of his apparatus for friction experiments. The spring D measures the

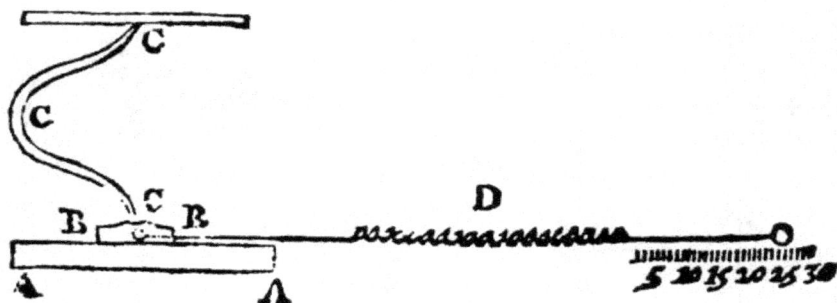

Figure 1.5: Amontons' sketch of his apparatus for friction experiments. The spring D measures the friction force during the sliding process between materials A and B. Spring C adjusts the normal force. (From[1])

friction force during the sliding process between materials A and B. Spring C adjusts the normal force.

1.4 Guillaume Amontons (1663-1705)

Two centuries after Leonardo's discoveries, the French physicist Guillaume Amontons considered the problem of friction again. In his experiments he used springs to measure lateral forces (see Fig. 1.5) and therefore he must have been able to measure both static and kinetic friction. However, we must conclude, that also Amontons wasn't aware of the difference of the two friction phenomena. Amonton postulated the following friction laws:

1. The resistance caused by rubbing only increases or diminishes in proportion to greater or lesser pressure (load) and not according to the greater or lesser extent of the surfaces.

2. The resistance caused by rubbing is more or less the same for iron, lead, copper and wood in any combination if the surfaces are coated with pork fat.

3. The resistance is more or less equal to one-third of the pressure (load).

Amontons found a material-independent friction coefficient of 0.33 and therefore also he believed in the existence of an universal friction coefficient.

1.5 Leonhard Euler (1707-1783)

The scientist Leonhard Euler, who was born in Basel
and later moved to St. Petersburg is famous for his work
in the field of mathematics. Less is known about his
important contributions in the field of friction physics.
He studied theoretically the mechanism of the sliding
motion of a block on an inclined plane. He adopted
the model of rigid interlocking asperities as the cause
of frictional resistance, shown in Fig 1.6. Euler considered Leonardo's experiments of the sliding block on the
inclined plane.

Leonhard Euler

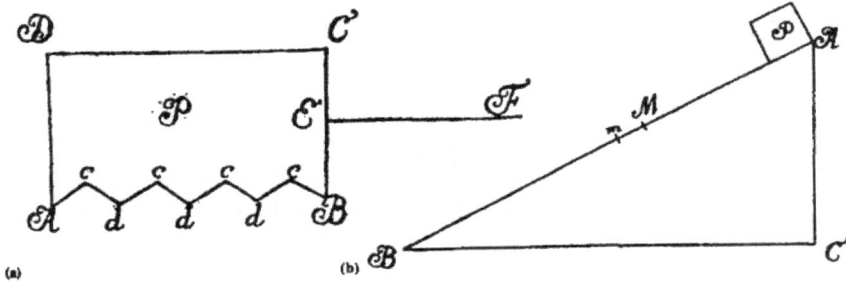

Figure 1.6: Interlocking asperities, corresponding to Euler's model. (From[1])

He assumed that the friction force results from gravitational forces, trying
to minimize the potential energy of the block. Within this model, sliding starts,
when the slope of the asperities gets horizontal. He pointed out, that the typical interlocking angle α of the asperities in respect to the macroscopic surface is
related to the friction coefficient by $\mu = \tan \alpha$. He found the relation $\mu = \tan \alpha$
between the friction coefficient and the inclination angle of an inclined plane,
where the block starts sliding. Assuming a velocity-independent friction coefficient, he found, that for the critical angle $\tan \alpha = \mu$ the acceleration of
the block should be exceedingly small, since gravity is nearly compensated by
kinetic friction. This result was against the experimental facts, where sliding started relatively fast. He concluded, that one has to distinguish between
static and kinetic friction and that static friction is always larger than kinetic
friction. With these assumptions he was able to describe the motion of a block
on an inclined plane. Euler was the first who distinguished between static and
kinetic friction.

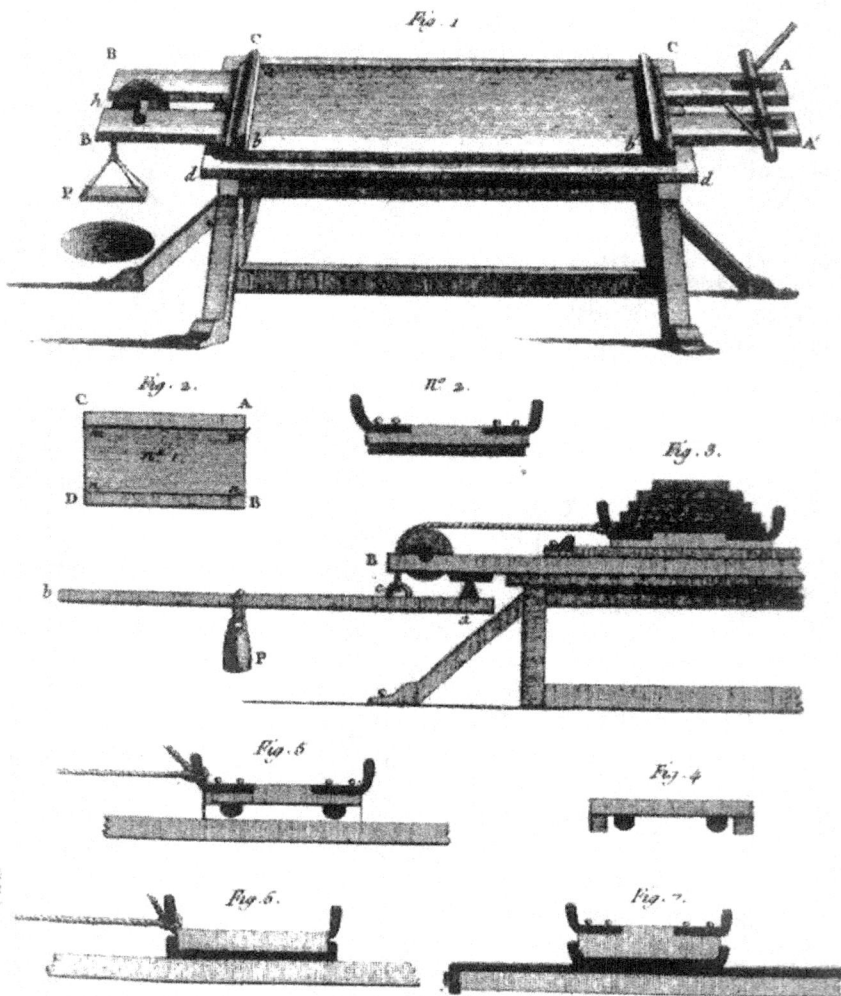

Figure 1.7: The apparatus that Coulomb used for his measurements. (From[1])

1.6 Charles Augustin Coulomb (1736-1806)

Coulomb learned about Amontons' work and got so interested in this daily life physics that he started making measurements himself. His was not only interested in friction coefficients, but also in the time dependence of the static friction force on the time of rest. He found an increase of the friction force with the time of rest and tried to find a mathematical description. Phenomenologically he found the following relation between the static friction $F_{L,stat}$ and time of rest t:

C. A. Coulomb

$$F_{L,stat} = \frac{A + mt^s}{C + t^s} \tag{1.2}$$

where A, B and m are material dependent constants. For the exponent s he found about $s = 0.2$. The friction force for $t = 0$ is $F_{L,stat} = A/C = 502$ and for $t \to \infty$ it increases to $F_{L,stat} = m = 2700$ measured for two pieces of well-worn oak lubricated with tallow [a].

He published his major results in an "Essai sur la théorie du frottement", namely the friction laws, often referred to as "Coulomb's laws of friction":

1. For wood sliding on wood under dry conditions, the friction rises initially but soon reaches a maximum. Thereafter, the force of friction is essentially proportional to load.

2. For wood sliding on wood the force of friction is essentially proportional to load at any speed, but kinetic friction is much lower than the static friction to long periods of repose.

3. For metals sliding on metals without lubricant the force of friction is essentially proportional to load and there is no difference between static and kinetic friction.

4. For metals on wood under dry conditions the static friction rises very slowly with time of repose and might take four, five or even more days to reach its limit. With metal-on-metal the limit is reached almost immediately and with wood-on-wood it takes only one or two minutes. For wood-on-wood or metal-on-metal under dry conditions speed has very little effect on kinetic friction, but in the case of wood-on-metal the kinetic friction increases with speed.

The second part of the fourth law, which describes the velocity independence of kinetic friction is nowadays well known as Coulomb's Law (See beginning of this chapter)

[a] The units are not SI-standard, but have to be viewed in the historical context.

1.7   Friction and wear

1.7.1   Ploughing

It is well known that sliding often damages the surfaces in contact and that mechanical energy is transformed into deformation energy on a small length scale. This mechanism is often referred to as ploughing and a more fundamental understanding has not been found yet although intuitively this process seems to be very easy.

1.7.2   Wearless friction

Wear is a possible origin for the dissipation in a friction process. However, if we compute the wear rate from the mechanical work, the wheels of a locomotive would be totally damaged after a few kilometers of travel. (Tomlinson 1929) From an atomistic point of view one would argue like: *"If we consider, for example a brake horse-power test of a 100 KW motor... .., we may quite safely conclude that only an extremely small proportion of those atoms taking part in contact are detached from their original position."*[b] This idea got lost due to the success of the plastic junction theory, which assumes every deformation to be totally plastic. Assuming a plastic contact, one obtains Amonton's law in a trivial manner. In 1961 J. F. Archard pointed out again, that plastic deformation would damage every machine after a few minutes of work. He postulated[18]:

"The analysis of wear experiments suggests that most of the events which occur in rubbing are contacts between protuberances which are deformed elastically and which separate without damage; an asperity encounter with damage is a relatively rare event." (Archard's principle)

Hence other dissipation mechanisms must exist which do not change the structure of the surfaces in contact. In this context one must be careful. The surfaces may change their structure, and for sure they undergo infinitesimal changes, but these changes do not remain after the separation of the surfaces. In such a process, the mechanical work is continuously transformed into heat.

In 1929 Tomlinson proposed a new dissipation mechanism, nowadays known as **Tomlinson's mechanism**[c] by taking into account the importance of mechanical adiabaticity and the role of instability: *"To explain friction it is necessary to suppose the existence of some irreversible stage in the passage of one atom past another, in which heat energy is developed at the expense of external work".*

---

[b]This assumption of Tomlinson has been confirmed experimentally with recent AFM measurements.
[c]see chap 4.2

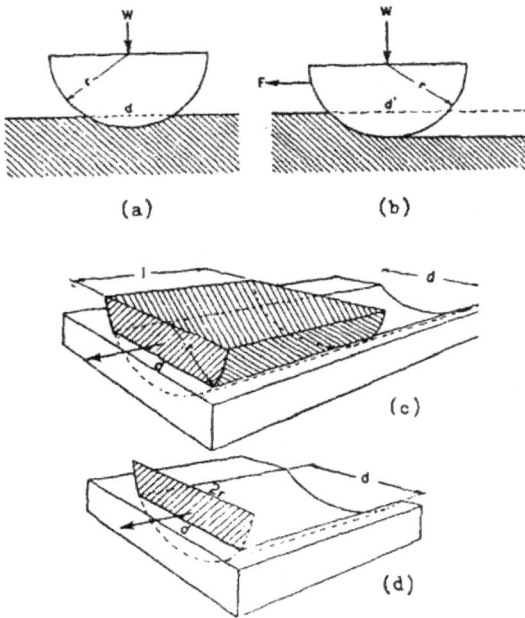

Figure 1.8: The mechanism of ploughing. (From[7])

The main achievement of Tomlinson's work was to solve the following problem: All known atomic potentials are conservative and therefore one might conclude in the limit of zero driving velocity, energy cannot be dissipated. His consideration was dealing with the existence of instable coordinates which become important in a friction mechanism (see chapter IV).

1.8   Friction on a macroscopic scale

Even a surface which appears to be flat on a millimeter scale may contain micrometer scale asperities i.e., the surface is rough. If we bring two surfaces in contact, only these asperities really touch each other. Friction is due to the interaction between the asperities of the different surfaces and the resulting energy dissipation is due to the interaction of these asperities. The real area of contact is therefore a few orders of magnitude smaller than the apparent area of contact. This important fact has to be taken into account while modelling a friction process.

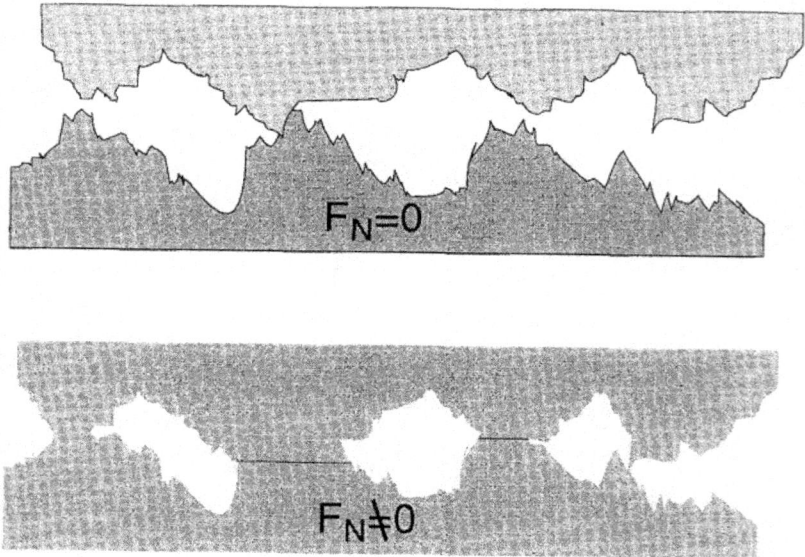

Figure 1.9: The length scale of interest depends on the normal load acting between the surfaces. A real contact area cannot be defined for zero load (a), whereas for finite normal force small asperities are destroyed (b).

1.9 The Bowden and Tabor adhesion model

In 1950 F. P. Bowden and D. Tabor produced a collection of knowledge on friction and lubrication, where it
must be noticed that most results given in the book
were obtained by themselves. The book "The friction and lubrication of solids" has become the standard work on friction and lubrication for a couple of
decades. Based upon their knowledge about friction,
Bowden and Tabor presented a simple model for friction on a micrometer scale: The Bowden and Tabor
adhesion model or plastic junction model. The model
assumes that friction is proportional to both the real
area of contact and a mean lateral force per unit area,
the so-called shear strength

F.P. Bowden

$$F_F = \sigma A_R \qquad (1.3)$$

where A_R denotes the real area of contact and σ is the shear strength.[d] Since
friction is proportional to the real area of contact as is adhesion, the model
may be called adhesion model. The energy loss in the friction mechanism is
described as plastic deformation of the asperities. Thus, it also may be called
plastic junction model. The understanding of friction at the micrometer scale
has been reduced to an understanding of two new quantities: shear strength
and area of contact.

In the adhesion model all changes of the asperities are assumed to be
plastic and therefore the energy loss due to friction is considered as plastic
deformation energy of the surfaces in contact.

1.10 The shear strength

The double number of asperities in real contact must produce the double lateral
resistance and hence double friction. It is therefore convenient to define the
lateral force per unit area, the shear strength σ. It has the dimension of a
pressure

$$\sigma = \frac{F_L}{A_R} \qquad (1.4)$$

where A_R is the **real** area of contact. The shear strength is a material constant.
We will see later, when discussing the real contact area, that experimentally

---

[d]In the historical context, σ was related to tensile experiments, where plastic deformation
in normal direction is observed. Bowden and Tabor assumed that it is identical to the
plastic deformation in lateral motion. Later, we will use τ for the shear strength in friction
experiments without wear.

we have good reasons to assume, that σ is independent of the applied load F_N, which seems to be counterintuitive. In the following chapters we will treat σ always as a constant. Only few is known about it. A more general assumption for the pressure-dependence of the shear strength is often used, dealing also with a linear dependence of σ on the applied load:

$$\sigma = \sigma_0 + \alpha P \qquad (1.5)$$

where σ_0 is a constant and P is the normal pressure (Normal force per unit area).

When P becomes very high, it follows, that

$$
\begin{aligned}
F_L = \sigma A_R &= (\sigma_0 + \alpha P)A_R && (1.6) \\
&= (\sigma_0 + \alpha F_N / A_R)A_R && (1.7) \\
&= \sigma_0 A_R + \alpha F_N && (1.8)
\end{aligned}
$$

For high normal pressure (P large compared to σ_0) equation 1.8 simplifies to Amonton's law of friction

$$F_L \approx \alpha F_N \qquad (1.9)$$

Therefore, the friction coefficient μ in Amonton's law is equal to α.

1.11   The real area of contact

When one neglects any dependence of σ on the normal force, the resolving experimental problem is to measure the real area of contact as a function of the applied load.

It is nearly impossible to determine the real contact area since it depends on the length scale of interest. It is conceptually the same problem as measuring the length of the coast of Great Britain. Let's do the following Gedanken experiment: In order to measure the length of the coast of Great Britain, a man walks along the shore, spanning a rope. He fixes the rope at the ground in equidistant distances of length l. It is now clear, that the length of the coast depends on this distance. To do his job correctly, he should fix the rope at every "coast atom". The coast of Great Britain is of fractal dimension and therefore a "length" cannot be defined.[e]

In order to measure the real contact area between two surfaces under zero normal load we need a "rule" of an atomic length scale, but also then we must be aware that we are neglecting subatomic effects. The problem changes when a finite normal force F_N is acting on the asperities. Due to elasticity of the materials, all structures on a length scale which are smaller than the

---

[e]For an introduction into the theory of fractals and chaos, see[8].

typical length scale $R_{rule} = \sqrt{DF}$[f]are destroyed due to either plastic or elastic deformation according to Hertzian contact theory which will be discussed in more detail below. The smallest "rule" has to be taken and is given by the length R_{rule}.

In regard of the complicated character of surface characterization the experimental results, measured under a well defined normal load, are astonishingly well reproducible.

1.11.1 Measuring the real contact area

We give a short introduction into the two most important techniques to measure the real contact area, which were available already for Bowden and Tabor. The recently developed nanotechniques already today make it possible to determine the real contact area with a much higher accuracy (see later Chap. IV).

Optical methods

The first and easiest approach to determine the area of contact is by observing the sample through an optical microscope in order to determine the size and shape of the asperities. However in most cases the result is wrong by orders of magnitude.

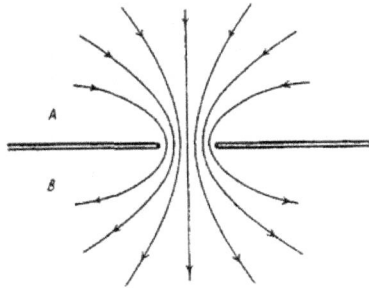

Figure 1.10: The current that flows through a small asperity is spread out before and after the asperity. (From[7])

---

[f]When the contact radius of the asperity is larger than the asperity itself, it is elastically destroyed. According to Hertz, the radius a of contact area for a perfectly spherical asperity of radius R is $a = (DRF)^{1/3}$. $a = R \Leftrightarrow R = \sqrt{DF}$

Load $[kg]$	A_{real} $[cm^2]$	$A_{real}/A_{app.}$	Resistance $[10^{-5}\Omega]$	Asperities in Contact	Radius a $[10^{-2}cm]$
500	0.05	1/400	0.9	35	2.1
100	0.01	1/2000	2.5	22	1.2
20	0.002	1/10000	9	9	0.9
5	0.0005	1/40000	25	5	0.6
2	0.0002	1/100000	50	3	0.5

Table 1.1: Real area of contact, resistance, number of asperities and radius a

Resistance or conductance

A simple method to measure the real contact area is to apply a voltage between two bodies and to measure the resulting current as a function of the applied load. The current will pass through the asperities in real contact. Of course there can be established a tunneling current between asperities which are only a few angstroems apart, but this contribution can be neglected in a first approximation. If the contact points are assumed to be cylindric, the resulting resistance of a pair of connected asperities would be:

$$R = \frac{\sigma h}{\pi a^2} \qquad (1.10)$$

where σ is the specific resistance of the material, h is the height and a is the radius of the cylinder. It has to be taken into account that the current coming out of a small cylinder in a very large bulk, is spread out and this results in an additional resistance $R = 1/(2a\lambda)$, which is called spreading resistance (Maxwell 1873), where λ is the conductivity of the material.

Hence, the total resistance is given by

$$R = \frac{1}{2a\lambda} + \frac{2\sigma h}{\pi a^2} \qquad (1.11)$$

where πa^2 denotes the real area of contact, σ is the specific resistance of the material and h is the height of the asperities.

For low pressure, the resistance of a contacting asperity can be written in the form:

$$R = R_0 + \frac{a}{2\lambda P} + O(\frac{a}{P^2}) \qquad (1.12)$$

Assuming, that the elastic properties of the asperities are the same than those of the bulk material, the real area of contact can be approximately determined. It is given by $A_{real} = F_N/Y$ where Y is the yield pressure of the material. On the other hand the real area of contact is given by: $A_{real} = n\pi a^2$, where n is the total number of asperities. Combining these equations with Eq. (1.12) we can compute the number n and radius a of the asperities for given Load F_N and Resistance R:

$$n = \frac{\pi Y}{4\lambda^2} \frac{1}{F_N R^2} \tag{1.13}$$

$$a = \frac{2\lambda}{\pi Y} R F_N \tag{1.14}$$

Results for steel on steel are presented in Table 1.1.

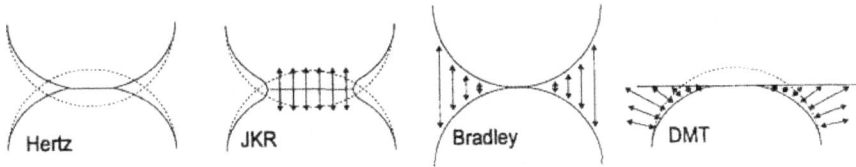

Figure 1.11: a) Hertzian contact, b) JKR contact, c) Bradley contact, d) DMT contact.

1.11.2 Single asperity contact

It is convenient to start our considerations of the real area of contact by a single asperity. We distinguish between plastic and elastic point contacts and additionally we consider the regime between plastic and elastic contact.

Fully plastic contact: The plastic junction theory

In Bowden and Tabor's model the deformation of the surface asperities is assumed to be totally plastic. Hence, when a normal load F_N is applied, the asperities are deformed until the pressure on the asperity becomes equal to the yield pressure of the asperity p_m^*. It is in this context important to note that the yield pressure of the asperity, here denoted p_m^*, can be essentially smaller than the yield pressure of the bulk material due to its small size. The resulting area of contact A_R is $A_R = F_N/p_m^*$ independent of the geometrical area of contact and proportional to the normal load. One can easily see that

$$F_F = \sigma A_r = \frac{\sigma}{p_m^*} F_N \tag{1.15}$$

and we recover Amonton's law. This simple proof of Amonton's law was the main reason why the Bowden Tabor model has been applied to many friction processes. However, it is not probable that friction in a normal machine is totally plastic, since after a while of action, the machine would be totally damaged. Hence, elastic deformation must play an important role in a friction contact.

Fully elastic: The Hertzian point contact

We now consider the opposite problem of a totally elastic interface within classical continuum mechanics. This setup has been calculated for the first time by Hertz in a general form with a contact between two spheres of radii R and R' respectively. Since, we are interested in the contact between a plane and a sphere we will always use the limit $R' \to \infty$. The full analysis for two spheres is given explicitly in[12]. The setup is shown in Fig. 1.11a.

The algebra is rather complicated and not easy to understand for people not being familiar with elasticity theory. Therefore, we directly give the results, which are relevant for the present problem. For symmetry reasons the area of contact between the sphere and the plane is a circle and its area is

$$A_R = \pi(DR)^{2/3} F_N^{2/3} \qquad (1.16)$$

Where R is the radius of the sphere, D describes the elastic moduli and Poisson's numbers of both the sphere and the plane (see Fig. 1.11a).[9] For the "penetration depth" h (see Fig. 1.11a) we obtain

$$h = \left(\frac{D^2}{R}\right)^{1/3} F_N^{2/3} \qquad (1.17)$$

The result $A_R \propto F_N^{2/3}$ is important since assuming that the shear strength σ is independent of the load, one obtains the relation $F_L \propto F_N^{2/3}$ which is in contradiction to Amonton's law.

Including adhesive forces: The JKR model

Adhesion plays an important role on the length scales we are interested in and hence it should be taken into account when small objects in contact are modelled. Johnson et al.[16] extended the Hertzian model by terms due to surface tension γ. The real area of contact is described by the equation:

$$A(F_N) = \pi(DR)^{2/3} \left(F_N + 3\pi\gamma R + \sqrt{6\pi\gamma RF_N + (3\pi\gamma R)^2}\right)^{2/3} \qquad (1.18)$$

where D and R are the same symbols used in the section before. It has to be pointed out, that in the JKR model, the real area of contact has the value $A(0) = \pi(6\pi D\gamma R^2)^{2/3}$. Hence, it has to be applied a negative load to break the contact.

---

[9]$D = 3/4((1 - \sigma_p^2)/E_p + (1 - \sigma_s^2)/E_s))$, where the indices p and s denote the plane and the sphere respectively. D^{-1} is the so-called "plane stress modulus".

In the limit of zero surface tension $\gamma \to 0$ and also for large forces compared to γR, the JKR result coincides asymptotically with the Hertzian result given in 1.16.

Experiments have been presented in[16], where the contact diameter of the interface between a rubber sphere of diameter 4.4 cm and a rubber flat has been measured using an optical microscope. The result is shown in Fig. 1.12. For small load the JKR model fits the experimental data very well, whereas the Hertz model is not able to reproduce the observed contact diameters. However, for loads smaller than -0.3 g, the jump-off-contact regime, also the JKR theory does not fit very well.

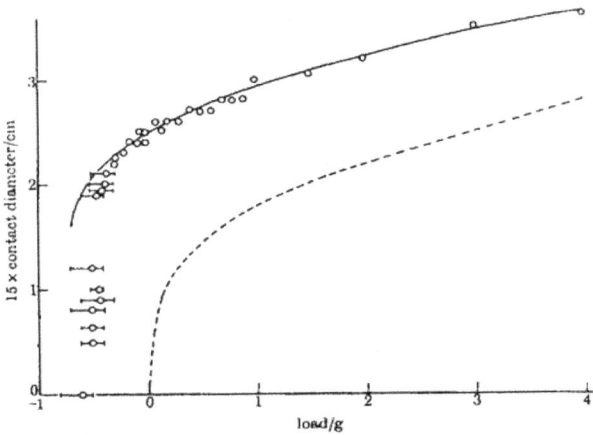

Figure 1.12: Contact diameter as a function of applied load between a rubber sphere (diameter 4.4 cm) and a rubber flat (o). The solid line represents the prediction from JKR theory, the dashed line refers to a Hertzian contact. (From[16])

Other approaches

The Hertzian theory and its extension, the JKR theory, are the most important contact mechanism theories. However, some more models have been developed to describe the contact between a plane and a sphere. Most of them are rather complex and therefore we content ourselves with a short overview.

Adhesion between rigid spheres: Bradley[13] Bradley considered only adhesion between two rigid spheres. No deformation is allowed.

Adhesion beyond the contact area: DMT[14] The DMT model assumes, that the deformation of the sphere, which is assumed to be significantly softer than the plane, can be described by the Hertz theory. Additionally the adhesion between the whole sphere and the plane give rise to an additional load. Contrary to the JKR model, where adhesion is assumed to be limited to the contact area, the whole adhesion is taken into account.

Validity of the above model contacts: The Maugis Dugdale theory

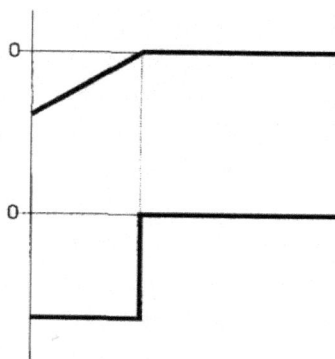

Figure 1.13: Schematics of the Dugdale model potential. Top: The potential shape. Bottom: The corresponding force law.

In a recent work by Maugis[15] a Dugdale potential is used to model the separation energy of a single asperity contact. The Dugdale potential is piecewise linear. It has a slope σ_0 for separations $d < h_0$ and is constant for $d > h_0$ (Fig. 1.13a). The resulting interaction force is constant with values $F = -\sigma_0$ for distances $d < h_0$ and vanishes for separations $d > h_0$ (Fig. 1.13b). The above described theories (Hertz, JKR, Bradley, DMT, Maugis) at first were thought to be competitive. Later, they were recognized to apply to the opposite ends of a spectrum of a non-dimensional parameter λ[10],

$$\lambda = 2\sigma_0 \left(\frac{9R}{16\pi w (E^*)^2} \right)^{1/3} \qquad (1.19)$$

which[7] is a measure of the magnitude of the elastic deformation at the point of separation compared to the range of surface forces. Thus, λ is small for hard materials and is large for soft materials. Here σ_0 denotes the depth of the potential well and $w = h_0 \sigma_0$ is the adhesion energy. R denotes the radius of the sphere and $E^* = 3/4D^{-1}$ is the combined elastic modulus.

The new insights due to the development of the Maugis-Dugdale theory made it possible to plot a two-dimensional phase diagram with axis load and elasticity parameter respectively. Different phases in this diagram correspond to the applicability of different contact mechanisms. It is shown in Fig. 1.14. The following regimes are important:

The Hertz regime

Although adhesive forces are present at all compressive contact loads, at a sufficiently high load the adhesive component may be neglected.

The JKR regime

The JKR theory is valid if the elastic deformation of the surface caused by the adhesive forces is large compared to their effective range of action.

The DMT regime

The DMT theory is nearly valid in the regime, where elastic deformation due to the surface forces is small compared to their range.

The Bradley regime

Valid, for stiff materials and light loads in order to neglect deformation.

The Maugis regime

It remains a regime where the Maugis theory must be used. The Maugis theory is valid on the whole spectrum, but it has the disadvantage, that it must be solved numerically.

1.11.3 Statistical ensemble of asperities: Apparently flat surfaces

We will extend our results obtained for the single asperity contact to a statistical ensemble of asperities.

The plastic contact

The same analysis as it has been done for the single asperity contact can be used, since we never used the condition that we have only one single asperity. Therefore we find again the result:

$$F_R = \sigma A_R = \frac{\sigma}{p_m^*} F_N \qquad (1.20)$$

The elastic contact

The Greenwood and Williamson model (1966)

Greenwood and Williamson[19] considered a surface with a random distribution of asperity heights but where every asperity has the same radius of curvature β at the summit, which is pressed again a plane. The separation of the reference lines is denoted d. We consider the deformation of an asperity with height z, then $w = z - d$ is the "penetration depth" of this asperity. Using Hertz's theory of a totally elastic contact we obtain the area of contact of this asperity:

$$A = \pi\beta w \qquad \text{if} \quad w \geq 0 \qquad (1.21)$$
$$A = 0 \qquad \text{if} \quad w < 0 \qquad (1.22)$$

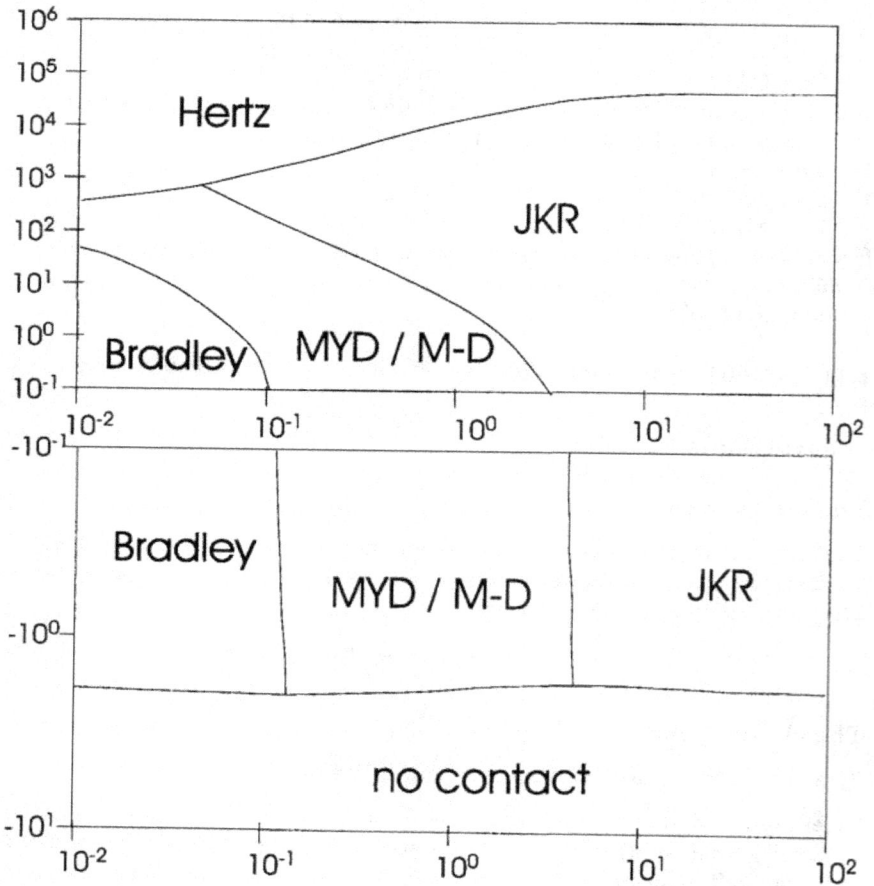

Figure 1.14: The Maugis-Dugdale Adhesion Map. In the two-dimensional Load vs. Elasticy-Parameter Diagram different regimes can be described by the corresponding contact mechanism theory. (From[10])

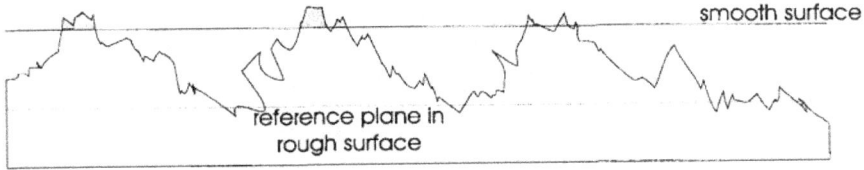

Figure 1.15: The overlap of a rough surface with a geometrical plane. According to the Greenwood and Williamson theory the asperities above the plane are deformed elastically according to Hertzian theory.

The load acting on the asperity after Hertz is

$$F_N = \frac{\beta^{1/2}}{D} w^{3/2} \tag{1.23}$$

where D is the inverse "plane stress modulus" discussed above. The probability, that a given asperity has a height between z and $z+dz$ will be $\Phi(z)dz$. The standard deviation of the height distribution is denoted σ_0, where the index 0 is in order to distinguish from the shear strength σ. After some simple algebra we get the real area of contact

$$A_R = \pi N \beta \int_d^\infty (z-d)\Phi(z)dz \tag{1.24}$$

Note, that the total number of asperities N is proportional to the nominal contact area $N = \eta A_G$ and therefore $A_R \propto A_G$. In the same manner and using (1.23) we find that the total load is

$$F_N = \frac{N\beta^{1/2}}{D} \int_d^\infty (z-d)^{3/2}\Phi(z)dz \tag{1.25}$$

The simplest Ansatz for the probability $\Phi(z)$ is $\Phi(z) = e^{-d/\sigma}$ and the integrals in (1.24) and (1.25) can be solved exactly:

$$A_R = \pi\beta\eta A_G \int_d^\infty (z-d)e^{-z/\sigma}dz = \pi\beta\eta A_G\sigma e^{-d/\sigma} \tag{1.26}$$

$$F_N = \frac{3\sqrt{\pi}}{4D}\eta\beta^{1/2}\sigma^{3/2}A_G e^{-d/\sigma} \tag{1.27}$$

The quotient A_R/F_N is :

$$\gamma := \frac{A_R}{F_N} = \frac{4\sqrt{\pi}}{3}D\sqrt{\frac{\beta}{\sigma}} \tag{1.28}$$

Thus, the real contact area is proportional to the load and independent of the nominal contact area. The Greenwood and Williamson model leads to Amonton's law as well as the Bowden and Tabor "plastic junction model", but it is important to note, that in the present Ansatz a totally elastic contact is used. Therefore, this result is consistent with Archard's principle

$$F_L = \sigma A_R = \sigma \gamma F_N \qquad (1.29)$$

Later, we will see that experimentally a Gaussian distribution of the asperity heights has been measured rather than an exponential distribution. The above integrals cannot be solved exactly for a distribution $\Phi(z) = e^{-x^2}$. However, approximatively one obtains again a linear dependence between area of contact and the applied load, shown in Fig. 1.16b.

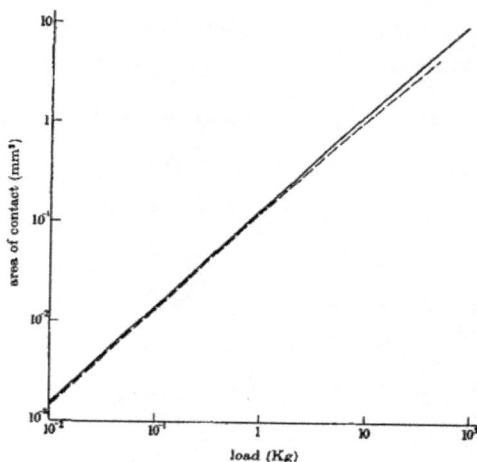

Figure 1.16: The relation between friction force and normal load within the Greenwood and Williamson model for a Gaussian height distribution $\Phi(z) = e^{-x^2}$. Amonton's law is recovered within a totally elastic approach.

Experimental observations of height distributions

The Greenwood and Williamson theory was a great success, since it was able to explain Amonton's law without the (unphysical) assumption, that every deformation between the asperities is plastic. However, there are two parameters working as input to the theory which are not easy to measure and even not easy to define. First, a well defined number of independent asperities N must be measured, which is a nearly impossible task to do, due to the existence of double peaks on one hand, and the self-similar structure of the asperities themselves on the other hand.

In order to image the height distribution of the surface, Greenwood and Williamson used a profilometer, which consists of a small tip which is scanned over the surface, while its movements perpendicular to the surface are recorded with a certain (finite) sampling rate. Within this method it is impossible to distinguish asperities of a smaller length scale than the sampling distance.

Nevertheless, Greenwood and Williamson measured the surface topography for various materials after different surface treatments. In Fig. 1.17 profile scans and cumulative height distribution for bead-blasted aluminium (a) and mild steel (b) are shown. Within the present model, an asperity is defined through a three point analysis. When the height of the middle point is larger than both the height of the left and the right sampling point, the point in the middle is recorded as an asperity. This rather trivial concept has been corrected by Whitehouse and Archard[20] and is discussed in the next section.

The height distribution, measured and interpreted as has been discussed, is quite different for various materials and strongly depends on the surface finish. For the aluminium sample a Gaussian distribution has been found, whereas for the steel sample it is definitely non-Gaussian. However, in the range between 20% and 100%, which is significant for tribological experiments, the steel sample shows a Gaussian behaviour as well. The results are shown in Fig. 1.17a and Fig. 1.17b. In the figure, the y-axis, showing the cumulative height distribution, is stretched corresponding to the error-function in order to get a linear dependence in the case of a Gaussian distribution.

The problem of correct counting and measuring asperities has been solved by Whitehouse and Archard[20] in 1970 using the concept of self-similarity (Fig. 1.18), first introduced by Archard in his pioneering work[17], where he was able to proof, that for a certain self-similar surface structure, according to Hertz theory, the exponent n in the relation $A_G \propto F_L^n$ goes to unity (Fig. 1.18).

Instead of counting asperities like Greenwood and Williamson, Whitehouse and Archard took an old concept of electronics. In analogy to the waveform of a random signal, the characteristics of the shape of the (one-dimensional) random surface is described by two parameters, a height distribution and an autocorrelation function. These two functions are described by a mean height σ_0 and a correlation distance β^*. This approach is much more reasonable, than assuming that all asperities have the same radius of curvature and additionally it is more convenient to fit these statistical parameters into a set of data points than counting single asperities with a three-point algorithm.

Within the model of Whitehouse and Archard, surface profiles have been analysed in order to learn about typical height distributions of rough surfaces. Again, they[20] found a Gaussian dependence, shown in Fig. 1.17c.

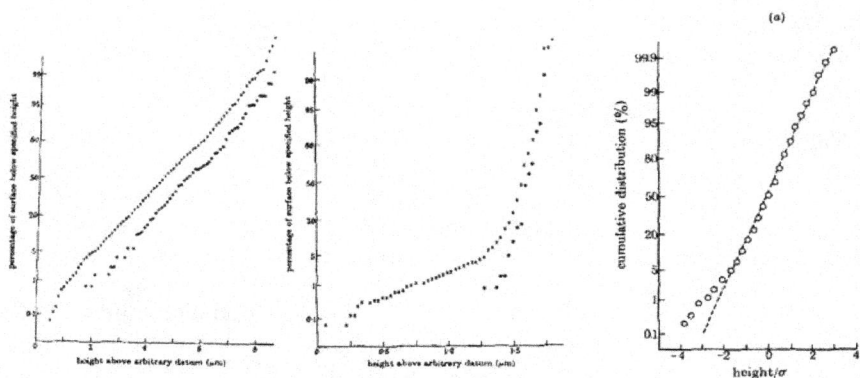

Figure 1.17: Cumulative height distributions measured with a profilometer. The data analysis has been done according to Greenwood and Williamson (a,b) and Whitehouse and Archard (c) respectively. The y-axis is deformed in a manner that a Gaussian distribution is represented by a straight line. (From[19](a, b) and[20] (c))

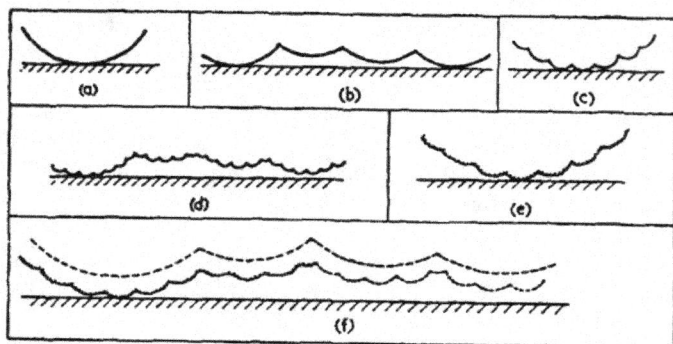

Figure 1.18: Models of surfaces containing asperities of different scales of size. When the deformation is elastic the relationships between area of contact A and normal load L converge to proportionality: a) $A \propto L^{\frac{2}{3}}$ b) $A \propto L^{\frac{4}{5}}$ c) $A \propto L^{\frac{8}{9}}$ d) $A \propto L^{\frac{14}{15}}$ e) $A \propto L^{\frac{26}{27}}$ f) $A \propto L^{\frac{44}{45}}$ (From[18])

In summary, we have found that the Bowden-Tabor adhesion model is valid for both plastic and elastic contacts. In the case of elastic contacts, some statistical analysis is needed to deduce the empirical Amonton's law.

1. D. Dowson, *History of Tribology*, Longman London (1979)
2. Leonardo da Vinci, Codex Atlanticus and other books
3. G. Amontons, *Mémoires de l'académie royale*, **A**, 257 (1706)
4. L. Euler, De Curvis elasticic, *Acta Acad. Pretropolitanae* (1778),
   L. Euler,Sur le frottement des corps solides, *Mém. Acad. Sci. Berl.* **4**, 122 (1748)
   L. Euler, Sur la diminuation de la résistance du frottement, *Mém. Acad. Sci. Berl.* **4**, 133 (1748)
   L. Euler, Remarques sur l'effet du frottement dans l'equilibre, *Mém Acad. Sci. Berl.* **18**, 256 (1762)
5. C. A. Coulomb, Théorie des machines simples, en ayant égard au frottement de leurs parties, et à la roideur des cordages, *Mém. Math. Phys.* 161 Paris (1785)
6. G. A. Tomlinson, *Phil. Mag.* **7**, (1929)
7. F. P. Bowden and D. Tabor, *The Friction and Lubrication of Solids*, Oxford University Press 1950
8. J. Gleick, *Chaos - Making a New Science*, Viking, New York (1978)
9. C. Maxwell, *Electricity and Magnetism* **1**, 308 (1873)
10. K. L. Johnson, A Continuum Mechanics Model of Adhesion and Friction in a Single Asperity Contact, in *Micro/Nanotribology*, 151, B. Bhushan (ed.) Kluwer Academic, Dordrecht, 1997
11. H. Hertz, *J. Reine Angew. Math.* **92**, 156 (1881)
12. L. D. Landau and E. M. Lifschitz, *Introducton into Theoretical Physics Vol. 7*, NAUKA, Moscow 1988
13. R. S. Bradley, *Phil. Mag* **13**, 853 (1992)
14. B. V. Derjaguin, V. M. Muller, Y. P. Toporov, *J. Coll. and Interface Sci.* **67**, 378 (1975)
15. D. Maugis, *J. Coll. and Interface Sci.* **150**, 243 (1992)
16. K.L. Johnson, K. Kendall and A.D. Roberts, *Proc. R. Soc. Lond.* **A324**, 301 (1971).
   K.L. Johnson, *Contact Mechanics*, Cambridge University Press, Cambridge, United Kingdom, (1985).
17. J. F. Archard, *Proc. Roy. Soc. (London)* **A243**, 190 (1957)
18. J. F. Archard, Single Contacts and Multiple Encounters, *J. Appl. Phys.* **32**, 1420 (1961)
19. J.A. Greenwood and J.B.P. Williamson, Contact of nominally flat surfaces, *Proc. R. Soc. Lond.* **A295**, 300-319 (1966).

20. D. J. Whitehouse and J. F. Archard, The properties of random surfaces of significance in their contact, *Proc. Roy. Soc.* **A 316**, 97 (1970)
21. J. A. Greenwood, Unified theory of surface roughness, *Proc. Roy. Soc.* **A393**, 133 (1984)

Chapter 2

Instruments

2.1 Introduction to instruments

In this chapter, the experimental tools, which are used in tribology are reviewed. The main focus will be on experiments, where physical properties are measured during relative motion of the interacting bodies. Surface science techniques, such as photoemission or Auger electron spectroscopy, ion mass spectroscopy or thermal desorption techniques will not be discussed. Eventhough, they are important tools for the characterization of surfaces before and after friction and wear experiments. E.g., the chemical composition of the wear track can be analyzed in this way, which is of importance for tribochemical investigations. For more details, the reader is referred to existing literature[1].

On a macroscopic scale, the classical tribometer experiments are described briefly, where friction is measured as a function of load, velocity or temperature. Extensions of tribometers, such as electrical and optical measurements of the contact zone, will be mentioned. Then, the tools of nanotribology are reviewed. Surface force apparatus experiments are found to represent single asperity experiments and give a rich information under well defined conditions. The quartz microbalance gives access to the shear properties at high frequencies. The resonant stick-slip of colloidal crystals shows that friction can couple into phonon modes and can cause surface melting. Finally, the basics of friction force microscopy will be given. This technique combines friction measurement with high lateral resolution. It also gives access to the phenomenon of atomic-scale stick slip.

2.2 Tribometer experiments

The aim of this section is to give a brief introduction into classical tribometer experiments. For a more complete description the reader is referred to existing literature. The book of Bowden and Tabor[2] is written in a comprehensive way and is also a pleasure to be read. The first issue dates from 1950. The concept to treat basic physical questions in great detail and to mention engineering questions only briefly, made it possible that this book could stand the test of time. Other more recent books are given in the reference list[3,4].

The classical tribometer experiment is to move a slider along a surface. The most common arrangement is the pin-on-disc experiment, where the pin slides over a rotating disc.

During the experiment, the lateral forces between the slider and the sample are measured with resistance strain gauges. Alternatively, forces are determined with induction, capacitance or beam deflection measurements. Usually, the lateral force is measured as a function of normal force. Typical normal forces are in the range of 10^{-3}N to 1000 N. In many cases, a linear function is found, which allows the user to determine the so-called friction coefficient μ.

$$F_L = \mu \cdot F_N \tag{2.1}$$

In macroscopic experiments, the normal force F_N is equalized to the externally applied normal force, also called loading L. This assumption is only valid when attractive surface forces, such as capillary forces or van der Waals are negligible. Thus, in a more general form the lateral force is given by:

$$F_L = \mu \cdot (F_A + L) \tag{2.2}$$

which can be simplified for small attractive forces F_A:

$$F_L = \mu \cdot L \tag{2.3}$$

It is common, that linearity is presumed and that the friction coefficient μ is just calculated by the ratio of lateral force F_L divided by normal force F_N.

$$\mu = \frac{F_L}{F_N} \approx \frac{F_F}{L} \tag{2.4}$$

Historically, the lateral force is called friction F_F. One has to keep in mind, that this lateral force has contributions from phenomena, such as wear-less friction, plastic deformation of asperities, lateral forces to move debris particles, viscous forces and ploughing terms.

2.2.1   Plastic or elastic deformation

Bowden and Tabor emphasized that the mean pressure acting on the asperities is close to the yield pressure. Thus, they assumed that the relevant asperities are fully plastically deformed.

Figure 2.1: Classical tribometer from[2]. Top: Picture of the whole tribometer. Lower left: View of the pin which is moved across the sample. Lower right: Schematical diagram of the instrument.

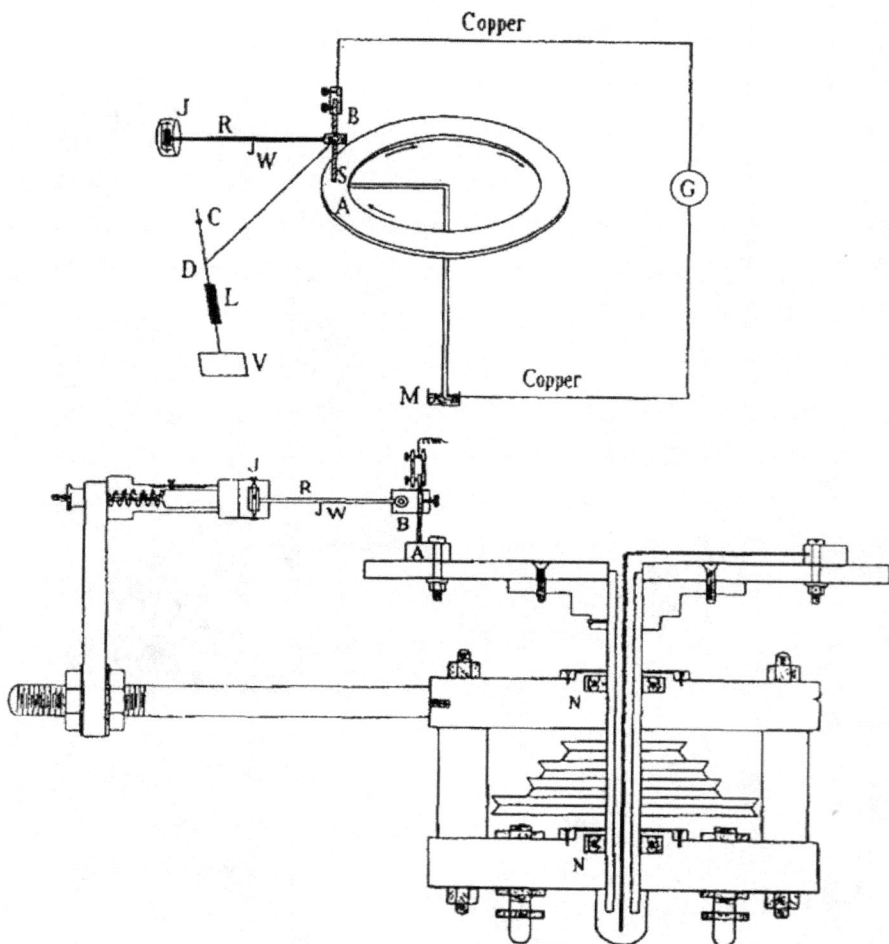

Figure 2.2: Classical tribometer from[5]. Top: Schematical diagram of the tribometer. Cylinder B is pressed on the rotating ring A. The cylinder B is attached to a light arm R which is carried on a gimbal J. Weights can be attached to W in order to vary the load. Bottom: Section of the apparatus. The rotating table which carries the metal ring A is mounted on a ball bearing N and can be driven at speeds from 4 cm/sec to 5000 cm/sec. When the metal ring rotates, the frictional force will tend to drag the cylinder. This movement is prevented by a fine wire which is attached to a pendulum. The deflection is measured by the movement of a spot of light reflected from the mirror C.

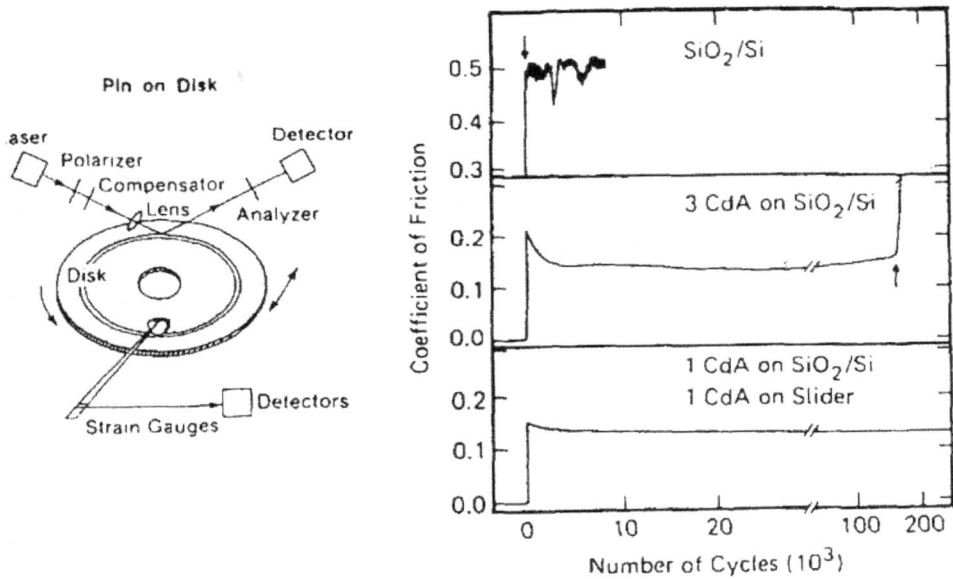

Figure 2.3: Left: Pin-on-disk set-up with in-situ scanning microellipsometry to profile wear of thin lubricant films. Sliding velocities are between 10 and 10^4cm/sec. The normal and lateral forces are measured both with a set of four semiconductor strain gauges in a Wheatstone bridge arrangement. Right: Dependence of the friction coefficient on the number of sliding cycles. (a) uncoated SiO_2. (b) SiO_2 with three Cd-arachidate layers and (c) SiO_2 with a Cd-arachidate layer on both slider and disc. The load was 150mN. From[6].

Material	Y	E	$F_{critical}$ $(R=1\text{cm})$	$F_{critical}$ $(R=1\mu\text{m})$
	10^8N/m^2	10^{11}N/m^2	N	10^{-8}N
Copper	3.1	1.2	4.5	4.5
Mild Steel	6.5	2	15	15

Table 2.1: Critical forces in a Hertzian contact for two radii of curvature, where plastic deformation starts $(p_m=1.1\ Y)$

It is characteristic for macroscopic experiments that plastic and elastic deformations are both present. Small asperities deform plastically at very low loads. On a macroscopic scale, plastic deformation starts at much higher loads. An example: The maximum pressure in a Hertzian contact is given by:

$$p_0 = (\frac{6 \cdot F_N \cdot (E^*)^2}{\pi^3 R^2})^{1/3} \tag{2.5}$$

Plastic deformation of a flat surface by a harder spherical surface starts at a critical pressure $p_m = 1.1 \cdot Y$, where Y is the yield strength and E^* is the effective Youngs modulus. The critical point P below the surface, where plastic deformation starts is indicated in Fig. 2.4. At a pressure of $p_m^* = 3 \cdot Y$ the whole region below the indenter is plastically deformed. A micron-sized asperity of copper will reach the yield point at a load of about $5 \cdot 10^{-8}$N, whereas the macroscopic sphere with a radius of curvature of 1cm, will deform plastically at a critical load of about 5N.

Thus, one has to keep in mind that different scales are relevant for macroscopic experiments. The surface roughness will determine the size of the asperities. The smallest asperities are plastically deformed at very low loads, whereas larger asperities are still in the elastic regime. It is not trivial at all to tell what percentage of the contact area is plastically, respectively inelastically deformed.

2.2.2   Velocity dependence

Typical velocities are between 0.0001m/s to 100m/s (e.g., 0-4000 r.p.m for commercial tribometers, with disc diameters of 5-10cm). The Coulomb law states that friction does not depend on velocity. However, it is found that this law is only approximately fulfilled for the case of dry friction. More accurate measurements have shown that dry friction decreases with increasing velocity. This decrease is attributed to frictional heating[2]. In the next section, we will

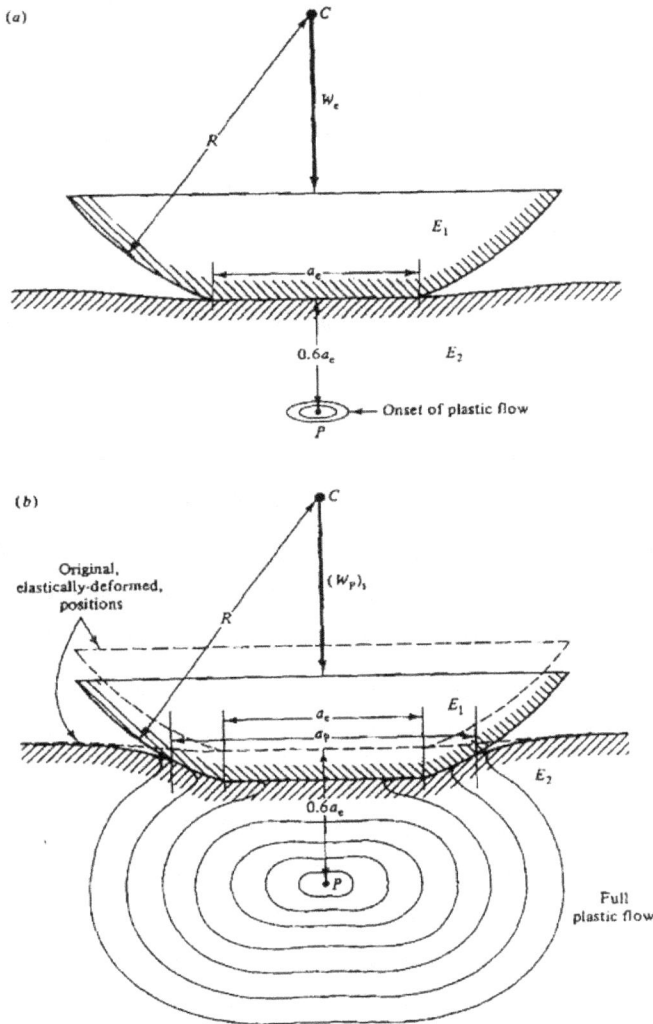

Figure 2.4: (a) Hertzian contact, showing the onset of plastic flow at a mean pressure $p_m = 1.1 \ Y$. (b) Full plastic flow of the sample region at a mean pressure $p_m = 3 \ Y$. From[4].

discuss the temperature dependence of friction. It will be shown, that friction decreases with increasing temperature in the case of dry friction.

In the presence of liquids, a characteristic velocity dependence is found, where three regimes are distinguished:

1. Elasto-hydrodynamic lubrication

2. Mixed lubrication

3. Boundary lubrication

This curve is called Stribeck-curve[8,9]. The regime of elasto-hydrodynamic lubrication is characterized by rather low friction coefficients (0.001-0.01). A fluid film prevents the surfaces from contacting each other. Friction can be calculated by elasto-hydrodynamic equations. Here, the Reynolds-equation is used to describe the hydrodynamics of the fluid. In addition, the influence of the elastic deformation of the solids and the increase in viscosity of the lubricant with pressure in highly stressed lubricated machine elements are taken into account[9,10].

The mixed lubrication regime is characterized by increased friction coefficients (0.01-0.1). The roughness is comparable with film thickness and some of the asperities will reach separations of molecular dimensions. This regime is a subject of current research[11].

Finally, the regime of boundary lubrication is described by contact separations of the solids which are of molecular dimensions. Metal-metal contact is only prevented by a few molecular layers of lubricant molecules. Typical friction coefficients are between 0.1 to 0.4. It was William Bate Hardy who first suggested that lubricant molecules might assemble with their long axis perpendicular to the surface[12]. Boundary lubrication is also a field of current research. There is much hope that nanotribology might help to give a better understanding. One of the first highlights was the orientation and distribution of lubricants that could be imaged with friction force microscopy[13].

2.3   Extensions of tribometers

2.3.1   Electrical contact resistance

The real area of contact cannot be easily calculated because of the mostly unknown size of the asperities. An elegant experimental procedure, which gives an estimate of the contact area, is to measure the electrical contact resistance.

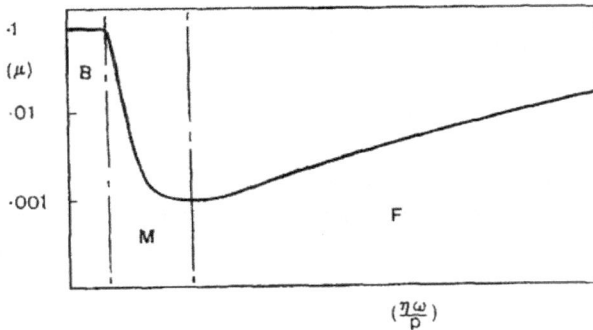

Figure 2.5: Stribeck-curve. The friction coefficient is plotted as a function of parameter $(\eta\omega/p)$, also called the Gumbel number, where η is the lubricant viscosity, ω the rotating speed and p the mean pressure on the bearing. Essentially, the Stribeck-curve shows the variation of the friction coefficient as function of speed and load, where 3 regimes are distinguished. The Elasto-hydrodynamic lubrication is rather well understood, whereas the mixed lubrication and the boundary lubrication are subjects of current research. From[9].

The resistance through a circular constriction of radius a_i is given by:

$$R_i = \frac{1}{2 \cdot a_i \cdot \lambda} \tag{2.6}$$

where λ is the electrical conductivity. The sum over all asperities gives then:

$$d_m = \sum_i a_i = \sum_i \frac{1}{2 \cdot R_i \cdot \lambda} = \frac{1}{2 \cdot R_{tot} \cdot \lambda} \tag{2.7}$$

where d_m is an approximate measure of the contact diameter. Roughly, the real area of contact is then given by: $A_R \approx d_m^2 = \frac{1}{4 \cdot R_{tot}^2 \cdot \lambda^2}$.

Experimentally, it is important to vibrate the contacts in order to break the oxide and to get well defined contacts. First, the experiments by Bowden and Tabor have shown that the real area of contact is much smaller than the apparent area of contact (e.g., 1:100 to $1:10^6$). The most important result was that the the square of the resistance is indirect proportional to the load. Thus, the real area of contact was found to be directly proportional to the load.

$$A_R = G \cdot L \tag{2.8}$$

where G is a constant that depends on the surface morphology and on the material properties (Youngs modulus, yield strength).

Recent experiments by Belin et al. from Lyon[14] have shown that the electrical resistance of MoS_2-lubricated tribocontacts can show variations, which are not directly related to friction changes. Two explanations might be relevant: (i) The asperities that cause friction are not identical with the asperities, where current flows through. (ii) Shortly before the lubricating MoS_2-film becomes worn away, an increase of conductance is observable. However, the film can still lubricate the system. More generally speaking, friction in the boundary regime is determined by molecularly thin films, whereas the current can flow through rather thick films.

2.3.2   Height measurements

If smooth surfaces, such as highly polished surfaces (e.g., silicon wafers or silicate glasses), are investigated, the height measurement of the slider gives information about wear processes. An accurate distance measurement set-up, such as an interferometer (for details about such sensors see appendix A), is positioned on the back of the slider. Ultimately, height variations as small as a monolayer can be observed. However, the method depends critically on the smoothness of the surfaces. In case, the roughness is comparable with the contact diameter, height variations will be dominated by the topography of the disc[14].

Several cases are observable: (1) Either the slider, the disc or both are weared off. A decreased height on the base of the slider is observed. (2) Particles (debris, dust...) are captured in the contact zone and are moved over the disc. First, the height increases. Later, it returns to its original value, when the particle remains left behind the moving slider. (3) Tribochemical reactions lead to the formation of a coating on slider, disc or both. A famous tribochemical reaction is silicon nitride, where the formation of silicon oxide is observed in the presence of humidity[15,16,17]. The reactions at the contacting asperities produce smooth surfaces. Thus, contact stresses are reduced and the wear rate is observed to decrease drastically. For a review in tribochemistry see[16,18].

2.3.3   Wear measurements

A possible technique to measure wear rates is to measure the weight loss or weight gain of the slider and/or the disc. However, it is more common to measure the shape of the slider with profilometry and to calculate the wear volume. In many cases, the onset of wear on lubricated tribocontacts is defined by a sudden increase of the friction coefficient (see e.g. Fig. 2.3).

2.3.4 Optical measurements

The observation of light emission out of the contact zone is of interest for the basics of friction and wear. Early examples were the observation of hot spots on glass surfaces[2] Other experiments are dedicated to characterize the optical properties of lubricants under contact or to measure the temperature of the contact zone[11].

The direct observation of the contact area is possible in the single asperity experiment of surface force apparatus (SFA) (see section II 4).

Micro-ellipsometry on the wear track

As shown in Fig. 2.3, a tribometer experiment can be combined with an ellipsometer with micrometer-sized focus, which gives the opportunity to measure the film thickness of a boundary lubricant during sliding. This technique is useful to observe the different stages of wear. Novotny et al. have observed that Cd-arachidate layers were removed after several hundreds of sliding cycles with the exception of the last monolayer next to the substrate, which retained its integrity for hundreds of thousands of cycles[6].

Optical interferometry and spacer layer imaging method (SLIM)

Optical interferometry proved to be a valuable tool in the study of elastohydrodynamic (EHD) lubrication. It was Gohar and Cameron[19], who first mapped the thickness of EHD films and confirmed theoretical predictions[20]. However, this method had a detection limit of 80-130nm. Therefore, classical EHD-films (> 130nm) could be studied, whereas the transition to the mixed lubrication film regime was not accessible (see section II 2 velocity dependence). Using silica spacer layers, which artificially augments the oil film, thicknesses as small as 20nm could be observed[21]. The group of Spikes improved the resolution with the help of fast image capture and analysis down to some nanometers[22]. Recently, they developed the spacer layer imaging method (SLIM), which gives the opportunity to map the spacer layer of thin films (down to some nanometers). Also, Play and Godet could directly visualize the contact zone[23].

In Fig. 2.6, the set-up of optical interferometry is shown. The contact is usually formed by a glass or sapphire disc, which is moved against a steel ball (rolling, sliding or a mixture of both). The underside of the glass disc is covered with a semi-reflecting chromium layer to give the necessary reflection for the interference. An additional silica layer on the chromium layer helps to lower the minimum detectable film thickness. The image from the contact is dispersed through a spectrometer and captured by a CCD camera. An intensity vs. wavelength distribution is measured, which yields the maximum constructive interference and then gives an average thickness of the film. The spacer layer imaging system is based on a high speed camera (shutter speed down to 0.00001sec). The fast shutter speed is necessary to get clear images of the moving contact. Then, the image processing separates color values into hue, saturation and intensity, where the hue reading is used for the thickness measurement[a].

Some results of SLIM are shown in Fig. 2.7[24]. In stationary contact, the oil film (synthetic poly-α-olefin) is squeezed out of the junction and the boundary appears smooth. Surface asperities conform to each other by elastic (plastic) deformation. The original roughness of the steel ball (R_a=0.012μm) is not visible. The observed irregularities in stationary contact are only 1-2nm. When the steel ball is moved, a finite film thickness is found, which grows with increasing speed. Interestingly, the surface roughness also increases with speed, because the asperities are getting less deformed elastically. The thickness and the profile of the lubricant film is in agreement with the classical Dowson and Hamrock equations[10] of EHL-theory. At higher speeds the film thickness is found to decrease again because of starvation, which means that there is insufficient lubricant to fill the inlet of the conjunction and an air/oil meniscus occurs close to the contact.

Nonlinear optical techniques

The structure of molecular films, which are confined in a tribological contact, are of interest for the understanding of boundary lubricants. Salmeron's group investigated monomolecular films between a glass lens and a glass flat[25]. Pressures of up to 50 MPa were applied over a contact area of 0.5mm radius. Only the flat window was coated by the monolayer. Non-linear optical techniques were applied to characterize the structure of molecular films under pressure. Second harmonic generation (SHG) and sum-frequency generation (SFG) were chosen, because these techniques were already adapted to study molecular films under static pressure to provide molecular orientation and conformation information[26,27]. In the experiment of Salmeron's group, a drastic decrease of

---

[a]Some explanations to colours: The colour sensation arises from radiation covering only a small part of the visible spectrum. Besides possessing luminosity (intensity relative to the eye's spectral sensitivity), colours have hue and saturation. Saturation is the degree to which a colour departs from white light and approaches a pure spectral colour. Hue is determined by wavelength. E.g., a pure continuous spectrum shows a continuous variation of saturated hues.

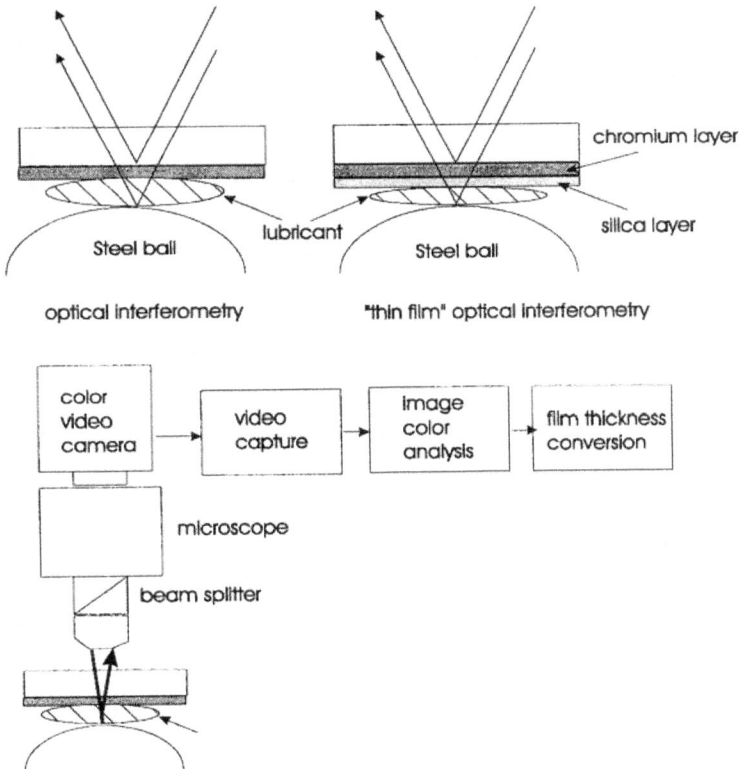

Figure 2.6: Top: Optical interferometric measurement of lubricant film thicknesses. The underside of a glass disc is coated with a semi-reflecting chromium layer to give the necessary reflection for the interference fringes. The minimum detectable film thickness is 80-130nm. In the thin film technique, shown on the right side, the chromium layer is overlaid by a silica film that augments the effective film thickness and removes the 100nm resolution limit. Bottom: Spacer layer imaging system. From[24].

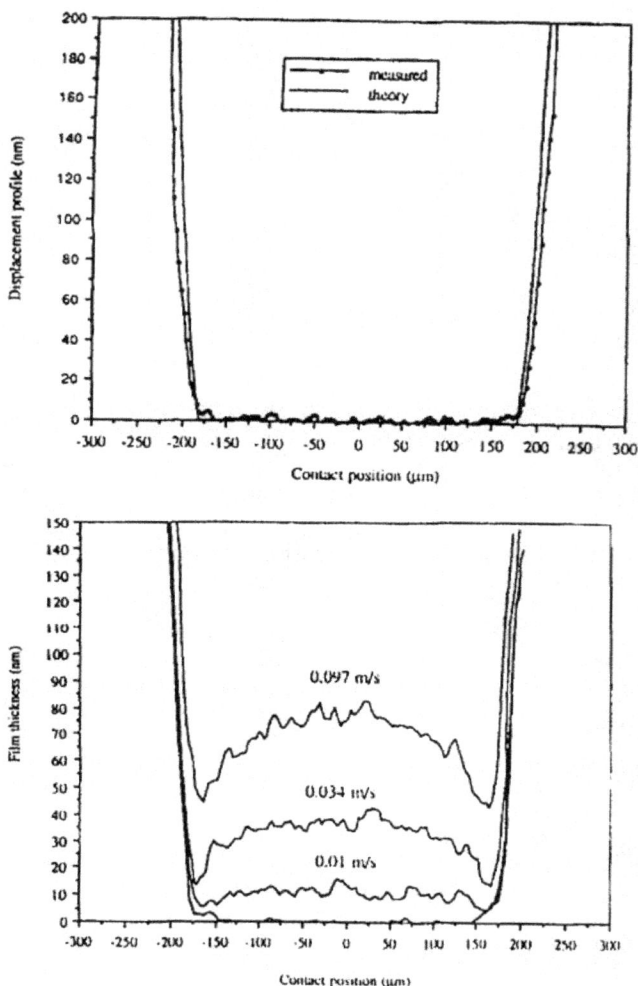

Figure 2.7: Top: Profile of a stationary Hertzian contact, acquired with SLIM. Surface asperities conform elastically to one another in the Hertzian region although some irregularities (1-2nm) can be seen. Bottom: Film thickness profiles (transverse to rolling direction) of a fluid film between a silica layer and a rolling steel ball. The film thickness increases with velocity. Remarkably, the observed roughness increases with speed as well (6nm at 0.034 ms^{-1}; 9nm at 0.097 ms^{-1}). The applied load was 17N. The viscosity of the synthetic poly-α-olefin oil was 0.056 PaS at 20°C. The theoretical curve was calculated with the Dowson and Hamrock equations[10]. From[24].

both the SHG and SFG signals were observed, from which they concluded that the molecules either lie flat on the substrate or that an increased number of Gauche defects is formed.

2.3.5   Temperature measurements

Frictional heat is created during sliding and will cause a change of the temperature field in the environment of the contact. Usually, one distinguishes between an average temperature of the contact zone and flash temperature. The latter is a localized increase of temperature at contacting asperities. The temperature of the contact zone can be measured with a thermocouple or thermoresistor, that is incorporated in the slider. Optical microscopy or infrared emission microscopy can be used for transparent samples. It is found that the temperature increases linearly with sliding speed and more slowly with loading, because the loading determines the real area of contact over which the friction power is dissipated[18,28,29,30].

The flash temperature is more difficult to be measured, because of the short period of time of such an event. Observation of light flashes with an optical microscope in a darkened room were described by Bowden and Tabor[2]. They found that with materials with low thermal conductivity, such as glass, flashes are observable. Alternatively, CCD-cameras can be used that are both sensitive to visible and IR-regime. These methods are only applicable for transparent samples or sliders.

In general, the friction coefficient of solid lubricants decreases with increasing temperature. This decrease of friction is consistent with the softening of the solids. A correlation of friction with the tensile strength of materials at high temperature is found[31,32]. There are many exceptions from this simple rule: Phase transitions from solid to liquid may lead to an abrupt drop of friction, e.g. boric oxide[33]. Sudden increases may be related to chemical reactions at the surfaces, e.g., the oxidation of MoS_2 at $500°C$[32] or decomposition of the lubricants, e.g. the failure of polymers or fatty acids above their melting temperature[2].

2.3.6   Triboscopy

A novel way to represent tribological data has been introduced by Belin et al. from Lyon. A tribometer is moved periodically over a sample at constant speed. During each cycle, physical data P_i, such as friction or electrical resistance are simultaneously measured. Data sets of $P_i(x, N)$, where x is the lateral position and N the cycle number, are acquired. After the acquisition, the data can be represented as grey scale images. Alternatively, histogram techniques can be used to improve the statistics and to visualize correlation between the different physical properties. These tribometer experiments provide spatial and temporal information, as it is common in other microscopic techniques. Therefore, the technique is called triboscopy[34].

Triboscopy can provide important information about the degradation of thin lubricating films. E.g, the location of the breakthrough of films can be observed[35]. The time evolution of the wear process is monitored at every stage. Correlation between friction and contact resistance show, that often the breakthrough of the films is accompanied by an increase of friction and a decrease of resistance. In certain cases, the resistance is found to drop earlier than the increase of friction. This phenomenon could be explained by the short-range nature of boundary lubrication, which is essentially only dependent on the last monolayer, whereas conductance can occur through much thicker layers.

2.3.7   Implementation into ultrahigh vacuum

The implementation of a tribometer into ultrahigh vacuum condition has been done by the group of Gellman at Carnegie Mellon. Large friction is found for clean Cu(111)-Cu(111) (μ_s=4.6±1.1) and is associated with wear. The coverage with lubricants, such as trifluoroethanol reduces friction significantly. The friction coefficient gradually decreases with increasing film thickness. A value of μ_s=0.38±0.07 is found for film thicknesses larger than 8ML. Other applications of this UHV-tribometer are indicated in the reference list[36]. The performance of friction and adhesion measurements in well-defined environment, where surface analytical and preparation tools are available, is rather novel and is promising for the fundamental understanding of tribological mechanisms.

2.4   Surface force apparatus

Tabor, Winterton and Israelachvili[37,38] developed the surface force apparatus (SFA) for measuring van der Waals forces in air or vacuum. Later, the instrument was adapted for measuring forces in liquids[39].

Two crossed cylinders are approached towards each other. The distance between the surfaces is controlled by the combination of micrometer screws, a differential spring mechanism and piezoelectric crystal transducer. The surface separation is measured with an accuracy of better than 1 Å by an optical technique using multiple beam interference fringes of equal chromatic order (FECO).

In SFA, forces are deduced from the deflection of the springs. The surface of the cylinders is usually made of mica. The special role of mica is explained by its smoothness. Mica is molecularly flat over hundreds of microns, which is quite unique for both natural crystals and synthetic ones. Mica can also be covered by surfactant layers or by metallic films. The main advantage of the original set-up of Tabor, Winterton and Israelachvili is the direct access to the contact area by optical visualization. Not only the area can be measured precisely, but also the shape of the contact, e.g., adhesive vs. non-adhesive, is observable[37]. Thus, the elastic deformation can be studied in detail, e.g., comparison with Hertz or JKR theories.

The group of Georges in Lyon has developed a SFA that measures the interaction of a highly polished sphere and flat plane[42]. Different spheres, such as cobalt coated boro-silicate glass spheres, were used, which shows that SFA is not necessarily limited to coated mica sheets. The stiffness of the apparatus is high $\approx 10^6 N/m$. Both lateral and normal forces are measured with capacitive force sensors. In addition, the motion of the sphere relative to the plane is measured with a 3-axial capacitance sensor. The electrical capacitance of the sphere-plane system yields the distance between sphere and plane, also called "electrical" distance. The damping as a function of normal displacement z gives the "hydrodynamic" distance. Extrapolation of these curves gives the origins O_e and O_η. The difference between these origins is twice the thickness of the immobile layer $(2D_s)$. The immobile layers are found to be of molecular thickness. E.g., for n-dodecane 0.3-0.5nm which is close to the molecular width of 0.392nm. For a collection of liquids, Georges et al. found a correlation between the bulk viscosity and the thickness of the immobile layer: For small molecules a linear relationship was observed. For larger molecules, the thickness increased much weaker with increasing viscosity $\sim (\eta^{0.10})$[44]. Georges et al. also reported that the elasticity of the confined liquid increases with decreasing distance. They conclude that the confined layer corresponds to a randomly organized immobile network of rigid molecules. Georges et al. also observed that the immobile layer is not formed on all substrates. E.g., n-dodecane forms immobile layers on cobalt oxide and steel surfaces, whereas no dodecane layer is observed on surfaces covered with stearic acid[43].

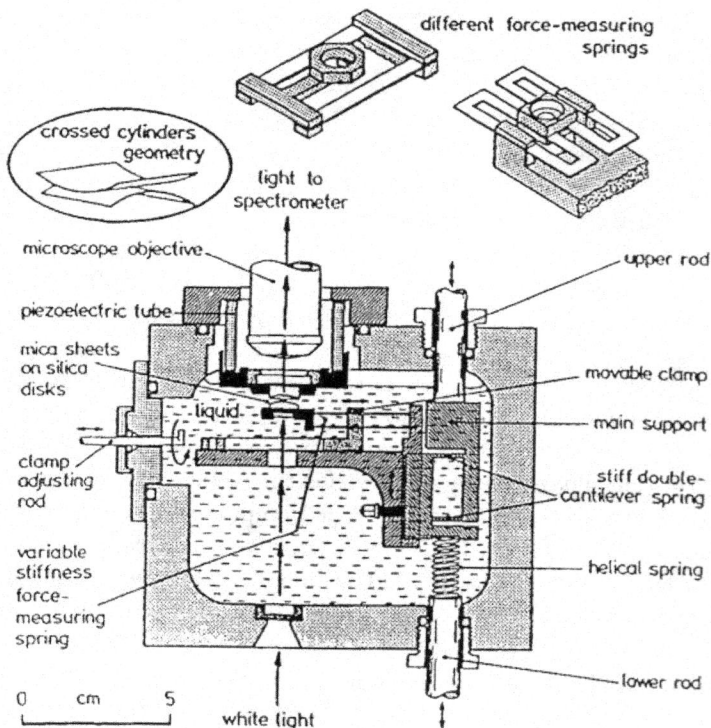

Figure 2.8: Surface force apparatus (SFA) from Israelachvili's group for measuring forces between two curved molecularly smooth surfaces (usually coated mica) in liquids or gases. Forces are measured from the deflection of a spring with adjustable stiffness. This apparatus has been used to identify and quantify fundamental interaction occurring between surfaces in various liquids and vapors, such as van der Waals and double-layer forces, solvation (hydration and hydrophobic) forces, adhesion and capillary forces, and the interactions between polymer-covered and surfactant-coated surfaces. From[40].

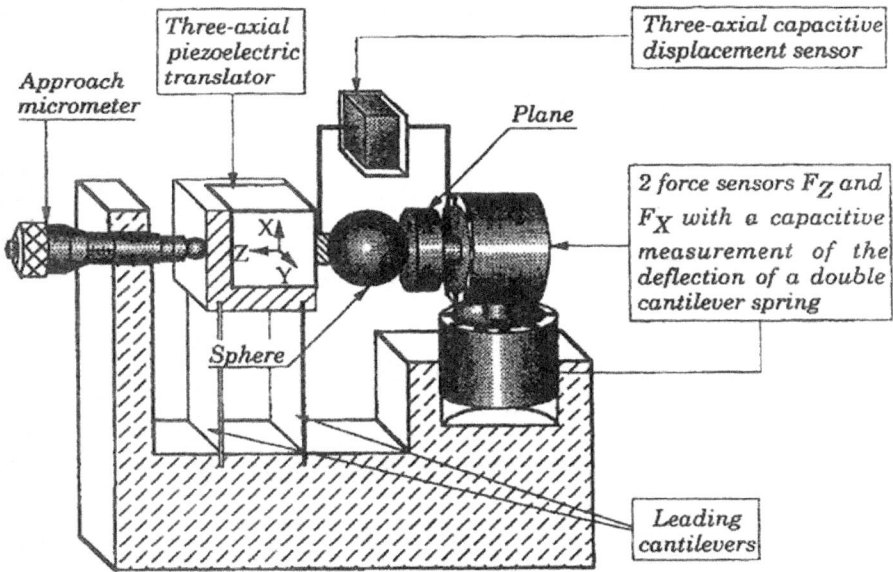

Figure 2.9: Three-axial surface force apparatus (SFA) from the group of Georges in Lyon for measuring forces between a sphere and a plane. Three capacitance sensors control the relative motion of the sphere with an accuracy of 10^{-2}nm. Normal and lateral forces are measured with 2 capacitive force sensors with a resolution of 10nN. The sample is mounted on a double-cantilever spring, which prevents unwanted rolling or shearing. The motion of the sample is achieved with a micrometer screw for coarse approach and piezoelectric crystals for small displacements in 3 orthogonal directions. From[43].

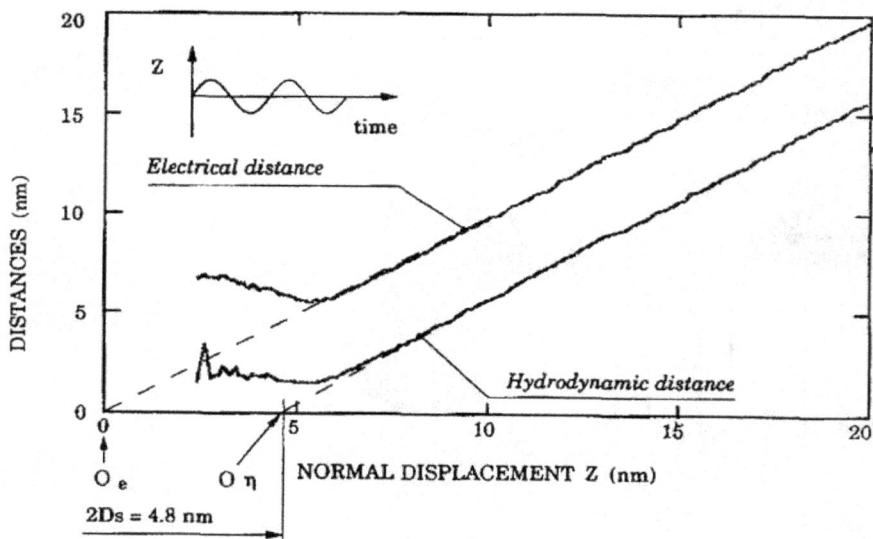

Figure 2.10: Plot of the electrical distance and the hydrodynamic distance. These curves are obtained thanks to dynamic measurements, where an ac-voltage is applied to the z-piezo. The difference between the origins O_e and O_η is twice the thickness of the immobile layer ($2D_s$). From[43].

2.4.1 Friction measurements with SFA

The first friction measurements with a SFA-apparatus were performed by Is-raelachvili's group[45,46]. Lateral springs were attached to the instrument. One of these springs acts as a frictional force detector by having four resistance strain gauges attached to it. Lateral motion is achieved with a motor-driven micrometer shaft which moves the translation stage. Typical sliding velocities are 0.1 to 20 μm·s^{-1}. Contact diameters are up to 100 μm and contact areas of 10^{-8}m^2. Externally applied loads are up to 0.5N down to negative loads. Contact pressures are 10-500MPa. The number of liquid layers separating surfaces during sliding ranges from 1 to 4, corresponding to gap thicknesses of up to 25Å. One of the most remarkable observations of Israelachvili's group[46] was the observation of friction without wear (so called "interfacial" or boundary friction) in contrast to friction with wear (called "normal" friction). In the case of boundary friction, the friction forces vs. normal load as well as the contact area vs. normal load were found to be in good agreement with the Bowden Tabor Adhesion model, meaning that the friction force F_F is directly proportional to the real contact area A_R, where the proportionality constant is the shear strength τ,

$$F_F = \tau A_R \qquad (2.9)$$

The loading dependence of the contact area was found to be in good agreement with the JKR-theory, which means that the contact is formed by a single asperity. In dry atmosphere a shear strength of 25MPa was determined. The shear strengths were found to depend on humidity (about 0.8MPa at higher humidities). In 10^{-2}M KCl solution very low friction forces were found ($\tau <$1MPa), showing rather linear loading dependence. In the presence of simple organic vapors or liquids such as decane, cyclohexane and OMCTS, high values of 10-12MPa were observed. The exact values were highly dependent on the number of liquid layers in the gap. For calcium stearate covered surfaces values of 3-4MPa were measured. For surfaces with high molecular adhesion, stick-slip was also observed. In the context of these experiments, Homola et al. proposed the cobblestone model for boundary friction.

The occurrence of damage could be observed in the FECO fringes, where discontinuities occur. A small mica flake is torn and protrudes from one of the surfaces. Once damage starts, it always propagates rapidly over the whole contact zone. Within seconds, and well before the sliding surfaces have traversed one full contact diameter, there is an abrupt transition to the "normal" friction, where a gap of wear debris (mica flakes) is formed. The friction is much reduced because of the reduced real area of contact. The loading dependence is now rather linear, corresponding to the classical Amontons' law: $F_F = \mu L$. Surprisingly, these friction coefficients were rather independent of humidity or liquid environment.

2.4.2   SFA-tribometry of ultrathin films

The question about the structure and the physical properties of confined ultrathin films is of importance for the understanding of lubricants. Isralachvili et al.[45] observed that dynamic friction was quantized as a function of the film thickness.

Van Alsten and Granick observed a dramatic increase of shear viscosity at close separations and high normal pressures[47], which was interpreted as a solidification of the confined hexadecane films. Georges et al. observed that the inverse of the damping function decreases slowly below a critical film thickness, which is quantitatively interpreted as an increase of the elasticity of the confined layer for small separations[44]. An elastic modulus of the confined layer of dodecane of 3MPa was determined, which indicates that this layer might consist of a randomly organized network of rigid molecules.

Yoshizawa and Israelachvili further studied the stick-slip regime in more detail[49]. It is found that the frequency of the stick-slip spikes increases with velocity and that above a critical velocity v_c stick-slip disappears. The authors suggest that the transition from the stick state to the slip state is a phase transition. Thus, the authors suggest that confined liquid films undergo melting and solidification cycles during the sliding. This type of stick-slip is novel and is currently investigated by further SFA-experiments.

The question whether the confined layers are in a glassy state or in a crystalline state has been addressed by Thompson and Robbins with molecular dynamics calculations[48]. These simulations show a solid-liquid phase transition, also called first order shear-melting transition, which is also observed in the bulk. In agreement with the experiments, they observed that the film remains in its critical state above a critical velocity v_c. They suggest that a glassy state might be possible for larger molecules with long relaxations times. Israelachvili suggested to call the states liquidlike and solidlike, because the geometrical arrangement is completely different from the bulk situation[49].

2.4.3   Adhesion vs. friction

Israelachvili's group performed systematic studies to investigate the role of adhesion and friction. Surfactant monolayers were studied at different temperature, relative humidity and velocity[50]. The velocity dependence of friction could be measured with (a) a micrometer drive, where low-speed sliding is achieved with a motor-driven translation stage and friction is determined with a resistance strain gauge attached to the lateral spring, and (b) with a piezo-electric drive, where faster motion is achieved with a piezoelectric bimorph and the friction is measured with a capacitor attached to the spring system. The authors find that the absolute magnitude of adhesion is not directly correlated with the friction values. E.g., they find that mica separated by one or two layers of cyclohexane exhibits high friction ($\mu > 1$) but a low adhesion energy of $\gamma \approx$1-5mJ/m^2. In contrast, two mica surfaces in humid air exhibit low friction ($\mu \approx 0.03$) but a high adhesion energy of $\gamma \approx$70mJ/m^2. Thus, there is no direct relationship between the magnitude of adhesion and friction. However, Israelachvili and coworkers suggest that the adhesion hysteresis is in close relationship with friction. Various surfactant coated surfaces were investigated. For the adhesion hysteresis the contact radius is measured as a function of normal force (r vs. F_N). Then, the JKR-theory is used to fit the data and to determine the surface energy γ.

$$a^3 = (R \cdot D)[F_N + 6\pi R\gamma + (12\pi R\gamma F_N + (6\pi R\gamma)^2)^{1/2}] \qquad (2.10)$$

where R is the radius of curvature, $D = 3/(4E^*)$ and E^* is the effective Youngs modulus. Both loading (advancing) and unloading (receding) curves are acquired and corresponding values γ_A and γ_R are determined. In addition, the maximum pull-off force is measured, which is related to the surface energy of the receding curve:

$$\gamma_R = F_{max}/3\pi R \qquad (2.11)$$

The measurement depends on the velocity. Thus, it is important that comparative measurements are done with the same speed. Also the friction measurements should be performed at similar speed. The authors argue that the front end of the contact can be thought of the advancing part, where the contact is continuously formed, whereas the trailing end corresponds to the receding part where the contact is separated again. In this perspective, the close relationship between friction and adhesion hysteresis becomes plausible.

Yoshizawa, Chen and Israelachvili then describe several measurements of friction and adhesion hysteresis at different temperature. They find two extreme behaviours: 1) Solidlike monolayers and 2) liquidlike monolayers. The solidlike films are characterized by stick-slip behaviour, where the frequency of the stick-slip spikes increases with velocity. At a critical speed the stick-slip disappears and friction proceeds smoothly. The liquidlike films exhibit lower friction and behave more like a viscous liquid. The friction increases with velocity. In contrast to bulk liquids, the values for the effective viscosity are higher.

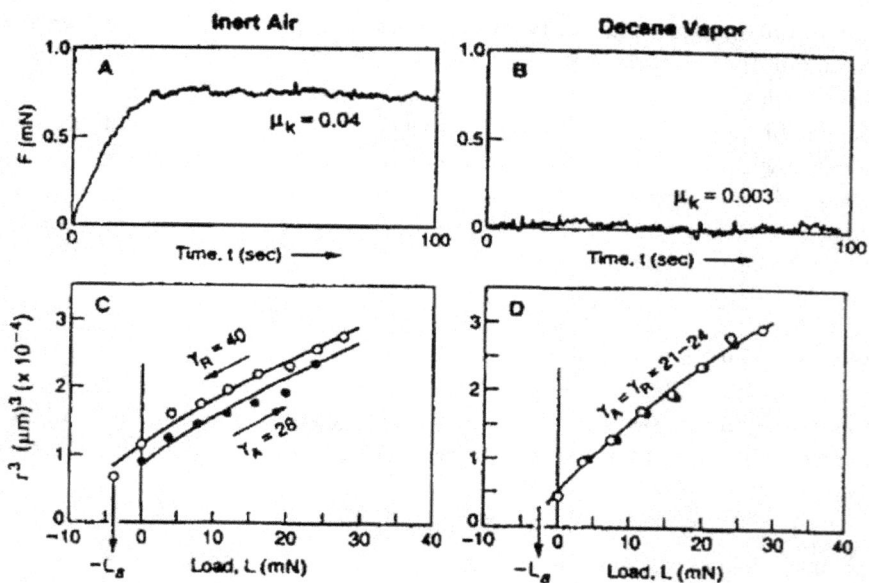

Figure 2.11: (A,B) Reproductions of friction traces for loading and unloading of two calcium alkybenzenesulfonate (CABS) surfaces, where the monolayers were in the liquidlike state. (C,D) Contact radius vs. load curves. Left: Surfaces exposed to dry air or nitrogen gas. Right: Surfaces exposed to nitrogen gas saturated with hydrocarbon (decane) vapor. From[50].

The liquidlike and the solidlike state are extremes. Other states fall somewhere inbetween. The transition regime from the solidlike to the liquidlike state is also called amorphous state and is characterized by the maximum friction. Thus, a phase diagram can be drawn as shown in Fig. 2.12. The maximum in this phase diagram is achieved, when maximum dissipation is achieved. This occurs when the sliding velocity v equals a characteristic molecular length δ divided by a characteristic relaxation time τ. This can be expressed by the ratio of relaxation time and the transit time $\tau_t = \delta/v$:

$$D_e = \frac{\tau}{\tau_t} \tag{2.12}$$

D_e is called Deborah number. When the Deborah number equals one $D_e \approx 1$, friction becomes maximum. Experimentally, this regime of maximum friction can be reached by changing the velocity. Alternatively, the temperature can be changed, which effectively changes the relaxation time $(\tau = \tau_0 e^{\Delta E/kT})$. The relative humidity makes the films more liquidlike, because the hydrophilic headgroups bind some water, which loosens the bond to the substrate and makes the molecules more mobile. In the perspective of the above mentioned phase diagram, the transition temperature of the film is lowered by the increased humidity. Thus, it depends on the starting and end positions in the phase diagram, whether friction is increased or decreased during the experiment.

On a molecular level, the authors suggest that friction occurs because of rearrangements of molecules or parts of molecules. Small interdigitations across the interface occur during sliding, respectively during the adhesion measurement. If the entanglement and disentanglement times are comparable with the transit time, maximum dissipation occurs. Finally, Yoshizawa et al. give an expression which relates adhesion hysteresis $\gamma_R - \gamma_A$ to the friction force:

$$F_R = \frac{\pi a^2}{\delta}(\gamma_R - \gamma_A) \tag{2.13}$$

where a is the contact radius, δ is a characteristic length scale. For $\delta \approx 5\text{Å}$ good agreement with the experiment is found. In this context, it should also be mentioned that the group of Nic Spencer[51] has found that friction force microscopy measurements as function of pH confirm the model of Israelachvili and coworkers in the sense that the friction force is proportional to the adhesion hysteresis.

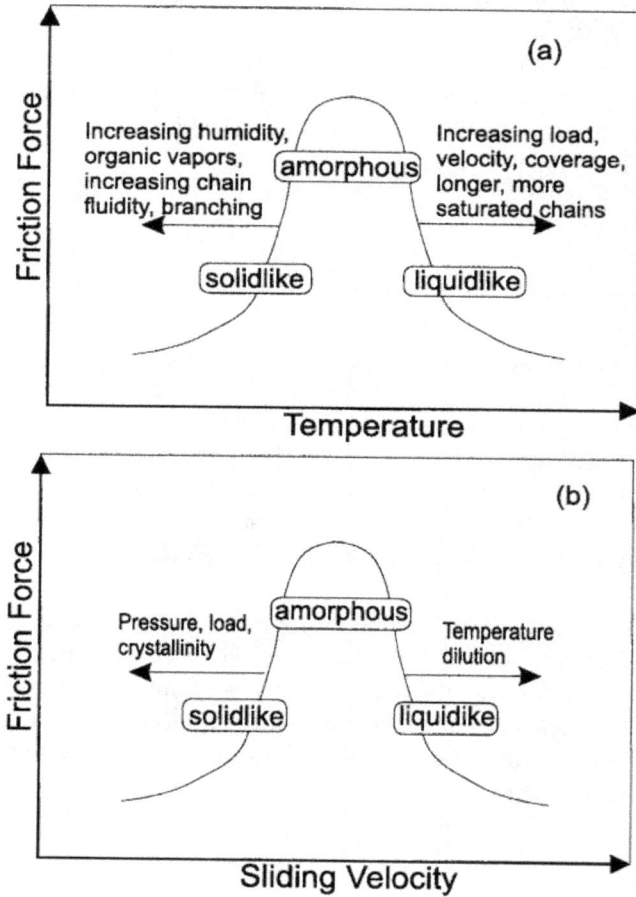

Figure 2.12: Phase diagram of friction (or adhesion hysteresis) as a function of a parameter, such as temperature (a) and velocity (b). The liquidlike state transforms continuously into the solidlike state via an amorphous states, where friction is maximum. Other parameters, such as humidity, organic vapors, load or molecular structure (chain branching, crystallinity) may shift the maximum of the curve. From[50].

2.5 Resonant stick-slip motion in colloidal crystals

Resonant stick-slip motion is important for common string instruments, such as the violin. Resin powder deposited on the bow modifies the tribological interaction between bow and string. When the periodicity of the stick-slip motion is synchronized with the string vibration modes, resonance occurs. Additional resonators of the instrument transform the string vibration into a characteristic set of harmonics and finally into acoustic waves. Another common example is the screech of chalk on a blackboard.

On a microscopic scale, stick-slip has been observed in surface force apparatus and friction force microscopy experiments. However, the coupling of stick-slip motion into resonances has not been observed so far. The chapter of K. Dransfeld describes a possible realization, where atomic-scale stick-slip motion of FFM is coupled into ultrasonic modes.

In this section, an experiment is discussed, where colloidal crystals are examined. This experiment has been performed by Palberg and Streicher[52]. Colloidal crystals are formed of sub-micron polystyrene spheres. The spheres are charged (dissociated carboxyl or sulphonate groups) and repel each other. This force is counter-balanced by attractive forces, such as van der Waals forces. In the solid phase, a lattice spacing is formed that is comparable with the wavelength of light. Due to the low shear modulus ($G \approx$ N/m^2) of such crystals, typical resonances frequencies of centimeter-sized crystals are in the range of 1 to 100Hz. Higher frequencies are strongly reduced due to viscous damping of the surrounding water. Palberg and Streicher have then sheared the colloidal crystal along the cell wall by applying external pressure. Simultaneously, the intensity of one of the Laue spots appearing under light illumination was monitored. For characteristic velocities, stick-slip motion is coupled into the internal vibration of the colloidal crystal, which is observed as a variation of the intensity of the Laue spot. The authors distinguish three regimes: 1) At low velocities, stick-slip motion excites resonances of the crystal. 2) Above a critical speed, the crystal is depinned and moves continuously without stick-slip. Probably, a premolten surface layer is formed that leads to floating without internal vibrations. 3) Finally, at a second critical speed, the whole crystal is shear melted, which is associated with the disappearance of the Laue spots. By varying the particle density or the salt concentration, the elastic constants of the crystals are altered, which leads to a shift of the internal resonance frequencies.

This experiment is the mesoscopic analoguon of a violin. In analogy to the violin, stick-slip motion couples into internal resonances. Variation of

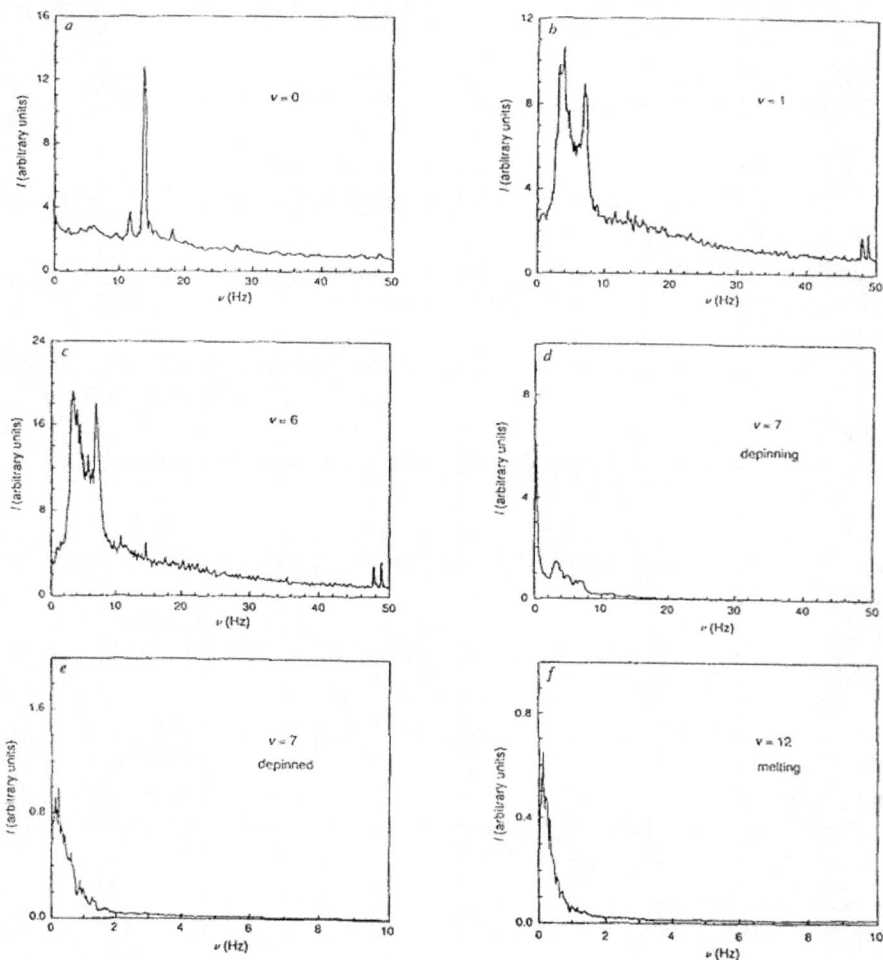

Figure 2.13: Laue diffraction pattern from a colloidal crystal. At low velocities, stick-slip excites resonances in the crystal. Above a critical speed, the crystal is depinned and moves continuously. Above a second critical speed, the crystal is shear melted. From[52].

the elastic constants lead to changes of the resonance frequencies. For future experiments, it would be interesting to observe the influence of the tribological properties of the interfaces between cell wall and colloidal crystal.

2.6 Quartz crystal microbalance

The quartz crystal microbalance (QCM) is common in thin film growth, where it is used for film thickness measurements. The QCM consists of a single crystal of quartz which oscillates in transverse motion with a high quality factor $Q \approx 10^5$. Metals or insulators can then be evaporated onto the electrode. Usually, the quartz crystals are water cooled for stabilization of the resonance frequency. In special cases, they can also be cooled to liquid nitrogen or helium temperature. Then, gas can be admitted to the vacuum chamber and adsorption occurs onto the cooled electrode. The frequency shift is a measure for the mass change of the crystal, which is commonly used for film thickness measurements. Jacqueline Krim and coworkers have extended this method to measure the tribological properties of thin adsorbed films[53]. Changes in the Q-factor are measured and can be related to a characteristic slip time σ_s of the adsorbed film.

$$\sigma_s = \delta(Q^{-1})/(4 \cdot \pi \cdot \delta f) \qquad (2.14)$$

where δf is the frequency shift and Q the Q-factor. The slip time σ_s corresponds to the time for the speed of the film to fall to $1/e$ of its original value, assuming that it had been moving at constant speed and was then stopped by friction. The relation between the slip time σ_s and the frictional force per unit area is:

$$\tau = F_F/A = \rho \cdot v/\sigma_s \qquad (2.15)$$

where ρ is the mass density of the adsorbed film and v the velocity. The calibration is made with helium gas, that does not adsorb at liquid nitrogen temperature. This gas will cause changes of the Q-factor due to well-known viscous damping.

Krim and coworkers have addressed the question, which contributions towards frictional energy losses are due to electronic or phonon processes. Ethane and ethylene were deposited on silver and oxygen surfaces. They found that the friction on the metallic surfaces is increased, which is attributed to the additional electronic friction on the conductive surface. The average slip time for ethane monolayers on silver, 9ns±1.3ns, is observed to be 1.8 times longer than that for ethylene on silver 5ns±1.7ns[54]. As theoretically expected[55], ethane has a longer slip time. However, theory predicts a factor of five, which is not found experimentally. At present, the uncertainties of theory are too large to make firm conclusions. However, it is remarkable that the role of oxygen in tribology could not only be to inhibit metal-metal contact, but also to reduce friction forces, due to the reduced electronic contribution.

An interesting combination of QCM and STM has been presented by Daly and Krim[56]. This combination allows one to measure the frictional force due to STM. It also gives the opportunity to perform fast lithography processes with the help of the moving quartz support. Even the dissipation and weight of mesoscopic structures, assembled by STM or AFM, might become accessible with this combined instrument.

2.7   Friction force microscopy

2.7.1   Introduction to friction force microscopy

With the invention of atomic force microscopy (AFM) by Binnig, Quate and Gerber[57,58] it became possible to study forces on a local scale. The force microscope where both normal and lateral forces are simultaneously measured is called friction force microscope (FFM), sometimes also called lateral force microscope (LFM)[60,61].

The force is a vector, thus in principle three components should be measured. In practice, where cantilevers with anisotropic force constants are used, only two components, the normal force F_N and the lateral force F_L, are measured. In order to measure normal and lateral forces simultaneously, several deflection sensors, which are shown in Fig. 2.15, were implemented in force microscopes. Neubauer et al. used two capacitance sensing plates located near the cantilever[62]. Marti et al.[63] and Amer et al.[64] have devised an optical beam deflection technique, where bending in normal direction and torsional motion of the cantilever are sensed simultaneously by a quadrant photodiode detector. A dual fiber interferometer was introduced by Mc Clelland[65]. This microscope uses two optical fiber interferometers of the type developed by Rugar et al.[66] to measure the cantilever deflection along two orthogonal directions angled 45° with respect to the surface normal. This orientation is more suited than the geometry, where one fiber is parallel to the surface, because of space requirements. The frictional and normal forces are measured from the difference and the sum of the two interferometer signals. Recently, Brugger et al.[67] and Kassing et al.[68] presented cantilevers, where two Wheatstone bridges with piezoresistive sensors are implemented at the base of the legs. The sum of both bridge signals gives the normal forces, where the difference signal is

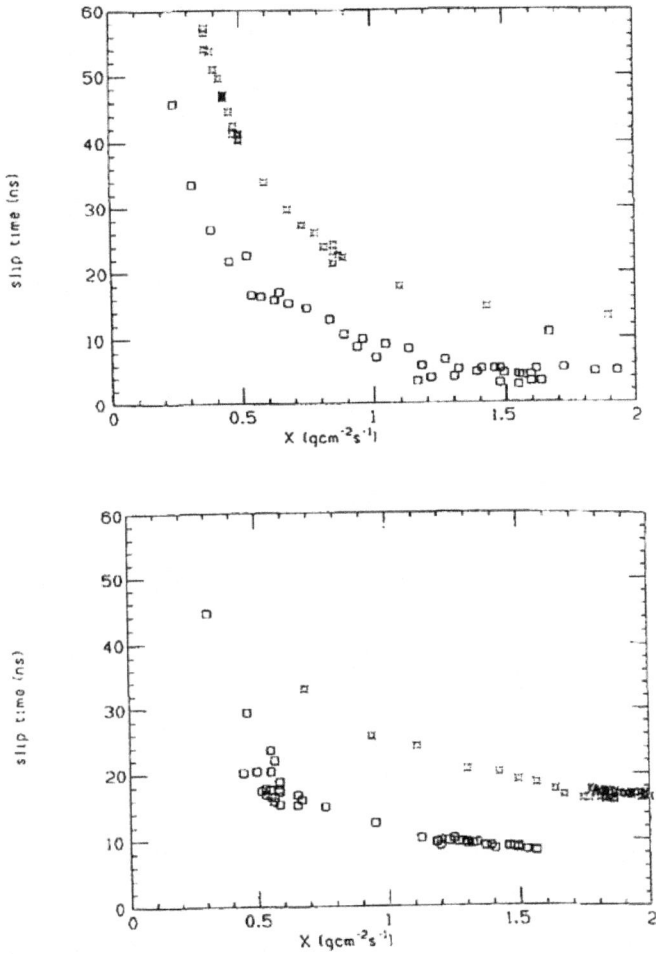

Figure 2.14: Quartz crystal microbalance. Top: Ethylene slip times before ($\square$) and after ($\star$) exposure of the surface to oxygen. Measurements were carried out at 77.4K. One monolayer corresponds to X=1.5g cm^{-2} s^{-1}. Bottom: Ethane slip times times before ($\square$) and after ($\star$) exposure of the surface to oxygen.

The average slip time for ethane monolayers on silver is observed to be longer than that for ethylene on silver. The slip times on the oxygen/silver surface, which is possibly less conducting, are longer. These observations indicate that electronic effects might be dominant. From[54].

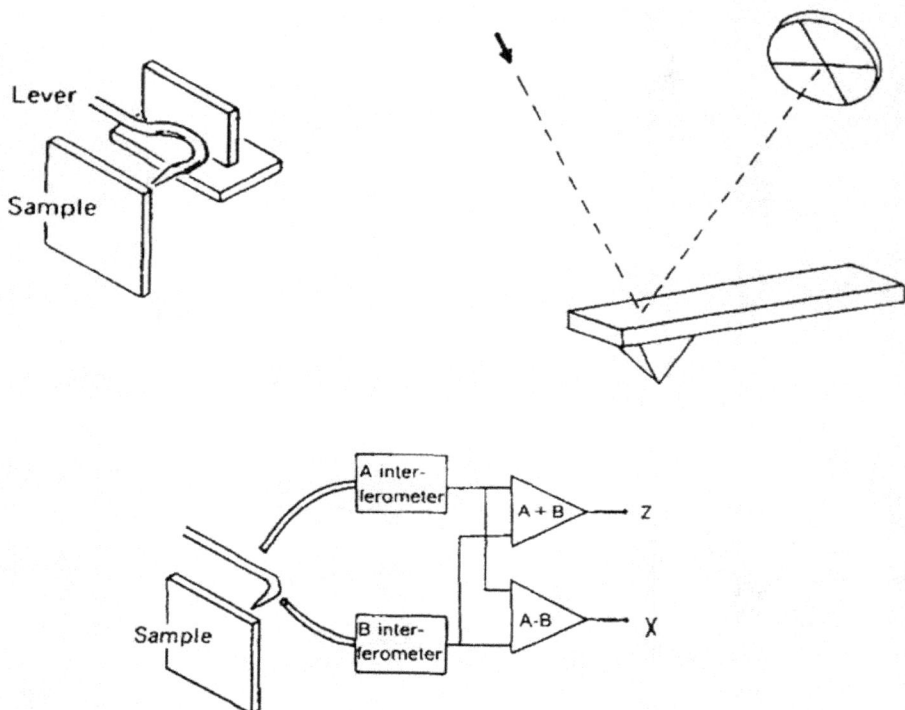

Figure 2.15: Schematics of friction force microscopes, which are designed for simultaneous measurements of normal and lateral forces. (a) Capacitance detection (b) Beam deflection method. (c) Dual fiber interferometry. From[65].

proportional to the torsion of the cantilever. A possible implementation into ultrahigh vacuum is shown in Fig. 2.16, which is based upon the beam deflection method. Spherical motors are used to position the incident laser beam on the the cantilever and the reflected beam on the photo-diode.

In 1987 Mate et al. demonstrated the first observation of friction on an atomic scale[59]. The frictional forces of a 300nm radius etched tungsten tip sliding on graphite in air was measured with a non-fiber interferometer[69]. This pioneering work has shown two major effects: The atomic-scale stick slip and the loading dependence of friction. These phenomena will be briefly discussed below:

Figure 2.16: A multifunctional Scanning Probe Microscope (SPM) is operated in Ultra High Vacuum (UHV). The instrument of beam deflection type is outfitted with a four quadrant photodiode (for friction force microscopy) and a fast photodiode (bandwidth 300 kHz, for dynamic applications). The optical beam deflection detector and the sample position can be adjusted by means of three compact inertial stepping motors. One advantage of this detection scheme is the relatively long working distance between the optics and the cantilever that facilitates in situ exchange of cantilevers. The whole AFM is mounted on a platform which is suspended by four springs and damped with efficient eddy current system.

2.7.2 Atomic-scale stick slip

Friction force loops showed hysteresis between the back- and forward scan, which is associated with the dissipation of energy (non-conservative forces). At low loads the hysterisis is barely visible, but increases with increasing normal force. A sawtooth pattern becomes visible, which varies on the atomic scale. By acquiring images where the tip is scanned in x- and y-direction, Mate et al. determined that the periodicity of friction is that of the atomic lattice of graphite. In contrast to the classical stick-slip, where kinetic friction is smaller than static friction, the atomic-scale stick slip is not dependent on velocity and the periodicity is given by the atomic structure. Typical velocities in force microscopy are 0.1-10μm/s, which corresponds practically to the static case. A surprising observation was that even at loads of 10^{-5}N, where continuum models suggest contact diameters of 100nm, atomic-scale stick slip is visible. Pethica[70] suggested that eventually graphite flakes were broken of the surface and adhered to the tip. Thus, friction between commensurate surfaces is observed. An alternative explanation was given by McClelland that the tip and surface make contact at only a few nm-scale asperities, so that the corrugation is not entirely averaged out. More recent measurements on non-layered materials in ultrahigh vacuum have shown that atomic-scale stick slip is limited to a rather low load regime for sharp tips[71]. A higher loads, plastic deformations of the sample or tip are observed. Thus, the observation of Mate et al. of atomic-scale stick slip at high loads might be restricted to layered materials or the presence of some lubricating contamination films. However, one has to emphasize that atomic-scale stick slip at low loads is observed on practically all materials and is probably a major source of dissipation. The mechanisms of this atomic scale slip will be discussed in more detail in chapter IV.

2.7.3 Loading dependence

Mate et al. also presented the normal force dependence of a tungsten tip on graphite[59]. They found a rather linear dependence with a friction coefficient of about 0.01. Although, linear dependences of friction vs. normal force are common in macroscopic experiments, this observation is not expected in microscopic, single asperity experiments. As explained in chapter I, the linear dependence in macroscopic contacts arises from the increase of contacting asperities with increasing load, which has been explained by statistical arguments. For a single-asperity contact a non-linear dependence is expected. A

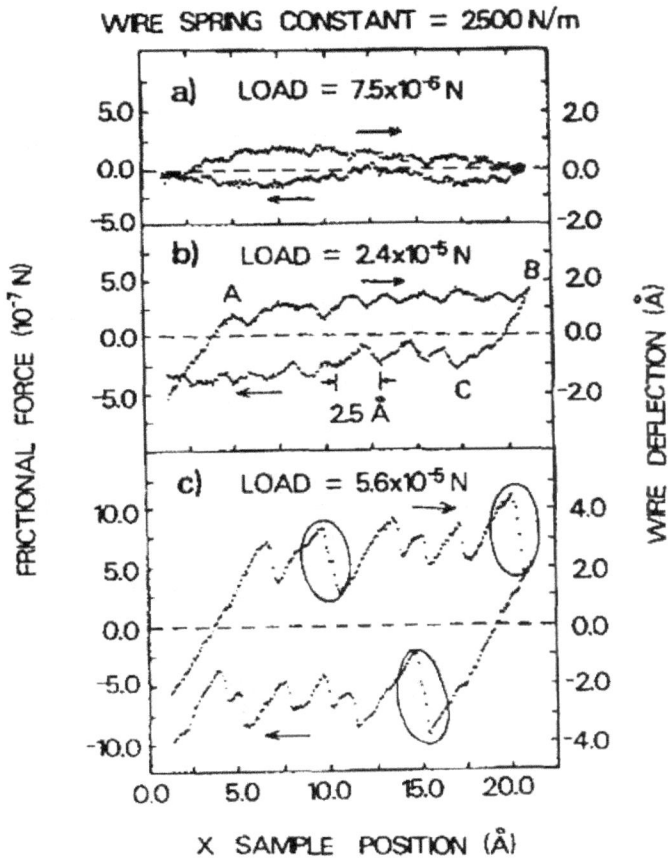

Figure 2.17: Frictional forces in ambient conditions between a tungsten tip and a graphite surface as a function of sample position for three different loads. Atomic stick slip is observed, where the spacing is given by the atomic lattice of graphite. From[59].

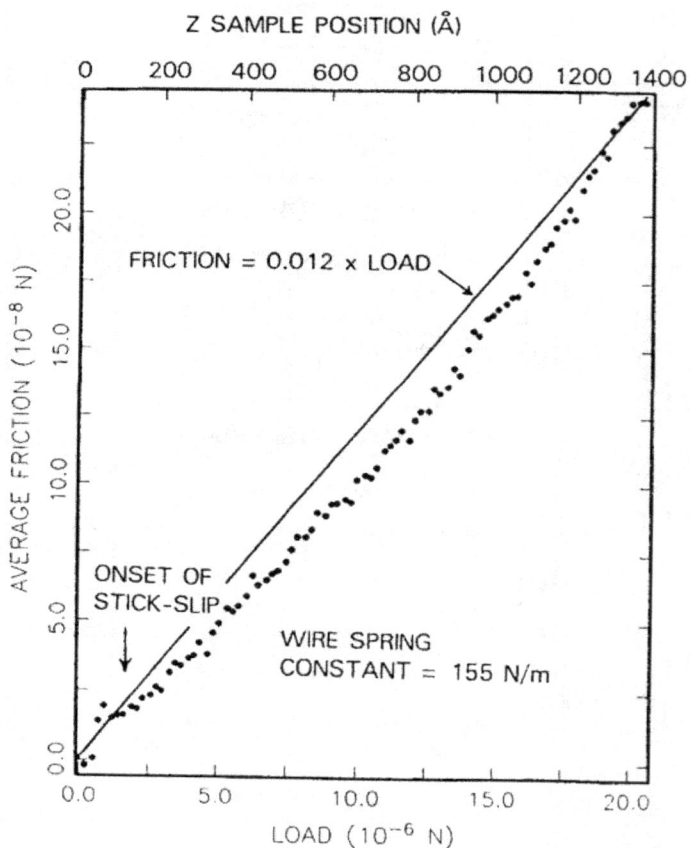

Figure 2.18: Frictional forces in ambient conditions between a tungsten tip and a graphite surface as a function of normal force. A linear dependence is found with a friction coefficient of 0.01. From[59].

simple Hertzian contact would result in a loading dependent contact area:

$$A = \pi (\frac{3RF_N}{4E_*})^{2/3} \qquad (2.16)$$

$$E^* = (\frac{1 - \nu_1^2}{E_1} + \frac{1 - \nu_2^2}{E_2})^{-1} \qquad (2.17)$$

where E_1 and E_2 are the Young's moduli of the sample and probing tip and ν_1 and ν_2 are the Poisson ratios. Assuming that the Bowden-Tabor adhesion model is valid, a loading dependence of friction of the form

$$F_F = \tau A = \tau \pi (\frac{3RF_N}{4E_*})^{2/3} \qquad (2.18)$$

is expected, where τ is the shear strength. Thus, the proportionality observed by Mate et al. may result from a multiasperity contact. The tip may have nm-scale roughness and the contacting asperities may increase with normal load as discussed in macroscopic experiments. Some evidence has been found by the Kaneko group[73], who found that the linear dependence of a silicon nitride tip on mica, observed in dry conditions, can change to non-linear dependence at high humidity. The explanation given by Putman et al. was that capillary condensation of water on the rough probing tip leads to a smoothening of the tip. With this smoothened tip, single-asperity behaviour is observed. Alternatively, the experiment of Mate et al. may also reflect a pressure-dependence of the shear strength τ. Briscoe[72] pointed out that the shear strength is given in first approximation by $\tau = \tau_0 + \alpha p$, where $p = F_N/A$ is the pressure and α the proportionality constant. Thus, in the framework of Hertzian deformation, a loading dependence of the form:

$$F_F = \tau A = \tau_0 \pi (\frac{3RF_N}{4E_*})^{2/3} + \alpha F_N \qquad (2.19)$$

Thus, for a large α-parameter a linear behaviour is observed. Whereas, small α-values lead to the non-linear dependence. This explanation found some support by measurements from Meyer et al.[76], which will be discussed in more detail in chapter IV, and Schwarz et al.[74], where both linear and non-linear behaviour were observed with the same tip on heterogeneous surfaces, consisting of islands of C_{60} on GeS in air. Linear behaviour was found on GeS and non-linear behaviour on C_{60}. Schwarz et al. interpreted their data with the Fogden-White model, which can be interpreted as an extended Hertz-model, where long-range capillary forces cause an additional attractive force, which shifts the normal force scale:

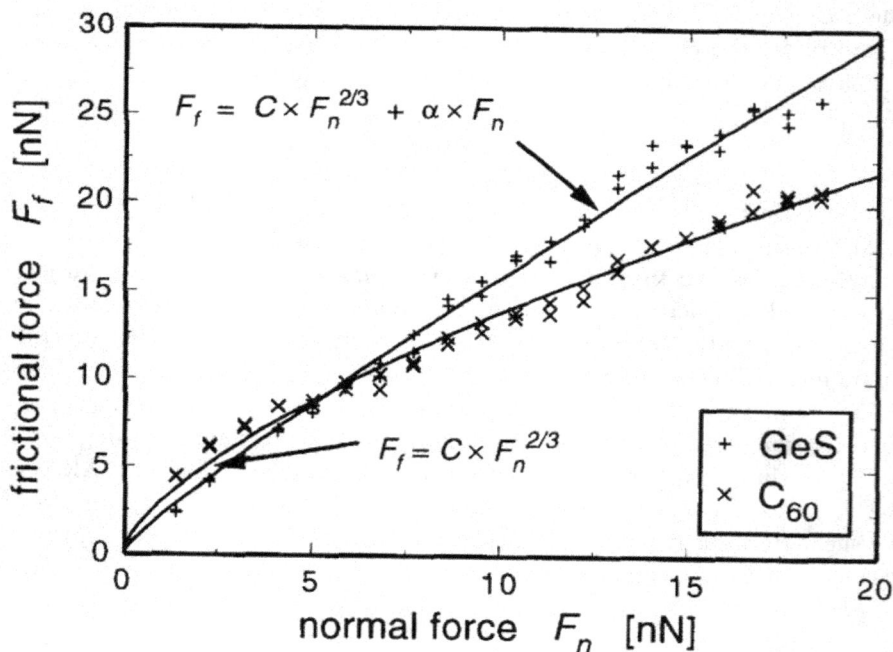

Figure 2.19: Frictional force vs. normal force. The data points from the GeS substrate are well approximated by a rather linear curve ($C=1.06\text{nN}^{1/3}$, $\alpha=1.08$). The C_{60}-data are approximated by a curve of the form $F_F \sim F_N^{2/3}$ ($C= 2.96\text{nN}^{1/3}$). Note the cross-over of the curves at about 5nN. From[74].

$$F_F = \tau A = \tau_0 \pi \left(\frac{3R(F_N + F_0)}{4E_*} \right)^{2/3} + \alpha (F_N + F_0) \qquad (2.20)$$

where $F_0 = 4\pi R \gamma (1 - D_0/2r_k)$, γ is the surface tension of the liquid-vapor interface, D_0 is the separation of the surfaces and r_k represents the radius of the meniscus. In the case of Schwarz et al., capillary condensation of water may occur in air. The experimental pull-off force of 6.7nN was consistent with a tip radius of $R \approx 7.8$nm. With this model, they found different α-values and τ_0-values for the two materials. However, Schwarz et al. also pointed out that deviations of the spherical geometry of the tip may explain the observed behaviour. They suggested that the contact area may follow a more general law:

$$A = C F_n^m \qquad (2.21)$$

where m is a parameter which varies with geometry: $m=2/3$ for sphere-plane geometry and $m = 1/2$ for pyramidal-shaped tip. With the fit parameters m and σ_0 ($\alpha = 0$) they found good agreement with the experiment. Thus, one has to make sure that the tip shape is spherical, in order to determine α and σ_0 accurately.

The effect of tip shape on the loading dependence of friction was studied by Carpick et al.[75]. The sliding of Pt-coated tips with different geometries on mica was investigated. The shape of the probing tip was characterized by imaging a stepped $SrTiO_3(305)$ surface. At the step sites the image is a convolution with the probing tip and allows one to determine the profile of the tip apex. The profiles were fitted with polynoms of the form $z \approx r^n$. Best fits for blunted tips were between r^4 to r^6. Analysis of the friction vs. normal force curves with an extended JKR-model, where the tip profile was included is shown in Fig. 2.20. Best agreement is found with the profiles r^6, which is in agreement with the independently measured tip profile. Thus, Carpick et al. find that the friction is proportional to contact area and that the shear strength τ is constant in first approximation for the case of Pt on mica in UHV. It is important to take into account deviations from the spherical geometry. Also, the results were applied to rather large radii of curvature (≈ 140nm).

2.7.4 2d-histogram technique

The 2d-histogram technique is a method to measure the loading dependence of friction as a function of normal force[76,77]. In contrast to conventional friction vs. normal force curves, the method is based upon the acquisition of images, so called friction force maps. During the acquisition of these data, the loading

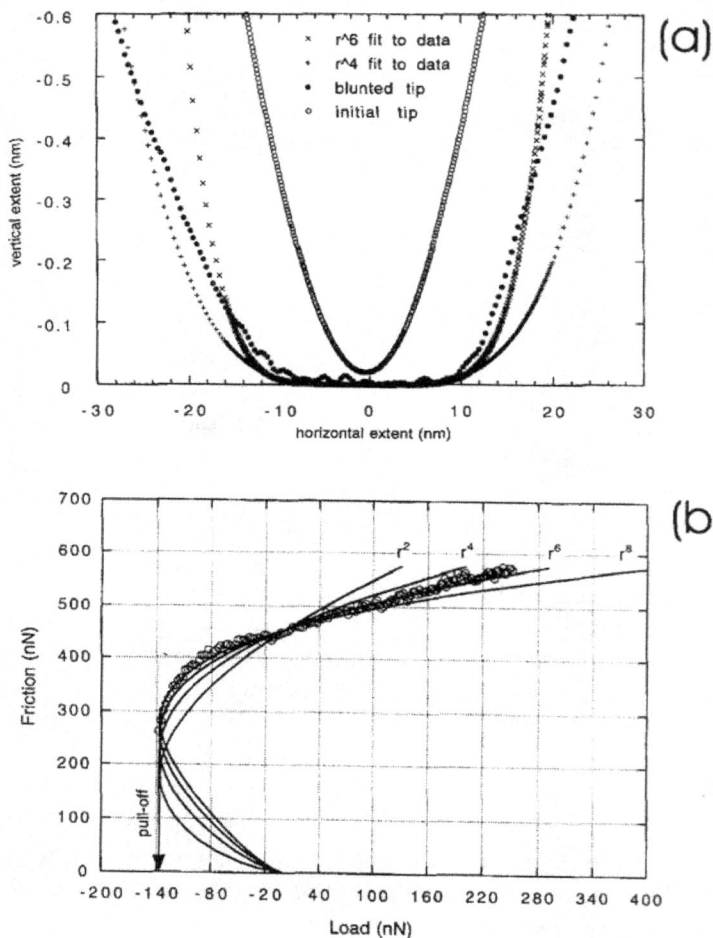

Figure 2.20: (a) Profile of the probing tip of a blunted tip compared to the original tip, as well as the fitted curves. (b) Friction vs. normal force for the blunted tip (see (a)). The four solid curves are calculated with the extended JKR-theory. The curve based on the r^6 tip gives the best fit and is also in agreement with the tip profile. From[75].

is increased or decreased. The probing tip can be either scanned on the same line or can change the position in y-direction after the acquisition of each friction loop. Forward- and backward scan images are then subtracted from each other and divided by two. For compensation of thermal and piezoelectric creep, the images can be shifted horizontally (along x-axis). Finally, the data are used to compute a 2d-histogram. The method has the advantage that good statistics can be achieved. It also does not presume any functional dependence between normal force and frictional forces. Each (F_F, F_N)-data point gives a contribution to the 2d-histogram. For weak or absent correlation, the data are randomly distributed in the (F_F, F_N)-plane. For strong correlation, the data are piling up in distinct regions of the (F_F, F_N)-plane, reflecting the functional or multifunctional dependence $F_F^i(F_N)$, where the index i represents different materials or inequivalent sites.

The first example in Fig. 2.21 shows how 2d-histograms are calculated. C_{60}-films deposited on NaCl(001) were investigated in this study[86]. First, friction loops are acquired (forward and backward scan). The y-position can be fixed or varied. After each friction loop, the loading is increased or decreased. Thus, two friction force maps are acquired (forward- and backward scan) with varying loading. Second, the two images can be shifted in x-direction to compensate thermal drift or piezoelectric creep, until the overlap between the two images is optimum. As indicated by the circles in Fig. 2.21c/d, the user will try to overlap both images in optimum way, in order to reduce artifacts at the boundaries between different materials. If the compensation is not well done, a large scatter in the 2d-histogram will be observed. Third, the two images are subtracted from each other and divided by two. Then, the difference image is used for the calculation of the 2d-histogram.

The second example shows the richness of information that can be gained from 2d-histogram technique. Thin films of AgBr(001) are grown epitaxially on NaCl(001). As shown in Fig. 2.22 islands of about 1-5nm, corresponding to 2-10 unit cell heights (a_0=5.77Å), can be observed. The 2d-histogram shows three different regimes, as can be recognized from the corresponding lateral force image: 1) wear-less friction. 2) wave-like structure 3) droplet-like structure. Regimes 2) and 3) show drastic changes of the morphology of the film during scanning and are therefore accompanied with wear processes. It is quite remarkable, that only the 2d-histogram technique allows us to distinguish clearly between wear-less friction and friction with wear. Spikes at normal forces above 13nN are therefore related to the transport of AgBr-material by the action of the probing tip. On the average, a rather linear behaviour is observed. The increase of the friction forces on AgBr compared to NaCl by

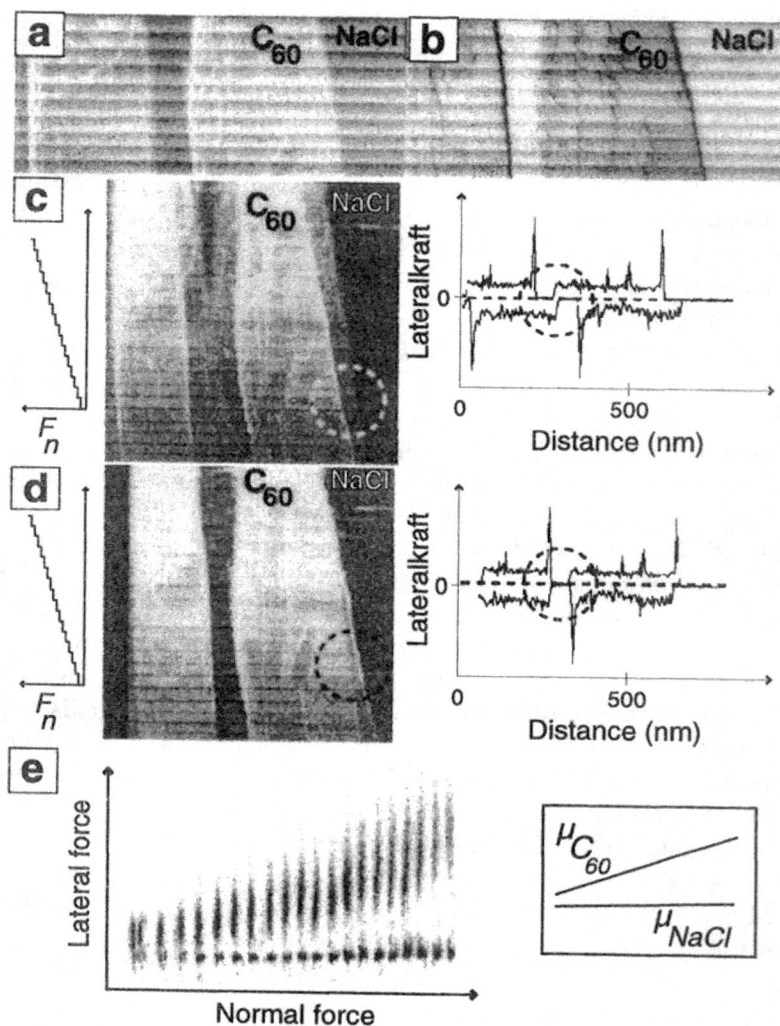

Figure 2.21: (a) to (e) shows the individual steps of the 2d-histogram technique. (a) Forward scan. (b) Backward scan. High friction is observed on C_{60} compared to NaCl(001). During the acquisition of the images, the normal force has been increased in discrete steps as indicated in (c). In (c) and (d) it iş shown that the difference image ((a)-(b))/2 has to be calculated in an interactive way. (c) shows a difference image without adequate correction of drift. (d) shows a well-compensated difference image. (e) The 2d-histogram as calculated from (d).

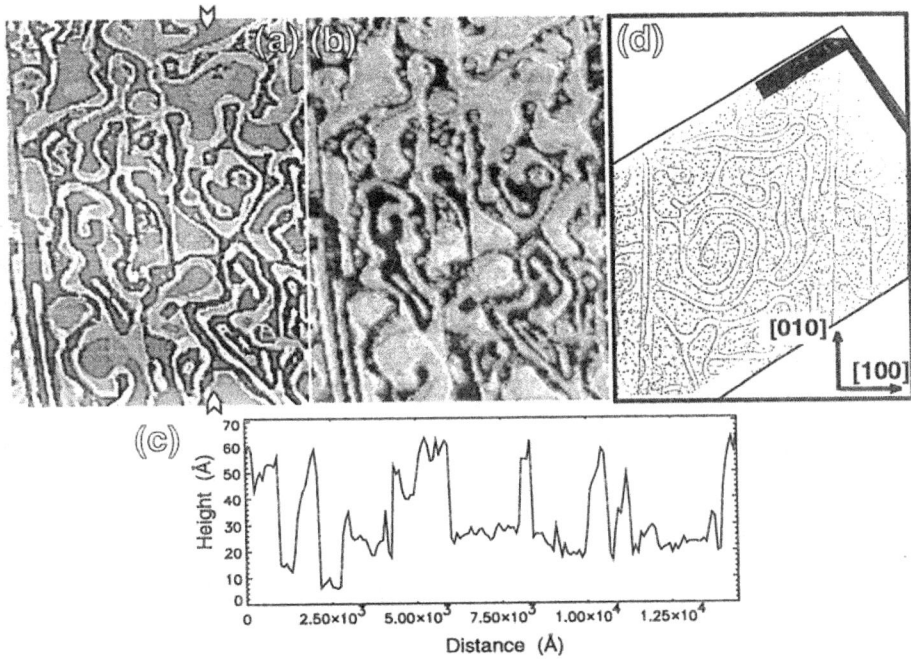

Figure 2.22: AgBr(001)-islands grown on NaCl(001). (a) Topography image. (b) Corresponding friction force map. High friction is observed on the AgBr(001) areas. (c) Profile as indicated in (a). The islands are 1-5 nm high. (d) Comparison with an scanning electron image of a replica, produced by metal decoration. The step structure is in good agreement with the structure observed by FFM.

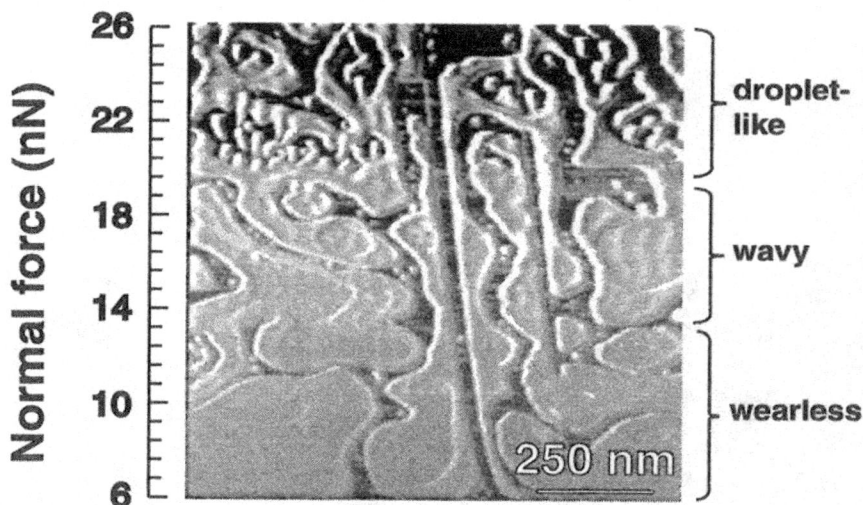

Figure 2.23: Friction force map acquired with variable normal force. Different regimes can be distinguished. The lower part (6-13nN) corresponds to wear-less friction. Between 13-20nN wave-like structure are visible, showing the first stage of wear. Above 20nN we observe droplet-like feature corresponding to strong deformation of the AgBr-islands.

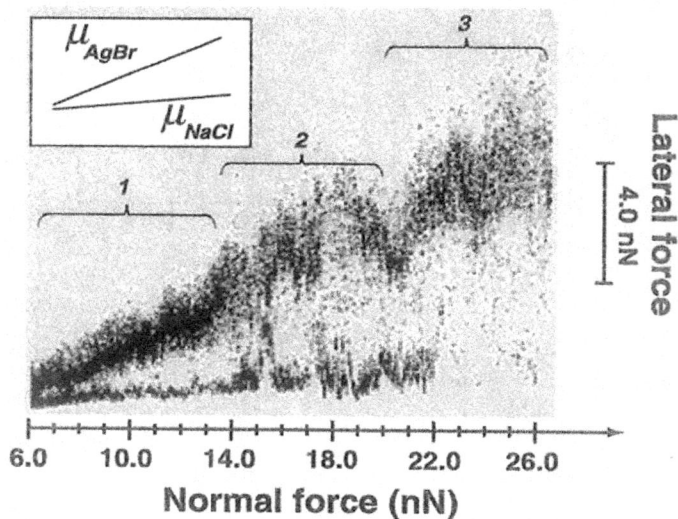

Figure 2.24: 2d-Histogram of the data shown in Fig. 2.23. The three regimes are indicated in the figure. Note that only regime 1 corresponds to the case of wear-less friction. The spikes on the histogram originate from the transport of AgBr-material.

a factor of 30 is quite drastic[77]. For comparison, the reduction of friction by an excellent boundary lubricant, such as Cd-arachidate, compared to a silicate substrate, is only a factor of ten. In literature, it is not well established that NaCl(001) is a reasonable lubricant under ultrahigh vacuum conditions. However, it is known from the field of extreme pressure lubricants (E.P.) that chlorides are excellent lubricants of steel at temperatures above 300°C where water can be excluded[2].

Although, both AgBr and NaCl are ionic crystals of similar structure, there seem to be fundamental differences. Several mechanisms may lead to this drastic difference in friction: 1) The Youngs modulus of AgBr is reduced ($E_{NaCl}=3.61 \cdot 10^{10} \text{N/m}^2$, $E_{AgBr}=2.53 \cdot 10^{10} \text{N/m}^2$) leading to an increased contact area for the same normal force. 2) AgBr is known for its high surface energy ($\gamma_{NaCl}=0.18 \text{J/m}^2$, $\gamma_{AgBr}=0.29 \text{J/m}^2$) which leads to increased adhesive force, increased net normal force and therefore also enlarges the contact area. Both contributions increase the contact area by a factor of $(\frac{\gamma_{AgBr}(1-\nu_{AgBr}^2)E_{NaCl}}{\gamma_{NaCl}(1-\nu_{NaCl}^2)E_{AgBr}})^{2/3}=1.63$. Thus, these contributions are not dominant.

Another possibility is surface diffusion: The surface diffusion coefficient of Ag^+-ions and Br^--ions is very high. Performing local scratch experiments with the AFM and observing the time evolution of the created structure (refilling of holes by surface diffusion) a diffusion coefficient of $9 \cdot 10^{-14} \text{cm}^2/\text{s}$ could be determined[101]. This increased surface diffusion might also be related to the contrast formation of friction. E.g., Ag-ions or Br-ions could be moved into interstitial positions. Even, a small ploughing term might be existent on the AgBr-surface, leading to the larger lateral forces. The small scratches may not be observable for small loadings, because they are refilled immediately after the passage of the tip. From the observation of wear processes above 13nN it becomes obvious that the mobility of AgBr is very high and that the suggested contrast formation is quite reasonable.

2.7.5 Resolution limits

The resolution limits of friction force microscopy are given by the resolution limits of force microscopy in contact. Contact-mode AFM is accompanied by the jump-in of the soft cantilever. When the condition

$$dF_a/dz < c_B \tag{2.22}$$

is met, an instability occurs, where dF_a/dz is the attractive force gradient and c_B the cantilever spring constant. The probing tip jumps towards the sample.

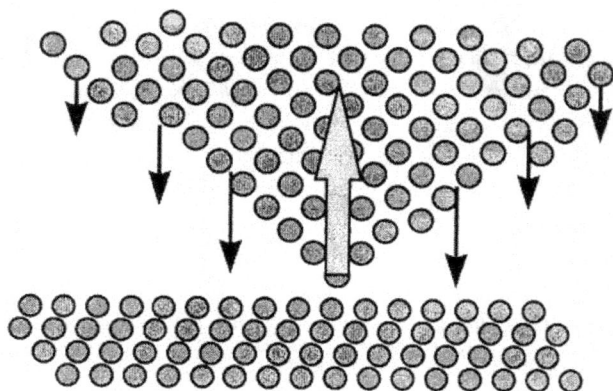

Figure 2.25: Long-range attractive forces have to be compensated by short-range repulsive forces.

Long-range attractive forces $F_{l.r.}$, such as van der Waals forces, capillary forces or electrostatic forces, have to be compensated by short-range repulsive forces on the foremost tip apex $F_{s.r.}$. After the instability, the operator will try to minimize the forces on the tip apex by compensating the long-range attractive force with the bending of the cantilever $F_B = c_B \cdot z_t$.

The equilibrium is given by:

$$F_{s.r.} = F_{l.r.} - c_B \cdot z_t \qquad (2.23)$$

which determines the force on the tip apex $F_{s.r.}$. The minimum force $F_{s.r._{min}}$ is achieved close to the jump-out of the contact. Experimentally, it is found that $F_{s.r._{min}}$ is a significant fraction of $F_{l.r.}$ (typically 10% - 50%). Thus, the compensation by the cantilever bending is not complete. This incomplete compensation is explained by local variations of the attractive force. These forces can change quite drastically above hillocks compared to valleys. During scanning, the tip will either jump out of contact above areas with low attractive forces or will experience high forces on the areas with high attractive forces. Thus, the minimum, experimentally achievable force on the tip apex depends also on the roughness and the scan area.

The consequences for the resolution of contact force microscopy become evident. Depending on the environment, the long-range forces $F_{l.r.}$ will vary significantly. In ambient pressure, capillary forces, originating from the formation of a liquid meniscus between probing tip and sample, will be dominant.

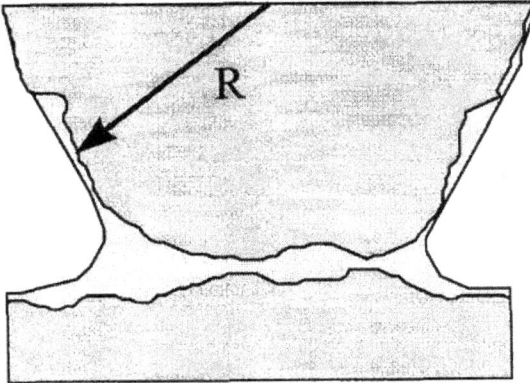

Figure 2.26: Formation of a meniscus between the probing tip and sample. Strong capillary forces arise.

With a tip radius R=100nm, the maximum attractive, capillary force is

$$F_{cap.} = 4\pi R\gamma cos(\theta) \approx 90nN \qquad (2.24)$$

where γ=0.07N/m^2 is the surface tension of water and θ is the contact angle. After the jump to contact this large force acts uncompensated on the tip apex and can deform the tip or sample. Even with optimum bending of the cantilever, the force will still be in the nN-regime.

Capillary forces can be eliminated by measuring in liquids or in vacuum conditions. In liquids, attractive forces can become very small. In best case, van der Waals forces become repulsive by choosing a suitable liquid[78]. Such a situation is met, when the refractive index of the liquid, n_l, is between the refractive index of the probing tip, n_t, and the refractive index of the sample, n_s: $n_s < n_l < n_t$. Ohnesorge and Binnig have shown that it is possible to achieve true molecular resolution on calcite in water, thus demonstrating that the contact diameter is of atomic dimensions[80].

In ultra-high vacuum conditions, van der Waals forces will always be present. Goodman and Garcia[79] have shown that typical van der Waals forces are between 1-10nN, where the exact values depend on the materials. A collection of their results is shown in table 2.2. A tip radius of R=100nm at a distance of z=1nm has been assumed.

The van der Waals force is given by

$$F_{vdW} = -\frac{B \cdot R}{z^2}\frac{1}{(1 + z/2R)^2} \qquad (2.25)$$

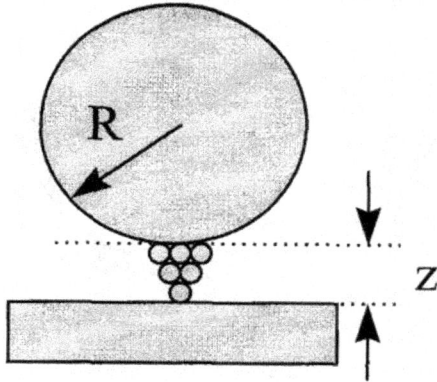

Figure 2.27: Calculation of van der Waals forces. The average radius of curvature R at the average distance z gives the van der Waals part.

where

$$B = \frac{3K}{4} \frac{(\varepsilon_s - 1)(\varepsilon_t - 1)}{(\varepsilon_t + 1)(\varepsilon_t + 2)} \qquad (2.26)$$

and K=1.41eV.

Thus, the attractive forces in ambient conditions are the largest: 1-100nN.

Materials	Force
Graphite-Graphite	8nN
Diamond-Diamond	17nN
Metal-Graphite	10nN
SiO_2-Graphite	1.2nN

Table 2.2: Estimation of van der Waals Forces

In ultra-high vacuum, van der Waals forces are always present and give values between 0.1-10nN. In liquids, van der Waals forces can become repulsive in suitable liquids and forces can become below 100pN. The Hertz model, gives us an estimate of the contact radius:

$$a_{min} = (\frac{3 \cdot R \cdot F}{4E^*})^{1/3} \quad , \qquad (2.27)$$

where $E^* = ((1 - \nu_1^2)/E_1 + (1 - \nu_2^2)/E_2)^{-1}$ and ν_i and E_i are the Poisson ratios and Youngs moduli of probing tip and sample. With typical parameters ($E_1 = E_2 = 1.7 \cdot 10^{11} N/m^2$, $\nu_i = 0.3$ and $R=90nm$), the contact diameter in ambient pressure is 2-10nm (1nN< F <100nN). In ultrahigh vacuum, we estimate contact diameters to be 1-4nm ($0.1nN< F <10nN$). Whereas, in liquids we expect to achieve true atomic resolution under best conditions ($a_{min} <1nm$). These rough estimates are confirmed experimentally: In ambient pressure, the resolution is typically in the range of 5-10nm. In ultrahigh vacuum, lateral resolutions of about 1nm could be achieved. E.g. Howald et al. could resolve a step edge on NaF(001) with 1nm width[83]. A similar resolution could be achieved on Si(111)7x7 with a PTFE-coated tip[84]. Fig. 2.28 and Fig. 2.29 show contact force microscopy images of C_{60}-islands on NaCl(001). The C_{60}-molecules could be resolved at the step edge both in normal mode and lateral force mode. Considering, that the distance between the C_{60}-molecules is 1nm, it becomes evident that the contact diameter is below 1nm. As mentioned above, true atomic resolution is possible in liquids[80]. Recently, it has been shown that force microscopy operated in the dynamic mode (oscillating the cantilever at its resonance and measuring frequency shifts), can achieve true atomic resolution[81,82,85].

In conclusion, the contact diameters in typical contact force microscopy are between 1-10nm. More sophisticated elasticity models for the calculation of the contact area are discussed elsewhere[87,93,95]. Generally, atomic-scale images have to interpreted with care. Even, when atomic-scale features are observed in normal or lateral force, the contrast originates from a multiple-atom contact. Exceptions from this rule are contact mode imaging in liquids with ultra-low forces ($<100pN$) and non-contact imaging in ultrahigh vacuum, where true atomic-resolution can be achieved.

2.7.6 Stiffness measurements: Ways to determine the contact area in FFM

The FFM is a useful tool to study friction, adhesion, lubrication and wear. Under certain conditions (low load, smooth tip shape, unreactive surfaces...) a single asperity contact is formed and wearless friction is observed. In this regime, several groups have observed that friction is proportional to the contact area. Thus, the Bowden-Tabor adhesion model seems to be valid even at scales of 1-100nm. Accordingly, the friction force, F_F is given by

$$F_F = \tau A = \tau \pi a^2 \qquad (2.28)$$

where τ is the shear strength and A is the contact area. One major problem for a quantitative analysis of friction data is the knowledge of the contact area. Unfortunately, the contact area is not directly measured in FFM. This is in contrast to the surface force apparatus, where the contact area is optically visible. Thus, the contact area has to be calculated with models, such

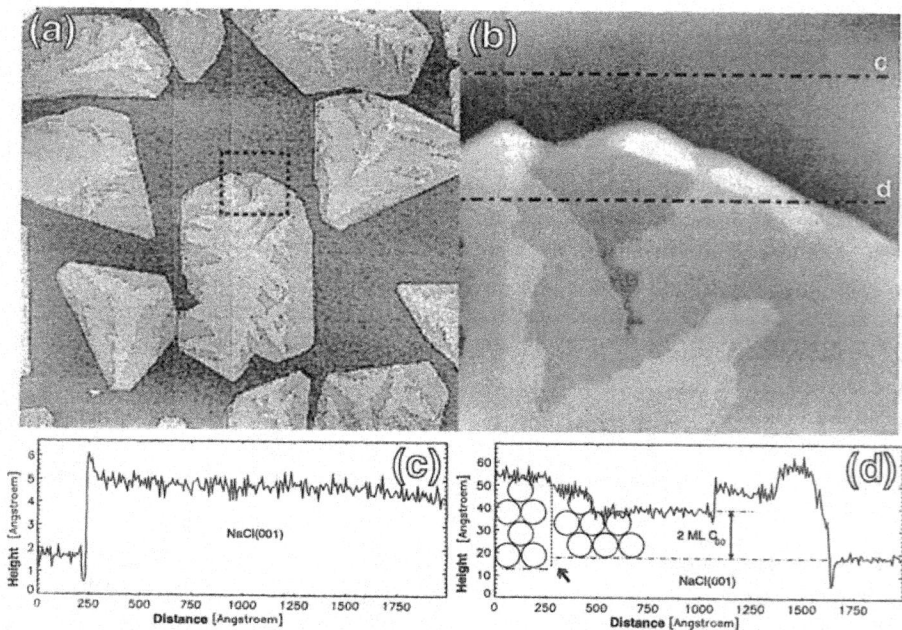

Figure 2.28: (a) Constant force image of C_{60}-islands grown on NaCl(001). The islands are 3-6 molecular layers high. (b) A zoom-in image of (a). The profiles (c) and (d) are indicated in (b).

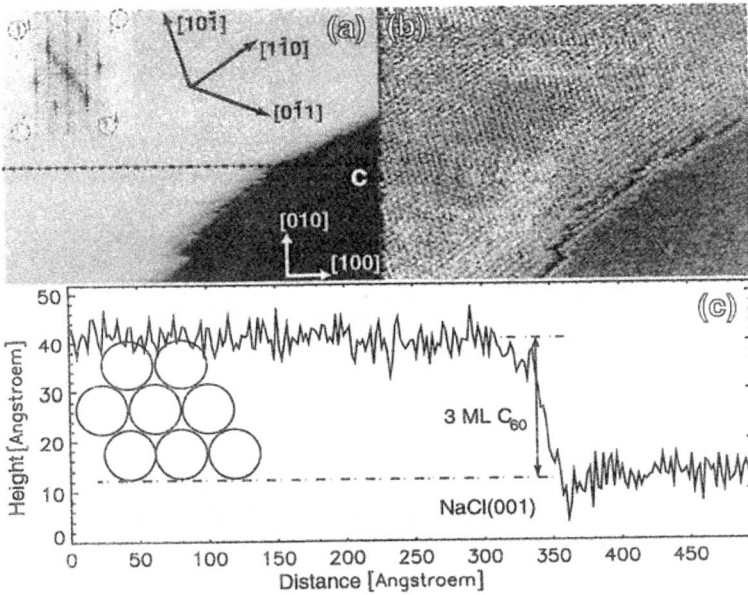

Figure 2.29: (a) High-resolution constant force image of C_{60} on NaCl(001). The inset shows the FFT-image, showing the spots from both the C_{60}-periodicity and the NaCl(001)-lattice. (b) Corresponding friction force map. Both the molecular structure on the C_{60}-terrace and the NaCl-lattice are visible. The observation of molecular structure at the step edge confirms that the resolution is about 1nm, which corresponds to the distance between C_{60}-molecules. (c) Profile as indicated in (a).

as continuum elasticity models (e.g., Hertz, Fogden-White, Johnson-Kendall-Roberts or Maugis-Dugdale). The particular choice of the model depends on the range of adhesive interactions, the load and tip shape. In addition, the shear strength depends also on the load (usually treated as a second order effect) and velocity (FFM is rather slow and can be treated as the quasi-static case). The relative motion and the lateral force may also affect the contact area, which is neglected in continuum elasticity models.

Normal contact stiffness measurements

One way to determine the contact area was suggested by the Pethica group[89]. They suggested to measure the contact stiffness, which is essentially the "spring constant" of the contact. According to the Hertz model, the normal contact stiffness $k^z_{contact}$ is given by

$$k^z_{contact} = 2aE^* \tag{2.29}$$

where a is the contact radius and

$$E^* = \left(\frac{1 - \nu_1^2}{E_1} + \frac{1 - \nu_2^2}{E_2} \right)^{-1} \tag{2.30}$$

and E_1 and E_2 are the Young's moduli of the sample and probing tip and ν_1 and ν_2 are the Poisson ratios. Thus, the contact area can be determined from the normal stiffness, when the elastic properties of sample and tip are known. However, one has to make sure that no plastic deformation occurs and that Hertz is applicable. The latter condition, is only approximately fulfilled, because surface adhesion is important in FFM-geometries. Models, such as Fogden and White[90], JKR[91] and Maugis-Dugdale[92], seem to be more realistic. A rather simple test of the applicability of Hertz, is to measure the loading dependence of the normal stiffness. Then, the contact radius should depend on the normal force:

$$a = \left(\frac{3RF_N}{4E^*} \right)^{1/3} \tag{2.31}$$

The normal contact stiffness can be measured by modulating the sample in z-direction Δz and measuring the elastic response of the cantilever Δz_{lever}. The elastic behaviour of the cantilever is modeled by two coupled springs in series. A normal displacement of the sample Δz is distributed between the two springs:

$$\Delta z = \Delta z_{contact} + \Delta z_{lever} \tag{2.32}$$

where $\Delta z_{contact}$ is the elastic deformation of the contact and Δz_{lever} the elastic deformation of the cantilever. The normal force acting on each of the springs is equal and the effective spring constant is given by:

$$F_{normal} = k_{eff}^z \Delta z \tag{2.33}$$

where

$$k_{eff}^z = (\frac{1}{k_{contact}^z} + \frac{1}{c_B})^{-1} \tag{2.34}$$

Thus, the normal contact stiffness can be determined from the measurement of the sample motion Δz, the deflection of the cantilever spring Δz_{lever} and the accurate knowledge of the normal spring constant c_B (see appendix for calibration procedures).

$$k_{contact}^z = \frac{F}{\Delta z_{contact}} = \frac{c_B \Delta z_{lever}}{\Delta z - \Delta z_{lever}} \tag{2.35}$$

For nanometer-sized contacts between common materials, such as metals or ceramics, the normal contact stiffness is in the range of $k_{contact}$=50-500N/m. On the other hand, typical normal spring constants are of the order of 0.01-1N/m. Thus, $k_{contact}^z$ cannot be easily determined with reasonable accuracy.

Jarvis et al.[89] constructed a modified force microscope, where a magnet is attached to the cantilever. An external coil is used to create an inhomogeneous magnetic field. Then, a magnetic force can be applied to the probing tip. The contact stiffness $k_{contact}^z$ is determined by modulating the applied magnetic force and measuring the displacement of the cantilever. Using this method, the contact diameter can be calculated with reasonable accuracy. However, the cantilevers are custom-made and are not commercially available at present. Therefore, this method is not applicable for standard operation of FFM.

Lateral contact stiffness measurements

An independent way to determine the contact area has been introduced by the Welland group in Cambridge[93,94] and the Salmeron group in Berkeley[96]. Instead of measuring the normal stiffness, where normal spring constants are commonly rather small, these authors suggested to measure the lateral contact stiffness, where typical spring constants are around 50-200N/m. Thus, the accuracy of the measurement is better than in normal contact stiffness measurements (at least for common cantilever geometries).

For a sphere-plane geometry, the lateral stiffness of the contact is given by:

$$k_{contact}^x = 8aG^* \tag{2.36}$$

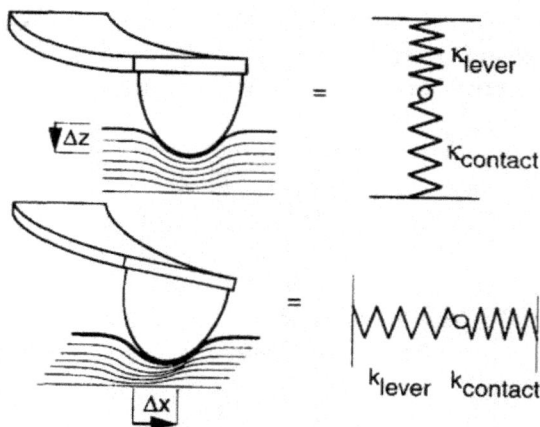

Figure 2.30: Schematic diagram of normal and lateral stiffness measurements. From[96].

where a is contact radius and

$$G^* = (\frac{2 - \nu_1^2}{G_1} + \frac{2 - \nu_2^2}{G_2})^{-1} \qquad (2.37)$$

and G_1 and G_2 are the shear moduli of the sample and probing tip and ν_1 and ν_2 are the Poisson ratios[95,97,98]. The equation is valid for various continuum elasticity models and does not depend on the interaction forces[95,96]. This is in contrast to the analogous equation of the normal stiffness, which is only valid for the Hertzian case. For other models, the equation of the normal stiffness 2.29 has to be modified.

A simple explanation has been given by Carpick et al.[96]: The lateral force $dF_{lateral}$ corresponds to a lateral stress $d\sigma$ which is distributed across the contact area A. This stress produces a strain $d\varepsilon \approx dx/a$, since a is the length scale of the stress distribution. Taking into account Hooke's law $d\sigma = G \cdot \varepsilon$, one can conclude that $dF_{lateral}/dx \approx G/a$. It has been assumed, that the lateral displacement does not change the contact area, which is reasonable for small displacements. However, in the case of the normal stiffness, the contact area changes with normal displacement, which explains that the normal stiffness is not generally proportional to the contact radius.

In close analogy to the treatment of normal stiffness, the elastic response in the experimental set-up is described by a series of springs. A lateral displacement of the sample Δz is distributed between three springs:

$$\Delta x = \Delta x_{contact} + \Delta x_{tip} + \Delta x_{cantilever} \qquad (2.38)$$

where $\Delta x_{contact}$ is the elastic deformation of the contact, Δx_{tip} is the elastic deformation of the tip and $\Delta x_{cantilever}$ the elastic deformation of the cantilever. The lateral force acting on each of the springs is equal and the effective spring constant is given by:

$$F_{lateral} = k^x_{eff} \Delta x \qquad (2.39)$$

where

$$k^x_{eff} = \left(\frac{1}{k^x_{contact}} + \frac{1}{k^x_{tip}} + \frac{1}{c_x} \right)^{-1} \qquad (2.40)$$

where $k^x_{contact}$ is the lateral stiffness of the contact, k^x_{tip} the lateral stiffness of the tip and c_x the lateral spring constant of tip (for most commercial cantilevers, the torsional spring constant is relevant!).

The experimental procedure to determine k^x_{eff} is as follows: Typical lateral force vs. lateral position loops, also called friction force loops consist of a sticking part, where the tip essentially stays at the same position, and a sliding part, where the tip starts to move and atomic-stick slip is observed. The slope of the sticking part corresponds to the effective spring constant:

$$dF_{lateral}/dx = k^x_{eff} \qquad (2.41)$$

If the piezoelectric scanner is calibrated, the lateral position can be measured accurately. Second, the lateral force scale depends on an accurate knowledge of the cantilever spring constant, c_x. Calibration procedures are described in the appendix. Having determined the effective spring constant, k^x_{eff}, the lateral stiffness of the contact can be determined according to equation 2.40. It was Lantz et al.[93], who pointed out that the lateral stiffness of the tip, k_{tip} can be important. Scanning transmission electron microscopy was used to determine the tip shape accurately. Finite element analysis was applied to determine the displacement at a given force. The lateral stiffness of commercial probing tips was found to be comparable to the lateral spring constant. For silicon levers, Lantz et al. found even a smaller lateral stiffness for the tip than for the cantilever $(k_{tip} < c_x)$[99].

Carpick et al. investigated the dependence of the lateral stiffness of the contact $k^x_{contact}$ as a function of loading for mica in humid air. They found good agreement with an extended Hertz model, also called Fogden and White

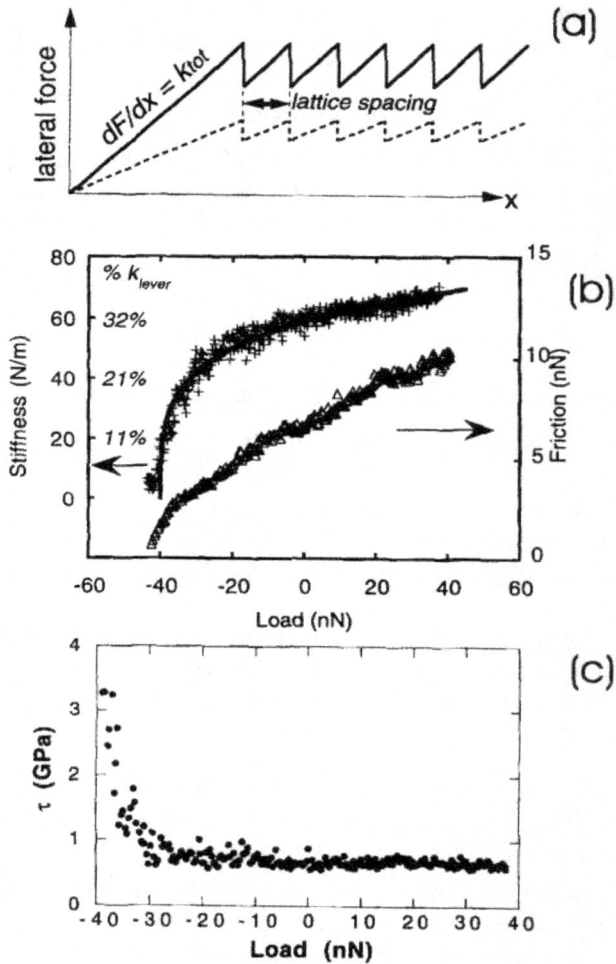

Figure 2.31: a) How to determine the lateral contact stiffness: Determine the slope dF/dx of the sticking part of a friction loop (lateral force vs. sample position x). Solid line: stiff contact. Dashed line: softer contact. b) Lateral stiffness vs. normal force. Solid line: a fit with a shifted Hertz model (Fogden and White). As the load increases, k_{eff}^x asymptotically approaches the lever spring constant $c_x \sim 190$N/m. c) Friction vs. normal force acquired shortly after the stiffness measurement. Bottom: The shear strength τ calculated from the data shown in the middle with equation 2.43. From[96].

model[90]. According to this model the loading axis has to be shifted by the pull-off force F_P. Thus, the contact stiffness is given by

$$k^x_{contact} = 8G^*(\frac{3R}{4E^*}(F_N + F_P))^{1/3} \qquad (2.42)$$

Lantz et al. applied the same method to NbSe$_2$ in ultrahigh vacuum conditions. They found that the contact radii calculated by the lateral contact stiffness measurement and independently by the Maugis-Dugdale theory (using a radius of curvature, determined from electron microscopy) were in good agreement (typical contact radius of 1-2nm). The loading dependence of the lateral contact stiffness and the corresponding contact radius was also found in good agreement with the Maugis-Dugdale theory[94].

One has to emphasize that the method of lateral contact stiffness measurements is a real extension to the more common friction measurements, which gives access to shear strength, when it is combined with normal friction measurements. Combining equations 2.36, 2.37, 2.28 the shear strength τ is given by:

$$\tau = \frac{64G^2 F_F}{\pi (k^x_{contact})^2} \qquad (2.43)$$

Thus, the lateral contact stiffness $k^x_{contact}$, combined with a friction force measurement F_F gives quantitative values of the shear strength, which depend less on the used continuum elasticity model than the normal stiffness measurement. The tip radius is not needed for the calculation. However, the tip shape has been assumed to be spherical or parabolic, which can be tested by scanning high aspect ratio features on surfaces or observed with an electron microscope. If the tip shape deviates, the prefactor in equation 2.36 has to be changed, accordingly[95].

2.8 Extensions of friction force microscopy: Nanosled experiments

An important disadvantage of common friction experiments using the FFM is the uncertainty about the real area of contact. The order of magnitude is about nm^2, but it is not simple to be calculated quantitatively. Methods such as normal and lateral stiffness measurements were described above, but are not yet routinely applied and are based upon continuum elasticity models. Therefore, measured friction forces after an exchange of probing tip can show rather large variations due to the different tip geometries, respectively, different contact areas.

An alternative technique is based upon the movements of an entire island by the action of the probing tip without destroying the island itself. However, this method is limited to weakly adsorbed islands, which exhibit the "nanosled" phenomenon. In other systems with stronger interfacial energies, the tip will start to scratch a line into the adsorbate island (step etching). The advantage of the nanosled technique is, that one knows the real area of contact between island and substrate, namely the area of the island itself, which is determined in the imaging mode. Second, one can measure the lateral force to move the island. Thus, the shear strength can be calculated.

The first example of nanometer-scale manipulation by FFM was the movement of a Cd-arachidate island on a Cd-arachidate-film[100]. E.g., the area of the island was $A = 4900nm^2$ and the force to move the island was F_{lat}=5nN, resulting in a shear strength of $\tau = F_{lat}/A$=1±0.2MPa. This value is compared to SFA-experiments, where monolayers of calcium stearate, deposited on mica sheets, were sheared across each other, yielding a shear strength of τ=1MPa.

The first example of nanometer-scale modification by FFM in UHV was the movement of C_{60}-islands on NaCl(001)[86,102], being several hundreds of nanometers in diameter. These islands can be rotated and translated in a controlled way. The lateral force is found to be proportional to the area of the island. A shear strength of $\tau = F_{lat}/A$=0.05-0.1MPa could be determined. This value is surprisingly low, even lower than the shear strength of boundary lubricants. Comparative experiments of C_{60} on graphite show that the interaction with this surface is stronger, being presumably related to the bonding of π-electrons. There, the islands cannot be moved by the action of the tip in a controlled way, but etching at step edges is observed. A second modification experiment is shown in Fig. 2.32 where an island is separated from another island (fracture experiment). The energy of 0.5keV±0.1keV divided by the number of molecules (3 layers times 125 molecules) gives an average energy of 1.5eV which corresponds well with the cohesive energy of C_{60}, as determined from sublimation experiments[103] and theoretical calculations[104].

In future, similar measurements with different materials should be performed. A direction dependence is expected for islands/substrate combinations with small mismatch. Thus, the influence of commensurability can be investigated on the submicron scale. Instead of pushing the islands from one side, the probing tip could be moved into the center of mass of the island, where rotation of the island cold be excluded.

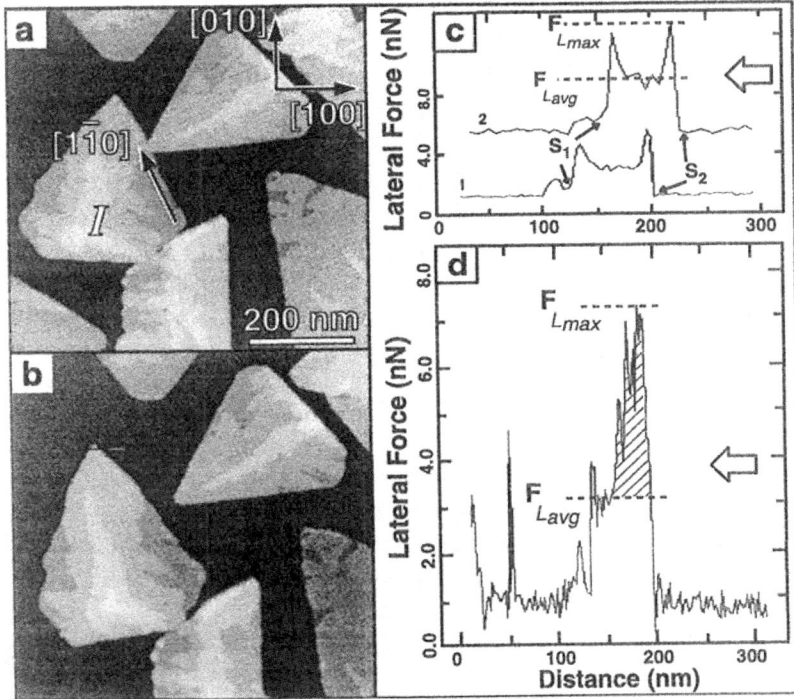

Figure 2.32: Nano-sled experiment. (a) and (b) show constant force images of C_{60}-islands deposited on NaCl(001). By the action of the probing tip, one island is moved over the substrate. (c) Lateral force during the nano-sled experiment. During the manipulation, it is observed that the lateral force is minimum on the substrate. Then, the lateral force increases to a maximum value $F_{l_{max}}$ until the island starts to move. During the motion of the island the lateral force equilibrates to an average value $F_{l_{av}}$. The lateral force increases again and the island stops. The tip experiences the frictional force on the C_{60}-island until it falls over the edge of the island and the minimum force on the substrate is observed again. (1) and (2) are two experiments with different islands. It is found that the lateral force is proportional to the island area A. Thus, the shear strength $\tau = 0.05$-0.1MPa could be determined. (d) Fracture experiment: Two C_{60}-islands, having a common boundary, are separated from each other. The energy needed to separate the islands is given by the striped area. In contrast to the movement of a free island, the shape of the curve resembles more to a triangle, showing that the force to break the islands decreases during the fracture-experiment.

2.8.1   Outlook

Friction force microscopy is a valuable tool for the characterization of heterogeneous surfaces. Material-specific contrast allows to identify different components. The contrast mechanisms are still not well understood.

However, it is possible to get more quantitative information about friction and wear. Calibration procedures are described, where the normal and
lateral spring constants are determined accurately. Even, the contact areas
can be well estimated from stiffness measurements and continuum elasticity
theory. It turns out, that long-range attractive forces are important, because
they have to be compensated by short-range repulsive forces. Thus, these
long-range forces determine the contact area and consequently the resolution.
In ambient conditions, capillary forces are dominant and limit the resolution
to about 5-10nm. In liquid environment, capillary forces are eliminated and
van der Waals forces can become repulsive: True atomic resolution is achievable. In ultrahigh vacuum, where van der Waals forces are always present, the
resolution in contact mode is about 1nm. In non-contact mode, true atomic
resolution can be achieved. The loading dependence of friction is investigated
with the 2d-histogram technique, where excellent statistics is achieved. In addition, comparison with the original images gives the opportunity to distinguish
between wear-less friction and wear processes.

Another type of experiment, that was performed with FFM, is the nano-
meter-sled experiment. Small islands are moved by the action of the probing
tip. Lateral forces are measured during the manipulation. Then, the shear
strength at the interface between island and substrate are measurable. The
following open questions might be of interest for future experiments: Is there
a dependence on the mismatch between island lattice and substrate lattice?
Are there angles, where friction disappears (incommensurate case) or does the
finite size of these objects make concepts of commensurability obsolete? What
happens when the size of the islands is shrinked below 10nm (single atom
limit)?

FFM is still a young technique. Many questions about the contrast mechanisms are unsolved. However, it opens a path to the fascinating world of
atoms and molecules. Today, microfabricated machines are already facing severe problems of high power dissipation. Adhesive forces become dominant at
those small scales. Boundary lubricants or novel materials are to be found,
that can be combined with microfabrication processes. FFM might be one

the instruments that yields information about the fundamentals of dissipation mechanisms in micro-mechanical systems and therefore can help to solve those problems in technology.

1. *Fundamentals of Friction*, edited by I.L. Singer and H.M. Pollock, p. 299-312, Kluwer Academic Publishers (1992)
   Terence J. Quinn, *Physical analysis for tribology*, Cambridge University Press (1991).
2. F.P. Bowden and D. Tabor, *Friction and Lubrication* , London, Methuen, revised edition (1967).
3. *Fundamentals of Friction*, edited by I.L. Singer and H.M. Pollock, Series E: Applied Sciences, Vol. 220, Kluwer Academic Publishers (1992)
   Physics of Sliding Friction, edited by B.N.J. Persson and E. Tosatti, Series E: Applied Sciences, Vol. 311, Kluwer Academic Publishers (1996)
   Micro/Nanotribology and Its Applications, edited by Bharat Bhushan, Series E: Applied Sciences, Vol. 330, Kluwer Academic Publishers (1997)
   Handbook of Micro/Nano Tribology, edited by Bharat Bhushan, CRC Series Mechanics and Materials Science, (1995)
4. Terence J. Quinn, *Physical analysis for tribology*, Cambridge University Press (1991).
5. F.P. Bowden and K.E.W. Ridler, Physical Properties of Surfaces, *Proc. Roy. Soc.* **152**, 640-656 (1935).
6. V. Novotny, J.D. Swalen and J.P. Rabe, Tribology of Langmuir-Blodgett Layers. *Langmuir* **5**, 485-489 (1989).
7. J.K. Lancaster, Solid Lubricants in CRC Handbook of Lubrication (Theory and Practice of Tribology), Volume II, Theory and Design, edited by E.R. Booser, CRC Press, Boca Raton, Florida, p. 269-290.
8. R. Stribeck, Die wesentlichen Eigenschaften der Gleit und Rollenlager, *Z. Ver. dt. Ing.* **46** No. 38, 1341-1348, 1432-1438; No. 39, 1463-1470 (1902).
9. D. Dowson in *Fundamentals of Friction*, edited by I.L. Singer and H.M. Pollock, p. 299-312, Kluwer Academic Publishers (1992)
10. B.J. Hamrock, Fundamentals of Fluid Film Lubrication, McGraw Hill, London, p. 474 (1994).
11. *Fundamentals of Friction*, edited by I.L. Singer and H.M. Pollock, p. 299-312, Kluwer Academic Publishers (1992).

12. W.B. Hardy and I. Doubleday, Boundary Lubrication - The Temperature Coefficient, *Proc. Roy. Soc.* **A 101** 487-492 (1922)
    W.B. Hardy and I. Doubleday, Boundary Lubrication - The Paraffin Series, *Proc. Roy. Soc.* **A 100** 550-574 (1922)
    W.B. Hardy and I. Doubleday, Boundary Lubrication - The Latents Period and Mixture of Two Lubricants, *Proc. Roy. Soc.* **A 104** 25-39 (1923)

13. E. Meyer, R.M. Overney, L. Howald, R. Lüthi, J. Frommer and H.-J. Güntherodt, *Phys. Rev. Lett.* **69**, 1777 (1992).

14. K.J. Wahl, M. Belin and I.L. Singer, A triboscopic investigation of the failure of MoS_2 in reciprocating sliding contact, *Thin Solid Films* (1997).

15. T.E. Fischer and H. Tomizawa, *Wear* **79**, 325 (1982).

16. T.E. Fischer, *Ann. Rev. Mater. Science* **18**, 303 (1988).

17. S. Jahanmir and T.E. Fischer, *Tribology Trans.* **31**, 32 (1988).

18. T.E. Fischer, in *Fundamentals of Friction*, edited by I.L. Singer and H.M. Pollock, p. 299-312, Kluwer Academic Publishers (1992)

19. R. Gohar and A. Cameron, Theoretical and Experimental Studies of the Oil Film in Lubricated Point Contacts, *Proc. Roy. Soc.* **A 291**, 520-536 (1966).
    R. Gohar and A. Cameron, The Mapping of Elastohydrodynamic Contacts, *ASLE Trans.*, **10**, 215-225 (1966).

20. C.R. Gentle, R.R. Duckworth and A. Cameron, Elastohydrodynamic Film Thickness at Extreme Presures, *ASME Journ. Lubr. Techn.*, **97**, 383-389 (1975).

21. F.J. Westlake, An Interometric Study of Ultra Thin Films, PhD Dissertation, London University, London (1970).

22. G.J. Johnston, R.C. Waynte and H.A. Spikes, The Measurement and Study of Very Thin Lubricant Films in Concentrated Contacts, *ASLE Trans.* **34**, 187-194 (1991).

23. D. Play and M. Godet, Visualization of chalk wear, in D. Dowson, M. Godet and C.M. Taylor (eds.), *The wear of non-metallic materials*, Mechanical Engineering Publications, London, p. 221 (1978).

24. P.M. Cann, H.A. Spikes, J. Hutchinson, The Development of a Spacer Layer Imaging Method (SLIM) for Mapping Elastohydrodynamic Contacts. *Tribology transactions* **39**, 915 (1996).

25. Q. Du, X.-d. Xiao, D. Charych, F. Wolf, P. Frantz, Y.R. Shen and M. Salmeron, Nonlinear optical studies of monomolecular films under pressure, *Phys. Rev.* **51**, 7456-7463 (1995).

26. Q. Du, E. Freysz and Y.R. Shen, *Science* **264**, 826 (1194).

27. S. Ong, X. Zhao and K.B. Eisenthal, *Chem. Phys. Lett.* **191**, 327 (1992).

28. J.P. Archard, *Tribology* **7**, 213 (1974).

29. D. Tabor, *Proc. Roy. Soc. London Ser.* **A 251**, 378 (1959).

30. M.F. Ashby, J. Abulawi and H.-S. Kong, *STLE Tribol. Trans.*, October (1991).

31. M.B. Peterson and S. Ramalingam in *Fundamentals of Friction and Wear of Materials*, edited by D. A. Rigney, p. 342, A.S.M., Metals Park OH, (1981)

32. I.L. Singer, in *Fundamentals of Friction*, edited by I.L. Singer and H.M. Pollock, p. 237-261, Kluwer Academic Publishers (1992)

33. M.B. Peterson, J.J. Florek and S.F. Murray, *Trans. ASLE* **2** 225 (1960).

34. M. Belin and J.M. Martin. *Wear*, **156**, 151 (1992).

35. M. Belin, J. Lopez and J.M. Martin. Triboscopy, a quantitative tool for the study o the wear of a coated material. *Surface and Coatings Technology*, **70**, 27-31 (1994).

36. C.F. McFadden and A.J. Gellman, Ultrahigh Vacuum Boundary Lubrication of the Cu-Cu Interface by 2,2,2-Trifluoroethanol, *Langmuir*, **11**, 273-280 (1995).
    A.J. Gellman, Lubrication by molecular monolayers at Ni-Ni interfaces, *J. Vac. Sci. Technol. A*, **10**, 180-187 (1992).
    Q. Dai an A.J. Gellman, *J. Am. Chem. Soc.*, **115**, 714-722 (1993).

37. D. Tabor and R.H.S. Winterton, *Proc. R. Soc. Lond.* **A312**, 435-450 (1969).

38. J.N. Israelachvili and D. Tabor, *Proc. R. Soc. Lond.* **A331**, 19-38 (1972).

39. J.N. Israelachvili and G.E. Adams, *J. Chem. Soc. Faraday Trans. I* **72**, 975-1001 (1978).

40. J. Israelachvili and P.M. Mc Guiggan, *Science* **241**, 795-800 (1988).

41. D. Rugar, H.J. Mamin, R. Erlandsson, J.E. Stern and B.D. Terris, Force microscope using a fiber-optic displacement sensor, *Rev. Sci. Instrum.* **59**,1045-1047 (1988).

42. A. Tonck, J.-M. Georges and J.L. Loubet, Measurements of Intermolecular Forces and the Rheology of Dodecane between Alumina Surfaces, *Journal of Colloid and Interface Science* **126**, 150-163 (1988).
    A. Tonck, F. Houze, L. Boyert, J.-L. Loubet and J.-M. Georges, Electrical and mechanical contact between rough gold surfaces in air, *J.*

Phys.: Condensed Matter **3**, 5195-5201 (1991).

J.P. Montfort, A. Tonck, J.L. Loube and J.M. Georges, Microrheology of High-Polymer Solutions, *J. of Polymer Science* **29**, 677-682 (1991).

43. J.-M. Georges, A. Tonck and D. Mazuyer, Interfacial friction and adhesion of wetted monolayers, in *Force in Scanning Probe Methods*, edited by H.-J. Güntherodt, D. Anselmetti and E. Meyer, p. 263, Kluwer Academic Publishers (1995).

44. J.M. Georges, S. Millor, J.L. Loube and A. Tonck, Drainage of thin liquid films between relatively smooth surfaces, *J. Chem. Phys.* **98**, 7345-7360 (1993).

45. J.N. Isralachvili, P.M. McGuiggan and A.M. Homola, *Science* **240**, 189 (1988).

46. A.M. Homola, J.N. Israelachvili, P.M. Mc Guiggan and M.L. Gee, Fundamental Experimental Studies in Tribology: The transition from interfacial friction of undamaged molecularly smooth surfaces to normal friction with wear, *Wear* **136**,65-83 (1990).

47. J. Van Alsten and S. Granick, Molecular Tribometry of Ultrathin Liquid Films, *Phys. Rev. Lett.* **61**, 2570-2573 (1988).

48. P.A. Thompson and M.O. Robbins, Origin of Stick-slip motion in Boundary Lubrication, *Science* **250**, 792-794 (1990).

    M.O. Robbins and P.A. Thompson, *Science* **253**, 916-918 (1991).

49. H. Yoshizawa, Y.-L. Chen and J. Israelachivili, Fundamental Mechanisms of Interfacial Friction 2: Stick-Slip Friction of Spherical and Chain Molecules, *J. of Phys. Chem.* **97**, 11300 (1993).

50. H. Yoshizawa, Y.-L. Chen and J. Israelachivili, Mechanisms of Interfacial Friction 1: Relation between Adhesion and Friction, *J. of Phys. Chem.* **97**, 4128 (1993).

51. A. Marti, G. Hähner, N. D. Spencer, Sensitivity of Frictional Forces to pH on a Nanometer Scale: A Lateral Force Microscopy Study, *Langmuir* **11**, 4632 (1995).

52. T. Palberg and K. Streicher, Resonant stick-slip motion in colloidal crystal, *Nature* **367**, 51-54 (1994).

    T. Palberg, R. Simon, M. Würth and P. Leiderer, Colloidal suspensions as model liquids and solids, *Proc. Colloid Polym. Sci.* **96**, 62-71 (1994).

53. J. Krim and A. Widom, *Phys. Rev. B* **38**, 12184 (1988).

    J. Krim, D.H. Solina and R. Chiarello, *Phys. Rev. Lett.* **66**, 181 (1991).

    J. Krim and R. Chiarello, *J. Vac. Sci. Technol* **A 9**, 2566 (1991).

E.T. Watts, J. Krim and A. Widom, *Phys. Rev. B* **41**, 3466 (1990).

C. Daly and J. Krim, Sliding friction of solid xenon monolayers and bilayers on Ag(111), *Phys. Rev. Lett.* **76**, 803-806 ((1996).

C. Daly and J. Krim, Sliding friction of compressing xenon monolayers, in *Micro/Nantotribology and Its Applications* edited by B. Bhushan, p. 311-316 , Kluwer Academic Publishers (1997).

54. C. Mak, C. Daly and J. Krim, *Thin Solid Films* **253**, 190 (1994).

55. B.N.J. Persson, Surface resistivity and vibrational damping in adsorbed layers, *Phys. Rev. B* **44**, 3277-3296 (1991).

56. C. Daly and J. Krim in *Atomic Force Microscopy/Scanning Tunneling Microscopy* p. 303, Ed. S.H. Cohen et al., Plenum Press, New York (1994).

57. G. Binnig, C.F. Quate and Ch. Gerber, Atomic force microscope, *Phys. Rev. Lett.* **56**, 930-933 (1986).

58. For an overview in force microscopy see: *Forces in Scanning Probe Methods*, Eds. H.-J. Güntherodt, D. Anselmetti and E. Meyer, NATO ASI Series E: Applied Sciences Vol. 286, Kluwer Academic publishers (1995).

59. C.M. Mate, G.M. McClelland, R. Erlandsson, and S. Chiang, Atomic-Scale Friction of a Tungsten Tip on a Graphite Surface, *Phys. Rev. Lett.* **59**, 1942-1945 (1987).

60. For recent reviews see e.g.:

R. Overney, and E. Meyer, *MRS Bulletin*, **18**, 26 (1993)

I.L. Singer in *Dissipative Process in Tribology*, Edts. Dowson, D., Taylor, C.M., Childs, T.H.C., Gopdet, M. and Dalmaz, G., Proceedings of the 20th Leed-Lyon Symposium on Tribology, Villeurbanne, 7-10 Sept (1993)

E. Meyer, R. Overney, and J. Frommer in *Handbook of Micro/Nanotribology*, Edt. B. Bhushan, CRC Press Inc. (1994)

O. Marti, Nanotribology: Friction on a Nanometer Scale, *Physica Scripta* **T49**, 599-604 (1993).

E. Meyer, R. Lüthi, L. Howald and H.-J. Güntherodt, p. 285 in *Forces in Scanning Probe Methods*, Eds. H.-J. Güntherodt, D. Anselmetti and E. Meyer, NATO ASI Series E: Applied Sciences Vol. 286, Kluwer Academic publishers (1995). J. Krim, *Comments Condens. Mater. Phys.*, **17** 263 (1995).

B. Bhushan, J.N. Israelachvili and U. Landman, *Nature* **374**, 607 (1995).

C.M. Mate, Force microscopy studies of the molecular origins of friction and lubrication, *IBM Journal of Research and Development*, **39**, 617-627 (1995).

R.W. Carpick and M. Salmeron, Scratching the Surface: Fundamental Investigations of Tribology with Atomic Force Microscopy, *Chemical Reviews* **97**, 1163-1194 (1997).

61. For instrumental aspects see e.g.:

    O. Marti, J. Colchero, and J. Mlynek, *Nanotechnology*, **1**, 141 (1990).

    G. Meyer and N. Amer, *Appl. Phys. Lett.* **57** 2089 (1990).

    L. Howald, E. Meyer, R. Lüthi, H. Haefke, R. Overney, H. Rudin, and H.-J. Güntherodt, *Appl. Phys. Lett.*, **63**, 117 (1993).

62. G. Neubauer, S.R. Cohen, G.M. McClelland, D.E.Horn and C.M. Mate, *Rev. Sci. Instrum.* **61**, 2296 (1990).

63. O. Marti, J. Colchero and J. Mlynek, Combined scanning force and friction microscopy of mica, *Nanotechnology* **1**, 141-144 (1990).

64. G. Meyer and N. Amer, Simultaneous measurement of lateral and normal forcs with an optical beam deflection atomic force microscope, *Appl. Phys. Lett.* **57**, 2089 (1990).

65. G.M. McClelland and J.N. Glosli, Friction at the atomic scale, in *Fundamentals of Friction: Macroscopic and Microscopic Processes* edited by I.L. Singer and H.M. Pollock, p. 405-425, NATO ASI Series E: Applied Sciences, Vol. 220, Kluwer Academic Publishers (1992).

66. D. Rugar, H.J. Mamin, P. Güthner, Improved fiber optic interferometer for atomic force microscopy, *Appl. Phys. Lett.* **55**, 2588-2590 (1989).

67. J. Brugger, J. Burger, M. Binggeli, R. Imura and N.F. de Rooij, Lateral force meaurements in a scanning force microscope with piezoresistive sensors, Proceedings ofthe 8th International Conference on Solid-State Sensors and Actuators and Eurosensors IX, Stockholm, Sweden June 25-29, 636-639 (1995).

68. R. Kassing and E. Oesterschulze, "Sensors for Scanning Probe Microscopy, in *Micro/Nantotribology and Its Applications* edited by B. Bhushan, p. 35-54, Kluwer Academic Publishers (1997).

69. G.M. McClelland, R. Erlandsson and S. Chiang, Atomic force microscopy: general principles an a new implementation, in *Review of Progress in Quantitative Non-Destructrive Evaluation*, edited by D.O. Thompson and D. E. Chimenti (Plenum, New York, 1987), Vol. 6B, pp. 1307-1314.

70. J.B. Pethica, Comment on Interatomic Forces in Scanning Tuneling Microscopy: Giant Corrugations of the Graphite Surface, *Phys. Rev. Lett.* **57**, 3235 (1986).

71. R. Lüthi, E. Meyer, M. Bammerlin, L. Howald, H. Haefke, T. Lehmann, C. Loppacher, H.-J. Güntherodt, T. Gyalog and H. Thomas. Friction on the atomic scale: An ultrahigh vacuum atomic force microscopy study on ionic crystals, *J. Vac. Sci. Technol. B* **14**, 1280-1284 (1996).

72. B. Briscoe and D.C.B. Evans, The shear properties of Langmuir-Blodgett layers, *Proc. R. Soc. London A* **380**, 389-407 (1982).
B.J. Briscoe and A.C. Smith, The interfacial shear strength of molybdenum disulfide and graphite films, *ASLE Transactions* **25**, 349-354.

73. C.A.J. Putmann, M. Igarshi and R. Kaneko, Single-asperity friction in friction force microscopy: The composite-tip model, *Appl. Phys. Lett.* **66**, 3221-3223 (1995).

74. U.D. Schwarz, W. Allers, G. Gensterblum and R. Wiesendanger, Low-load friction behaviour of epitaxial C_{60} monolayers under Hertzian contact, *Phys. Rev. B* **52**, 14976-14984 (1995).

75. R.W. Carpick, N. Agrait, D.F. Ogletree adn M. Salmeron, Measurement of interfacial shear (friction) with an ultrahigh vacumm force microscope, *J. Vac. Sci. Technol. B* **14**, 1289-1295 (1996).

76. E. Meyer et al. in *Physics of Sliding Friction*, edited by B.N.J. Persson and E. Tosatti, Series E: Applied Sciences, Vol. 311, Kluwer Academic Publishers (1996).

77. R. Lüthi, E. Meyer, H. Haefke, L. Howald, W. Gutmannsbauer, M. Guggisberg, M. Bammerlin and H.-J. Güntherodt. *Surf. Sci.* **338**, 247 (1995).

78. J.N. Israelachvili, *Intermolecular and Surface Forces*, Academic Press, London (1985).

79. F.O. Goodman and N. Garcia, Roles of the attractive and repulsive forces in atomic-force microscopy, *Phys. Rev. B* **43**, 4728 (1991).

80. F. Ohnesorge and G. Binnig, *Science* **260**, 1451 (1993).

81. F.J. Giessibl, Atomic Resolution of the Silicon(111)7x7 Surface by Atomic Force Microscopy, *Science* **267**, 68-71 (1995).

82. Y. Sugawara, M. Ohta, H. Ueyama and S. Morita, Defect Motion on an InP(110) Surface Observed with Noncontact Atomic Force Microscopy, *Science* **270**, 1646 (1995).

83. L. Howald, H. Haefke, R. Lüthi, E. Meyer, G. Gerth, H. Rudin and H.-J. Güntherodt, *Phys. Rev. B* **49**, 5651 (1994).

84. L. Howald, R. Lüthi, E. Meyer, and H.-J. Güntherodt, Atomic force microscopy on the Si(111)7x7, *Phys. Rev. B* **51**, 5484 (1995).

85. R. Lüthi et al., Atomic resolution in dynamic force microscopy across steps on Si(111)7x7, *Z. Phys. B* **100**, 165-166 (1996).

86. R. Lüthi, H. Haefke, E. Meyer, L. Howald, H.-P. Lang, G. Gerth and H.-J. Güntherodt, *Z. Phys. B.*, **95**, 1 (1994).

87. J.B. Pethica and A.P. Sutton. Nanomechanics: Atomic Resolution and frictional energy dissipation in atomic force microscopy, p. 353 in *Forces in Scanning Probe Methods*, Eds. H.-J. Güntherodt, D. Anselmetti and E. Meyer, NATO ASI Series E: Applied Sciences Vol. 286, Kluwer Academic publishers (1995).

88. G.J. Germann, S.R. Cohen, G. Neubauer, G.M. McClelland and H. Seki, Atomic scale friction of a diamond on diamond(100) and (111) surfaces, *J. Appl. Phys.*, **73**, 163-167 (1993).

89. S.P. Jarvis, A. Oral, T.P. Weihs and J.B. Pethica, *Rev. Sci. Intstrum.* **64**, 3515 (1993).

90. A. Fogden and L.R. White, *J. Colloid Interface Sci.* **138**, 414 (1990).

91. K.L. Johnson, K. Kendall and A.D. Roberts, *Proc. Royal Soc. London* **A 324**, 301 (1971).

92. D. Maugis *J. Colloid Interface Sci.* **150**, 243 (1992).

93. M.A. Lantz, S.J. O'Shea, A.C.F. Hoole and M.E. Welland, Lateral stiffness of the tip and tip-sample contact in frictional force microscopy, *Appl. Phys. Lett.*, **70**, 970-972 (1997).

94. M.A. Lantz, S.J. O'Shea, M.E. Welland and K.L. Johnson, Atomic force microscope study of contact area and friction on $NbSe_2$, *Phys. Rev. B*, **55**, 10776 (1997).

95. K.L. Johnson, *Contact Mechanics*, Cambridge University Press, Cambridge, United Kingdom, (1985).

96. R.W. Carpick, D.F. Ogletree and M. Salmeron, Lateral stiffness: A new nanomechanical measurement for the determination of shear strengths with friction force microscopy, *Appl. Phys. Lett.* **70**, 1548-1550 (1997).

97. J. Colchero, M. Luna and A.M. Baro, Lock-in technique for measuring friction on a nanometer scale, *Appl. Phys. Lett.*, **68**, 2896-2898 (1996).

98. J. Colchero, M. Luna and A.M. Baro, Energy dissipation in scanning force microscopy - friction on an atomic scale, *Tribology Letters*, **2**, 327-343 (1996).

99. Lantz et al. found for a silicon tip: $c_n=1.1N/m$; $c_x=110N/m$; $k_{tip}=84N/m$ and for a Si_3N_4-tip: $c_n=0.6N/m$; $c_x=8.2N/m$; $k_{tip}=39N/m$

100. E. Meyer, R.M. Overney, L. Howald, R. Lüthi, J. Frommer and H.-J. Güntherodt, Friction and Wear of Langmuir-Blodgett Films Observed by Friction Force Microscopy, *Phys. Rev. Lett.*, **69**, 1777-1780 (1992).

101. E. Meyer, L. Howald, R. Overney, D. Brodbeck, R. Lüthi, H. Haefke, J. Frommer and H.-J. Güntherodt, *Ultramicroscopy* **42-44**, 274 (1992).

102. R. Lüthi, E. Meyer, H. Haefke, L. Howald, W. Gutmannsbauer and H.-J. Güntherodt, *Science* **266**, 1979 (1994).

103. C. Pan, P. Sampson, R. Chai, H. Hauge, and J.L. Margrave, *J. Phys. Chem.* **95**, 2945 (1991).

104. Ph. Lambin, A.A. Lucas, and J.-P. Vigneron, *Phys. Rev. B* **46**, 1794 (1992).

Chapter 3

Normal forces at the atomic scale

3.1   Important forces between atoms and molecules

There are four known fundamental forces in nature: the strong, weak interaction, electromagnetic and gravitational interactions. The strong and the weak interaction are very short ranged ($\approx 10^{-15}$m) and describe the behavior of protons, neutrons and other elementary particles. The gravitational interaction is long ranged and deals with the motion of planets, moons, satellites and causes phenomena like tides and the weight of earth's creatures. Gravitational interaction is about forty orders of magnitude weaker than the electromagnetic interaction. This is the reason why it is less important on a microscopic scale. In the world of atoms and molecules the electromagnetic interaction is dominant and leads to a rich variety of forces.

Ionic bonds

These are simple Coulombic forces which are a result of the electron transfer. For example in lithium fluoride the lithium transfers its 2s-electron to the fluorine 2p-state. Consequently the shells of the atoms are filled up, but the lithium has a net positive charge and the fluorine a net negative charge. These ions attract each other by Coulombic interaction which stabilizes the ionic crystal in the rocksalt structure.

Covalent bonds

The standard example for a covalent bond is the hydrogen atom. When the wavefunction overlap is considerable, the electrons of the hydrogen atoms will be indistinguishable. The total energy will be decreased by the "exchange energy", which causes the attractive force. The characteristic property of covalent bonds is a concentration of the electron charge density between the two nuclei. The force is strongly directed and falls off within a few Å.

Metallic adhesive forces

The strong metallic bonds are only observed when the atoms are condensed in a crystal. They originates from the free valency electron sea which holds together the ionic cores. A similar effect is observed when two metallic surfaces approach each other. The electron clouds have the tendency to spread out, in order to minimize the surface energy. Thus a strong exponentially decreasing, attractive interaction is observed.

Van der Waals forces

Van der Waals forces exist between all kind of atoms or molecules and can be divided into three groups:

- *Dipole-dipole force:* Molecules having permanent dipoles will interact by dipole-dipole interaction.

- *Dipole-induced dipole forces:* The field of a permanent dipole induces a dipole in a non-polar atom or molecule.

- *Dispersion forces:*

  Due to charge fluctuations of the atoms there is an instantaneous displacement of the center of positive charge against the center of negative charge. Thus at a certain moment a dipole exists and induces a dipole in another atom. Therefore non-polar atoms (e.g. neon) or molecules attract each other.

The attractive van der Waals force between the atoms is proportional to $1/r^7$, where r is the distance between the atoms. The van der Waals forces are effective from a distance of a few Å to several hundreds of Å. Due to the finite velocity of light the dispersion forces are retarded for distances more than 100Å, which leads to a different power law index ($\approx 1/r^8$).

3.2 Important forces between probing tip and sample

3.2.1 Van der Waals forces

For distances above a few Å to hundreds of Å, van der Waals forces are significant. They can be used to measure topography with a resolution of a few nm[1]. Whether atomic resolution can be achieved with these forces is still unclear, because these forces are smeared out on the atomic scale and have shown no atomic structure up to now. The interaction between different geometries, such as plane against plane, sphere against plane and crossed cylinders, can be calculated by integration, assuming additivity, simple power laws result and the Hamaker constants specific for the materials can be defined. For example, the force between a sphere and a plane is $F(D) = -AR/6D^2$ where R is the radius, D the distance between the top of sphere and the plane, and A the Hamaker constant. The assumption of the additivity ignores the existence of multiple reflections. Multiple reflections occur when atom A induces a dipole in atom B. At the same moment the field of atom A polarizes also another atom C. The induced dipole of atom C, influences atom B. Therefore the field of atom A reaches atom B directly and via reflection from atom C. The Lifshitz theory has overcome the problem of additivity[2]. It is a continuum theory which neglects the atomic structure. The input parameters are the dielectric constants and refractive indices. The interpretation of the Lifshitz theory goes beyond the scope of this book. The main point we want to emphasize here is that the results of the Lifshitz theory are in qualitative agreement with the results deduced by simple pairwise integration. The power laws remain the same, but the Hamaker constants are different. A detailed treatment has been performed by Israelachvili[3] with the surface force apparatus. Moiseev et al.[4], Hartmann et al.[5] and Hutter et al.[6] performed AFM experiments.

3.2.2 Magnetic forces

Magnetic microstructure plays an important role in science and technology. All of the data storage of present computers is based on magnetic recording media. Therefore much effort has been made to investigate magnetic forces in AFM. This kind of force microscopy has developed into an independent field and is called magnetic force microscopy (MFM). Basically, a ferromagnetic tip probes the domain wall structure of a ferromagnetic sample. At present most of the measurements have been performed at a distance of a few hundred of Ångstroms. Therefore we can neglect most of the other forces, which decrease more rapidly. Due to the superposition principle, the magnetic forces can be summed up and a theory has been developed which explains most of the experimental results.

3.2.3   Electrostatic forces

Similarly to magnetic forces, we can apply the Maxwell theory to the treatment of the interaction between a charged tip and the charged regions of an insulator. Stern et al.[7] studied the dissipation of a localized charges on a polymer surface, which has been created by a voltage pulse. A deviation of the rate of charge dissipation from macroscopic values was found. Terris et al.[8] deposited charges by mechanical contact (without voltage pulse) between tip and sample and measured the charge distribution as well as the ratio of positive and negative charge as they sat side by side on the sample. Another application is the observation of ferroelectric domain walls by electrostatic forces[9]. The transport of single charges was studied by Schönenberger and Alvarado with contact electrification experiments on PMMA[10]. They could demonstrate that the force sensitivity is sufficient to detect electrostatic forces of single charge carriers.

3.2.4   Capillary forces

It is well known that microcontacts act as nuclei of condensation. In air water vapor plays the dominant role. If the radius of curvature of the microcontact is below a certain critical radius a meniscus will be formed. This critical radius is defined approximately by the size of the Kelvin radius $r_K = 1/(1/r_1 + 1/r_2)$ where r_1 and r_2 are the radii of curvature of the meniscus. The Kelvin radius is connected with the partial pressure p_s by the equation

$$r_K = \frac{\gamma V}{RT log(p/p_S)}$$

where γ is the surface tension, R the gas constant, T the temperature, V the mol volume and p_S the saturation vapor pressure[11]. The surface tension γ of water is 0.074N/m (T=20°C) which gives the parameter $\gamma V/RT$=5.4Å. Therefore we obtain for p/p_s=0.9 a Kelvin radius of 100Å. For smaller vapor pressures the Kelvin radius gets comparable to the dimensions of the molecules and therefore the Kelvin equation is not applicable anymore. Now the question arises of what will happen in the case of STM and AFM. Typical tips having radii of less than 1000Å are possible nuclei of condensation. If such a meniscus is formed an additional capillary force acts on the tip. A simple estimation is given by the equation:

$$F = \frac{4\pi R\gamma \cos\Theta}{1 + D/(R(1 - \cos\phi))}$$

where R is the radius of curvature, Θ the contact angle, D the distance between tip and sample and ϕ the angle of the meniscus[3]. The maximum force is given by $F_{max} = 4\pi R\gamma \cos\Theta$. For a tip radius of 1000Å we obtain a force $F_{max} = 9.3 \cdot 10^{-8}$N, which is a rather high value for typical forces in AFM. Typical force vs. distance curves in ambient conditions reveal forces of the order of 10^{-8}-10^{-7}N, which mainly originates from capillary forces. These adhesion forces limit the minimum force which acts on the outermost tip region, because these attractive forces have to be equilibrated with the repulsive force in this small contact region. Therefore capillary forces can play an essential role in STM and AFM measurements in air. At least in humid air, the presence of a meniscus has to be taken into account. A reduction of capillary forces is achieved by covering hydrophilic sample and probing tip with amphiphilic molecules that render the surfaces hydrophobic. Then, the surfaces are not covered by thick water films and the capillary forces are reduced. Immersing the whole tip-sample area under liquids can completely remove capillary interaction.

3.2.5   Short-range forces

Magnetic, electrostatic, van der Waals and capillary forces have in common that they are quite long ranged. Therefore a large part of the tip will contribute to the force. The definition of this "effective tip" is somewhat arbitrary, but is usually done with a sphere of a radius of a few hundreds of Å. This sphere corresponds to a homogeneously magnetized, polarized or charged part of the tip (domain). The size of this sphere is chosen to be closely related to the observed AFM resolution.

In the case of the short ranged forces the "effective tip" is smaller. To get real atomic resolution the size should be comparable to the atomic dimensions. For distances below a few Å the tip and sample are in contact. It is assumed that the force microscope follows closely the contours of the total charge density. Thus the images can be identified with the topography. We have to bear in mind that the assumption is only valid when the surface is homogeneous enough that the influence of the local variations in elasticity is not strong. An assumption which has held up to now for many applications on a nm-scale. On the atomic scale several effects seem to contribute to the image contrast.

1. The *Pauli repulsion and ionic repulsion* plays an important role. These forces hinder the tip from penetrating the sample, and give a simple explanation for the observed atomic corrugation, which arises from the variations of total charge density.

   The repulsive forces have two origins: First, the strong overlap of the two electron clouds leads to incomplete screening of nuclear charges, which causes a Coulombic repulsion. Second, according to the Pauli exclusion principle, two electrons having the same quantum numbers cannot occupy the same place. Thus the electrons can only overlap when the energy of one electron is increased, which yields the repulsive force. The repulsion forces are very short ranged ($\approx$Å) and can be described by a power law $F = const/r^n$ or by an exponential function, where n is greater than 8.

2. *Physisorption and chemisorption:* Bonds can be formed between tip and sample. In some materials this will give an additional attractive force which varies on an atomic scale.

3. *Metallic adhesion:* This short ranged interaction has been studied in the most detail so far by force microscopy. This interaction becomes important when two metals come so close to each other that their individual electron wave functions overlap and electron exchange becomes important. This is a quantum mechanical effect and leads to an attraction which is exponentially dependent on distance. Metallic adhesion has been theoretically studied extensively by Ferrante and co-workers[12] who found a universal behavior of the adhesion energy for different metals. Chen has postulated that a close relationship between tunneling conductance and force should exist[13]. With a force microscope-type set-up which involved a rigid STM tip and a sample mounted on a cantilevered beam, Dürig et al. were able to measure metallic adhesion directly for different tip and sample metals[14].

4. *Friction:* The component of the force parallel to the surface, called frictional force, can be quite large. It has been shown by Mate et al.[15] that these frictional forces vary on an atomic scale. See chapter II and IV.

5. *Elasticity:* Elastic deformations can play an important role in the image contrast. The theory of Tomanek et al.[16] predicts that variations in the local elasticity can be observed by AFM on layered materials, such as intercalated graphite.

6. *Plastic deformation:* Landman et al. studied the tip sample interaction in AFM- and STM-type geometries by molecular dynamics calculations[18]. Their simulations show strong plastic deformations for some metal-metal and silicon-silicon arrangements, often accompanied by complete collapse of the original tip and sample surface geometries. These results indicate that there are limits on suitable tip materials in applications to specific sample materials. These calculations provide one of the most valuable theoretical tools for understanding the processes in AFM and STM on an atomic scale.

The above classification is very useful to describe qualitatively the observed phenomena. In the next section we will expand on these topics from a more fundamental point of view, allowing a more quantitative description.

3.3 Microscopic description of the tip-sample contact

The aim of this section is to give examples of the theoretical description of the tip-sample interaction. The simplest way to describe an interaction is the application of empirical potentials such as the Lennard-Jones or the Stillinger-Weber potential. Although the actual choice of the potential is rather arbitrary, the parameters can be determined by fitting them to certain experimental data or by adjusting them to the results of *ab initio* calculations. These empirical potentials will be described in more detail. If an empirical potential has been chosen, it can be used for molecular dynamics (MD) calculations. In MD-calculations the Newtonian equations of motions are solved. For large systems, including 100 to 1000 atoms, the calculations need sophisticated numerical methods. Some MD-calculations will be discussed. A different approach has been shown by Tomanek et al.. They developed a continuum elasticity theory for the special case of layered materials. The elastic constants were determined by *ab initio* calculations, which do not rely on empirical potentials and have no adjustable parameters. They essentially start with the Schrödinger-equation and inherit the full quantum mechanics. In Sect. III 3.4 some examples of *ab initio* calculations will be discussed.

3.3.1 Empirical potentials

The interaction energy between N identical atoms is described by a potential function $\Phi(1,\ldots N)$. Φ can be split up in one-body, two-body, three-body, etc.

contributions:

$$\Phi(1,\ldots N) = \sum_i v_1(i) + \sum_{i,j;i<j} v_2(i) + \sum_{i,j,k;i<j<k} v_3(i,j,k) + \ldots + v_N(1,\ldots N)$$

The single particle interaction v_1 describes the interaction with external fields and v_2 is the usual pair potential, v_3 describes the three-body interaction, which will be important for the following discussion. A priori it is not clear that the higher terms will decrease with increasing order, but it has been shown that for insulators, semiconductors like Si, Ge and semi-metals like graphite the lower terms dominate. For metals, where we have strongly delocalized conduction electrons, the situation is different. Probably the most famous two-body potential is the Lennard Jones (LJ) potential:

$$v_2(r) = -3\varepsilon \left(\left(\frac{\sigma}{r}\right)^{12} - \left(\frac{\sigma}{r}\right)^6 \right)$$

the parameter ε is the minimal potential energy at the equilibrium distance $2^{1/6}\sigma$[19]. The LJ potential has the common $1/r^6$ van der Waals power law for large distances and describes the ionic repulsion reasonable. It has been applied to describe phase transitions of noble gases (Ar, Kr, Xe). Simple two-body interactions, such as the LJ potential, are not adequate for covalent materials, because they do not take into account the directionality of forces (bond-angle forces). A striking evidence for this inadequacy is the prediction of two-body forces, that all solids should have a close-packed structure, whereas other structures like the diamond structure would be unstable. Several authors have proposed empirical potentials which take into account three-body forces. Among these are the Keating potential[20] and the Stillinger-Weber (SW) potential[21,22]. Chelikowsky et al.[23] introduced an improved potential which prevents discontinuities of the angle dependence. The rest of this section will describe the SW potential in more detail, because it will be used in the molecular dynamics calculations. The Stillinger-Weber potential takes into account two and three-body forces. It has been introduced to investigate the melting process of Silicon[21]. Similar to the LJ potential the parameter ε, σ are introduced, which are chosen to give v_2 the depth $-\varepsilon$ and to minimize v_2 at the distance $r_{ij}=2^{1/6}\sigma$. The two body part v_2 is

$$v_2(r_{ij}) = \varepsilon \cdot f_2(r_{ij})/\sigma.$$

the reduced function $f_2(r_{ij})$ is

$$f_2(r_{ij}) = \begin{cases} A(Br_{ij}^{-p} - r_{ij}^{-q})\exp((r_{ij} - a)^{-1}), & r_{ij} < a \\ 0, & r_{ij} \geq a \end{cases}$$

where A, B, p and q are positive real numbers. This function cuts off at $r_{ij}=a$ without discontinuities. The three body part is

$$v_3(\vec{r}_i, \vec{r}_j, \vec{r}_k) = \varepsilon \cdot f_3(\vec{r}_i/\sigma, \vec{r}_j/\sigma, \vec{r}_k/\sigma)$$

with the reduced function

$$f_3(\vec{r}_i, \vec{r}_j, \vec{r}_k) = h(r_{ij}, r_{ik}, \Theta_{jik}) + h(r_{ji}, r_{jk}, \Theta_{ijk}) + h(r_{ki}, r_{kj}, \Theta_{ikj})$$

where Θ_{jik} is the angle between $\vec{r}_j$ and $\vec{r}_k$ originated at the vertex i. The function h has two parameters λ and γ and has the form

$$h(r_{ij}, r_{ik}, \Theta_{jik}) = \lambda \exp(\gamma(r_{ij} - a)^{-1} + \gamma(r_{ik} - a)^{-1}) \cdot (\cos(\Theta_{jik} + 1/3)^2$$

for r_{ij}, $r_{ik} < a$, otherwise h vanishes. The part with the angle dependence clearly favors bonds having tetrahedral angles Θ_t, because $\cos(\Theta_t)=-1/3$. There are several ways to determine the parameters $A, B, \lambda, \gamma, p, q$. One way is to fit the potential to certain available experimental data like the lattice constants, several elastic constants, cohesive energies, melting temperature, etc. Another way would be to use ab initio calculations to determine those parameters.

3.3.2 Molecular dynamics

Abraham et al.[24] have investigated the relaxation of a Si(001)-(2x1) surface in the presence of a silicon tip by using molecular dynamics (MD) based calculations. Fig. 3.1 shows some of the results of the total normal forces which act on several types of tips (monomer T_1, dimer T_2, trimer T_3 and tetramer T_4)) placed at a vertical distance of ≈ 2Å. The corresponding forces are of the order of $\approx 10^{-9}$ N. Clearly the tip geometry changes the occurrence of the AFM images. The multiatom tip image can be produced by linear superposition of single atom tip images, if no surface relaxation is allowed. When the surface is allowed to relax the things get more complicated and linear superposition is not valid anymore. The authors found qualitatively the same appearance of the images, but the forces were reduced by the relaxation. It was found that with forces of the order of 10^{-9}N the AFM image represents approximately the undisturbed surface, whereas with large repulsive forces the surface relaxes significantly and the observed image deviates from the real structure.

Landman et al.[25] have applied MD calculations to investigate the interaction between a silicon tip and the unreconstructed Si(111) surface. The input for these calculations are the initial positions of the atoms and certain parameters for an interatomic potential. In this case, the Stillinger-Weber(SW)

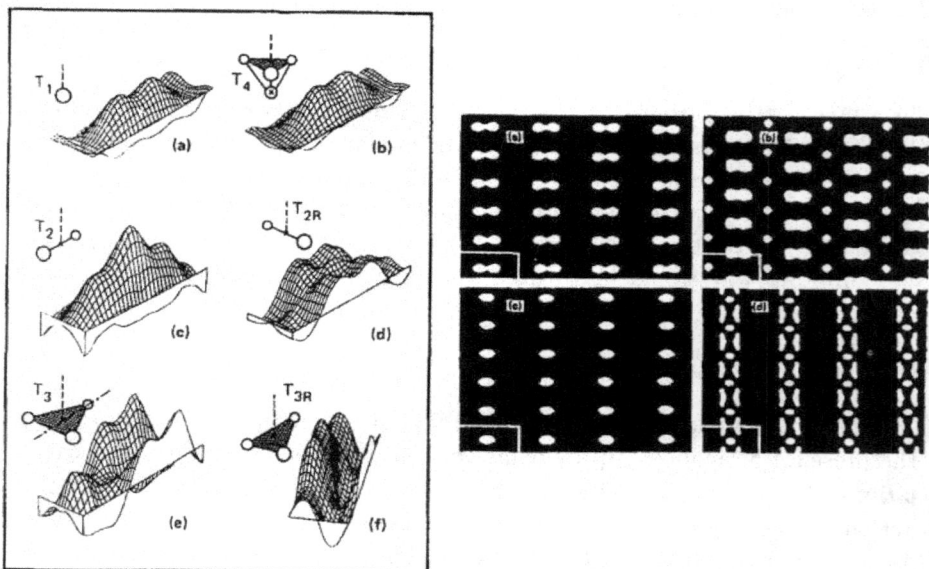

Figure 3.1: (I) Influence of the tip geometry on the AFM contrast. For each tip ($T_1...T_3$) the corresponding force dependence on the Si(100)2x2 is reproduced. The relaxation of the sample was not considered in this case. (II) Black-white AFM-picture for several tip geometries (a) T_1 (b) T_3 (c) T_2 (d) T_{2R} From[24].

potential has been used. In MD the atoms follow the Newtonian laws of motion, which are solved numerically. Typical time steps are $\approx 10^{-15}$sec and the system is equilibrated in a few hundred steps. The sample consists of 4-8 layers of about 100 atoms per layer. Several kind of tips have been employed such as a cluster of 1-4 Si atoms, a more realistic, ordered tip consisting of about 100 atoms and a disordered tip of 100 atoms.

In chapter IV some of the results from Landman are illustrated. Constant height scans were simulated with the 4 atom tip in the $\langle 1\bar{2}1 \rangle$ direction. The components of the forces, which are acting on the rigid tip, are plotted for all three directions. Although the tip is in contact with the surface, which would have expected to yield a strong repulsion, attractive as well as repulsive forces are observed. The attractive force is connected with the creation of surface interstitials. Thus the tip can induce a rearrangement of surface atoms, which results in a different recorded force. When the tip advances, the defect anneals again. Constant force simulation are shown for the ordered and disordered 100 atom tips. In this case the tip atoms were allowed to relax. An interesting stick and slip behavior is observed. The tip atoms remain in energetically favored positions, until the forces of the moving rest of the tip exceeds the forces of the sample. Then the tip atoms jump approximately one unit cell. This stick-slip behavior is very well observed in the lateral force F_x.

Abraham et al.[26] have applied a similar theory for the case of graphite. SW-potentials have been used to calculate the relaxation of a graphite layer in the presence of a single carbon atom. For a force of $\approx 2\cdot 10^{-8}$N the first layer is deformed 2Å and the fourth layer is still deformed 0.5Å. For forces above $\approx 5\cdot 10^{-8}$N the tip punctures the graphite layer. The authors conclude that atomic resolution for experiments with forces of the order of 10^{-7}N, cannot be explained with monatomic tips. They suggest that probably a more complex tip, such as a graphite flake, will yield such atomic scale features.

In summary we take the following conclusions:

- MD calculations are able to simulate the dynamics of the scan process and to include temperature effects.

- For forces of the order of 10^{-9}N the AFM image reflects the surface topography (contours of total charge density). Tip induced atomic relaxations lead to a reduction of the corrugation.

- Forces of the order of 10^{-8}N can lead to a rearrangement of the tip or substrate atoms. The creation of interstitials in silicon are observed, which leads to a different contrast. For the layered materials, such as graphite, a force of $2 \cdot 10^{-8}$N is sufficient, that the tip penetrates this layer.

- Ordered as well as disordered tips, having realistic dimensions, lead to force variations on an atomic scale, but the image can be altered by multiple tip effects. Multiple tips can explain the high loadings, which were measured experimentally.

- Single defects can be observed, but during the scan process complicated rearrangements of the sample atoms are going on.

- For layered materials such as graphite, a significant elastic deformation of the layers have been calculated, whereas for the non-layered silicon no long-ranged elastic deformations have been found.

- The force components in x,y,z directions are calculated. In certain cases the lateral components can be compared to or even exceed the normal force variations in the z direction. When the tip atoms are allowed to relax, pronounced stick-slip behavior is observed, which corresponds to variations of the frictional force on an atomic scale.

On the other hand we have to add the following critical remark to these calculations:

- Due to the limited computer power only time scales of the order of 10^{-12}sec can be investigated. The experimental time scale is of the order of seconds. The question arises if the simulated ultrafast scans of the theory can be readily compared to the quasistatic experimental scans?

- Tip and sample consist of the same material, whereas in experiments the tip is usually built of a hard material like diamond, amorphous SiO_2 or Si_3N_4. Therefore the simulations which allow the relaxation of the tip correspond to a more reactive tip than in the experiments. On the other hand the simulations which do not allow the relaxation of the tip, overestimate the rigidity of the tip. Probably the truth lies somewhere in between.

- Up to now the calculations are based on empirical potentials and the Newtonian equations of motion. Therefore it has to be shown if *ab initio* calculations yield the same results. Car and Parinello[28,29] have introduced a new method, which combines MD calculations with ab initio calculations of the interaction potentials. An application to the case of AFM might give new insights.

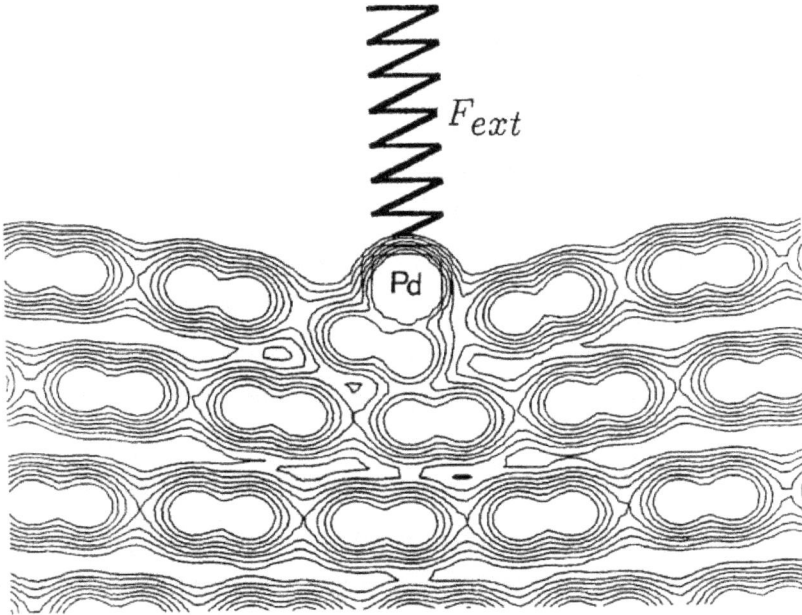

Figure 3.2: By means of continuum elasticity the deformation of the graphite surface has been calculated. The force is 10^{-8}N. The contour lines of the charge density has been calculated by linear superposition. The contour line correspond to following values of charge density: $1.00 \cdot 10^{-3}(1)$, $1.25 \cdot 10^{-2}(2)$, $5.00 \cdot 10^{-2}(3)$, $1.00 \cdot 10^{-1}(4)$ und $2.00 \cdot 10^{-1}(5)$ in units of $e/(a.u.)^3 = 1.2 \cdot 10^{-18}$C/Å^2 (Courtesy of G. Overney).

3.3.3 Continuum elasticity theory

Tomanek and G. Overney et al.[16,17] have presented a theory of deformable surfaces being suitable to calculate AFM forces on graphite with or without intercalant. The layers of graphite are substituted by elastic plates which are described by elastic constants. Essentially three elastic constants are sufficient namely the flexural rigidity D, the transversal rigidity K and the compressibility G along the c-axis. The vertical displacement $w_1(\vec{r})$ of the first layer as a function of the force distribution $F_1(\vec{r})$ is given by the differential equation

$$(D \bigtriangledown^4 - K\bigtriangledown^2)w_1 + G(w_1 - w_2) = F_1(\vec{r})$$

The distortion of the lower layers is described by

$$(D \bigtriangledown^4 - K\bigtriangledown^2)w_n + G(-w_{n-1} + 2w_n - w_{n+1}) = F_n(\vec{r})$$

for $n \geq 2$. The authors have indicated an analytical solution for the distortion $w_n(\vec{r})$ provided that $F_1(\vec{r})$ is a δ-function and $F_n(\vec{r})=0$ for $n \geq 2$ (cf. Fig. 3.2). Furthermore the model has been extended to cylindrical tips. By means of ab initio calculations the elastical constants have been determined. Experimentally small variations of the tip position as a function of the force can be measured, whence the local elastical constants can be determined. Especially in the vicinity of defects such as steps or intercalants It is expected that a change of the rigidity D occurs. These changes of the rigidity should be measurable by AFM. Therefore a new physical property, the local elasticity, can be determined by AFM.

3.3.4   Ab initio calculations

The previous models rely on empirical potentials. These potentials are used in combination with classical concepts like the Newtonian equations of motion or continuum elasticity theory. This approach has the big advantage that the simplified mathematics allow the treatment of complex systems. Furthermore temperature effects and dynamical processes can be included.

Ab initio calculations have proven to be applicable to determine band structures, to investigate surface reconstructions[27] and to derive certain parameters for empirical potentials[16]. In the field of STM the charge density distributions of several surfaces have been calculated[30,31]. In combination with the Tersoff-Hamann theory[32,33] the current densities for the STM could be derived. One important conclusion which can be drawn from recent AFM and certain STM experiments (giant corrugation on graphite[34], atomic resolution on Au(111), Al(111)[36]) is that the interaction of the tip with the surface can lead to significant disturbances in the electronic and atomic structure. It will be shown that tip-induced electronic states can be calculated and that the forces for certain distances between a monatomic tip and the sample can be derived.

Present ab initio calculations are based on density functional theory introduced by Hohenberg and Kohn[37]. They showed that the ground state total energy of a many-electron system in the field of the ions can be written as a functional of the electron charge density $\rho(\vec{r})$. This functional has a minimum value for the correct charge density $\rho(\vec{r})$, provided that the total number of electrons is conserved. Furthermore Kohn and Sham[38] have shown that the many-electrons system can be mapped exactly onto a system of non-interacting electrons moving in an effective potential. Consequently the problem is reduced to the solution of the Kohn-Sham equations:

$$\left(\frac{-\hbar^2}{2m} \Delta + V_{ion} + V_H + V_{xc} \right) \psi_i = \varepsilon_i \psi_i$$

where V_H is the Hartree potential, V_{ion} is the ionic potential, ψ_i are the one-electron eigenstates and ε_i the eigenvalues. Unfortunately the function $E_{xc}(\rho)$, which determines V_{xc} by $V_{xc} = \delta E_{xc}(\rho)/\delta \rho$, is unknown. A common approximation, the local density approximation (LDA), gives good results for the total energies. In the local density approximation the exchange correlation potential is calculated by assuming that the electrons between locally like a uniform electron gas.

The Kohn-Sham equations must be solved self-consistently. The potential of the ionic cores V_{ion} are commonly described by pseudopotentials[39]. For the expansion of the pseudowavefunctions plane waves or Gaussian-type fuctions are used. In order to describe the tip-sample system in AFM a repeated slab model has been introduced. Each slab consists of several layers of the sample (typically 1-4 layers) and is separated by a vacuum region. As shown in Fig. 3.3 the slab is extended in such a way that the tip is part of the system.

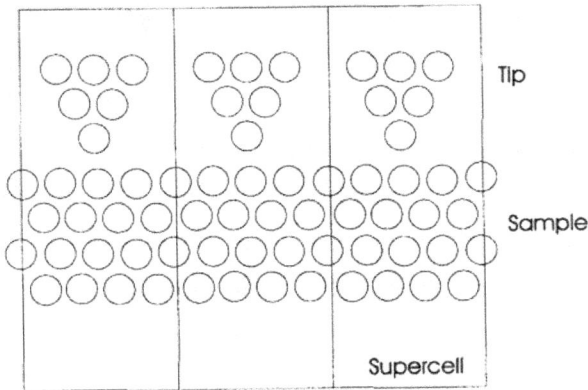

Figure 3.3: For the AFM calculations a supercell is needed, consisting of a several layers of the sample, a vacuum gap and a tip.

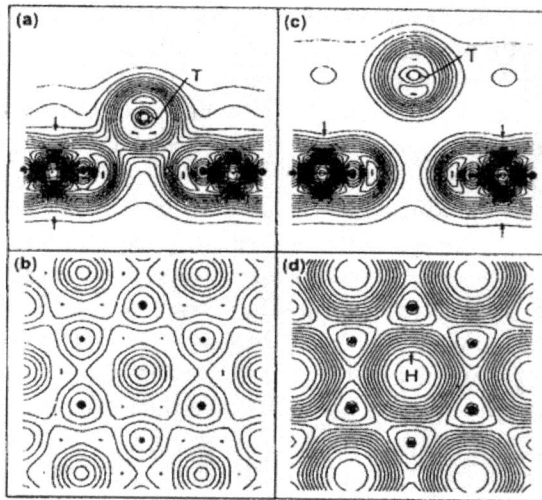

Figure 3.4: Contour plot of the charge density of a tip atom being positioned at two distance above the H site of graphite. (a) and (c) are cuts perpendicular (b) and (d) parallel to the layer (From[40]).

The main disadvantage of this model is caused by the artificial interaction between the adjacent tips. This interaction can be reduced by increasing the size of the supercell. Batra and Ciraci[40] have applied these calculations to the case of a monatomic carbon tip in close proximity to the graphite sample. The 3-layer slab with a vacuum gap of 10.05Å and a single carbon tip at several heights has been investigated. The distance between the carbon tips is 2.46Å. The close separations of the tips and the corresponding interaction between the tips limits the validity of the calculations to distances less than ≈ 3Å. Fig. 3.4 shows the contour plots of the total charge density of the tip-surface system at two distances (h=1.43Å, 2.49Å). The forces acting on the tip on top of different sites have been calculated and some predictions about the z-dependence of the corrugation could be made. Recently calculations with enhanced accuracy and larger supercells have been performed by Ciraci, Baratoff and Batra[41] for a Al-tip on graphite, an Al-tip on Al-surface and by Overney et al.[17] for a Pd-tip on graphite. Furthermore a theory has been proposed[44], which uses *ab initio* calculations to explain atomic scale friction. Friction coefficients have been derived which agrees with experimental values[15].

In summary the following conclusions can be drawn:

- With ab initio calculations selected atomic configurations of the tip-sample system can be investigated. Out of this the distance dependence of the force between tip and sample can be determined. But we have to bear in mind that the ab initio calculations determine only the ground state. The influence of finite temperature and the dynamical processes are neglected.

- Although the calculations are rather involved numerically they inherit a rich variety of physical properties, such as the band structure, charge density distributions and the total energy of the ground state, which can be applied for different purposes. The calculations are specially useful in combinations with empirical models in order to determine certain parameters which are not available by experiments.

3.4   True atomic resolution with normal forces

It was Ohnesorge and Binnig, who were able to get the first true atomic resolution on calcite in contact mode. They immersed the surface under water, which reduced attractive forces, such as capillary forces and van der Waals forces, down to the order of 10pN[46]. Atoms were observed on the terrace and at the step edges. Thus, one could conclude that atomic resolution is achieved in liquids due to the reduced attractive forces. However, atomic resolution in air was not achieved, because of the presence of capillary forces and van der Waals forces. In vacuum, van der Waals are always attractive. Estimations of Goodman and Garcia gave values of 1-10nN, which is too large to get a single point contact[47]. The attractive forces will cause a pressure of the order of GPa in the contact zone. Thus, the tip or sample will be deformed elastically or plastically in order to increase the contact area and to reduce the pressure. Only in liquids, van der Waals forces can become small or even repulsive[3] and a single atom contact appears to be possible in contact mode. See also chapter IV for resolution limits with lateral forces.

In contrast to the contact-mode AFM, where atomic-lattice imaging was achieved rather early, non-contact AFM, also called dynamic force microscopy was not promising at the beginning. The first attempts by Martin et al.[1], McClelland et al.[48] and Nonnenmacher et al.[49] gave resolutions in the 10nm range. The contrast was related to van der Waals forces, which were expected to be smeared out on the atomic scale. It was Giessibl, who could first achieve atomic resolution on the Si(111)7x7 with the nc-AFM[50]. Rather high amplitudes of about 300 Å were used for the cantilever oscillation and the cantilever resonance was detected with FM-detection[51]. The microscope was operated in the constant frequency mode. The presented image showed part of the unit cell. Thus, true atomic resolution of this complex surface structure was achieved. However, the operation conditions were found to be unstable, which was related to tip changes. Günthner presented true atomic resolution on Si(111)7x7, where an average tunneling current[52], $\overline{I_t}$, was used as input for the the feed-back loop (dynamic lever STM-mode). The unstable behaviour in the constant frequency mode (nc-AFM) was related to variations of frequency shift on the atomic scale not allowing stable frequency shift imaging. Lüthi et al. presented frequency shift vs. distance curves[53], where three regimes are observed:

Regime I: Long-range attractive forces, such as van der Waals and electrostatic forces, cause frequency shifts at distances as far as a few hundred Ångstroms.

Regime II: Short-range forces cause frequency shifts over a distance of approximately a nanometer. A minimum of the frequency shift is observed. True atomic resolution can be achieved with these short-range chemical forces.

Regime III: Repulsive contact forces become dominant. The conditions of

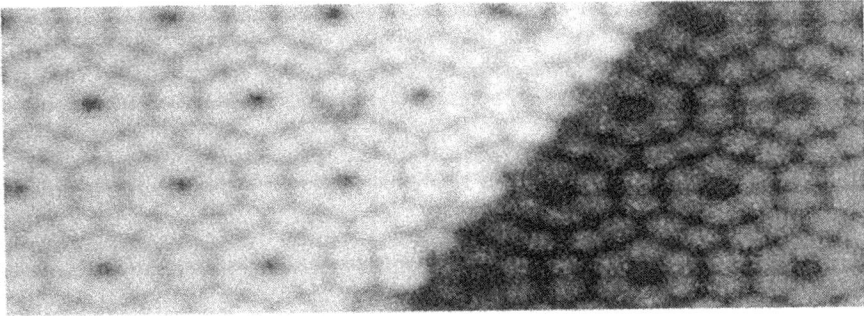

Figure 3.5: High resolution resonance frequency shift image of the Si(111)7x7 (8 x 3.5 nm^2) surface reconstruction recorded in the vicinity of a monatomic step. A time averaged tunneling current of 25pA was used as feedback input. From[53].

the tip/sample geometry are not stable and plastic deformation is observed. This regime corresponds to the more conventional contact mode.

Using dynamic lever STM-mode, Lüthi et al. could acquire frequency shift images across a step site. Larger frequency shifts (more attractive) were observed on the lower terrace. The transition regime between upper and lower terrace was found to be atomically sharp. The explanation of these frequency shift differences between upper and lower terrace (FREDUL) were related to long-range forces[54]. On the lower terrace, the average interaction volume is bigger compared to the upper terrace. It becomes obvious, that it is rather difficult to run the feed-back in the constant frequency shift mode, because of the local changes across step sites. However, on small areas, without step sites, true atomic resolution could be achieved in the constant frequency mode (nc-AFM-mode)[55]. With the appropriate choice of a small frequency shift on the lower terrace, it is also possible to run nc-AFM across a step site without touching on the upper terrace.

In contrast to the dynamic lever STM-mode, the frequency shift in nc-AFM-mode above the adatoms was found to be stronger than above the corner holes. The observation, which is made in nc-AFM mode, is in agreement with molecular dynamics simulations of Perez et al.[56]. Thus, the strongest attraction is above the adatom sites, where the silicon tip forms a weak chemical bond with the sample. Similar observations were also made by Erlandsson et al., where the contrast of nc-AFM was related to the local reactivity of the sample[57]. Erlandsson et al. observed also a difference between the adatoms close to corner holes compared to center adatoms, where the nc-AFM was op-

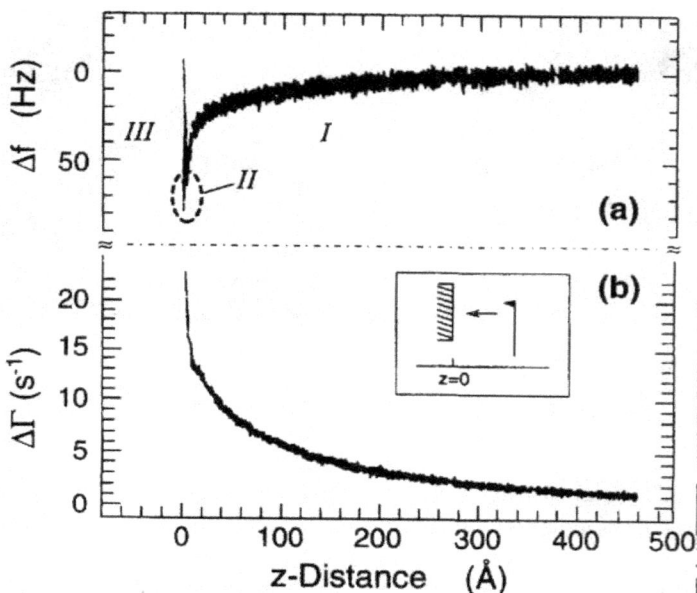

Figure 3.6: Distance dependence of frequency shift (a) and damping rate (b) of a silicon probing tip above NaCl(001). Long-range forces, such as van der Waals and electrostatic force determine regime I. Chemical forces are dominant in regime II, where true atomic resolution can be achieved. Repulsive contact forces take over in regime II, where plastic deformation of both probing tip and sample can occur. From[59].

erated with slope detection (fixed excitation frequency and varying amplitude) and a tungsten tip was used. The difference between nc-AFM mode (most attractive force above adatoms) and dynamic lever STM-mode (most attractive force above corner holes), observed by Lüthi et al.[53] and Guggisberg et al.[54] is not so trivial to be explained. First, STM might be responsible due to changes of interaction distances (STM depends on electronic effects and does not keep the distance constant). Recent experiments have shown that this effect is not the primary reason. However, it is found that STM works in the branch of the frequency shift vs. distance curve with negative slope (repulsive part), whereas nc-AFM can be operated in the part with the positive slope (attractive part). Thus, nc-AFM with true atomic resolution is operated at even lower forces than typical STM operation.

Other applications of nc-AFM on semiconductors were presented by Sugawara et al.[58] on InP(001), where point defects were observed. The first application of nc-AFM on an insulator was presented by Bammerlin et al.[59]. The

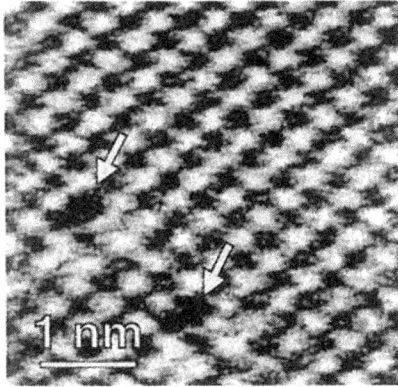

Figure 3.7: Non-contact force microscopy on NaCl(001). Two point defects are observable. The spacing between the periodically arranged bright spots corresponds to the distance between equally charged ions. From theoretical calculations, it is concluded that the contrast originates from the sodium ions. From[59].

unreconstructed surface of NaCl(001) was observed. Only one ionic species was visible. From theoretical studies, it was concluded that the more attractive force is above the Na^+. The probing tip is a silicon tip (argon sputtered), which is assumed to be negatively charged (dangling bond). Then, the most attractive force is above the positive sodium ions. In addition, a pair of defects was observed, exhibiting stronger attraction than the rest of the imaged surface. The pair of defects was found to be stable on the time scale of minutes. After 80 minutes, the defects were found to jump one atomic spacing, which is either related to thermal motion or due to the action of the probing tip. The high stability of the defects at room temperature, the asymmetric shape and the occurence in pairs, makes it plausible that the defects are OH^--centers. Recently, Landman et al. presented molecular dynamics simulations, where the NaCl-surface was predicted to play an important role in the catalytic reaction with water[60].

With the achievement of true atomic resolution, nc-AFM has made a real break-through. Insulators, semiconductors and metals are accessible to the application of nc-AFM. The comparison of STM and AFM on conductive surfaces might be important for the understanding of STM-operation. The surface physics of insulators is still at the beginning. Possible fields of studies are: Unknown surface reconstruction of insulators, the physics and chemistry of color centers, catalytic reactions and thin oxide films of semiconductors.

1. Y. Martin, C.C.Williams and H.K. Wickramasinghe, Atomic force microscope-force mapping and profiling on a sub 100-Å scale, *J. Appl. Phys.* **95**, 4723-4729 (1987).

2. L.D. Landau and E.M. Lifshitz, *Statistical Physics*, Volume 1. Pergamon, Oxford (1980).

3. J.N. Israelachvili, *Intermolecular and Surface Forces*, Academic Press, London (1985).

4. Yu.N. Moiseev, V.M. Mostepanenko, V.I. Panov and I.Yu. Sokolov, *Phys. Lett. A* **132**, 354 (1987).

5. U. Hartmann, *Phys. Rev. B* **42**, 1541 (1990).

6. J.L. Hutter and J. Bechhoefer, *Rev. Sci. Instrum.* **64**, 1868-1873 (1993).

7. J.E. Stern, B.D. Terris, H.J. Mamin and D. Rugar, *Appl. Phys. Lett.* **53**, 2717 (1988).

8. B.D. Terris, J.E. Stern, D. Rugar and H.J. Mamin, *Appl. Phys. Lett.* **63**, 2669 (1989).

9. F. Saurenbach and B.D. Terris, *Appl. Phys. Lett.* **56**, 1703 (1990).

10. C. Schönenberger and S. Alvarado, *Phys. Rev. Lett.* **65**, 3162 (1990).

11. A.W. Adamson, *Physical Chemistry of Surfaces*, John Wiley Sons, (1976).

12. J. Ferrante and J.R. Smith, *Phys. Rev. B* **31**, 3427 (1985).

13. C.J. Chen, *J. Phys. Cond. Matter* **3**, 1227-1245 (1985).
    C.J. Chen, *Introduction to Scanning Tunneling Microscopy*, Oxford Series in Optical and Imaging Sciences, Oxford University Press (1993).

14. U. Dürig, O. Züger and D.W. Pohl, *Phys. Rev. Lett.* **65**, 349 (1990).

15. C.M. Mate, G.M. McClelland, R. Erlandsson, and S. Chiang, Atomic-Scale Friction of a Tungsten Tip on a Graphite Surface, *Phys. Rev. Lett.* **59**, 1942-1945 (1987).

16. D. Tománek, G. Overney, H. Miyazaki, S.D. Mahanti and H.-J. Güntherodt, Theory for the atomic force microscopy of deformable surfaces, *Phys. Rev. Lett.*, **63**, 876-879 (1989).

17. G. Overney, W. Zhong and D. Tomanek, Theory of elastic tip-surface interaction in AFM, *J. Vac. Sci. Techn. B* **9**, 479-482 (1991).

18. U. Landman, W.D. Luedtke, N.A. Burnham and R.J. Colton, *Science* **248**, 454 (1990).

19. J.E. Lennard-Jones, *Trans. Faraday Soc.* **28**, 334 (1932).

20. P.N. Keating, *Phys. Rev.* **145**, 637 (1966).

21. F.H. Stillinger and T.A. Weber, *Phys. Rev. B* **31**, 5262 (1985).

22. Zi Jian, Zhong Kainig and Xie Xide, *Phys. Rev. B* **41**, 12915 (1990).

23. J.R. Chelikowsky, J.C. Phillips, M. Kamal and M. Strauss, *Phys. Rev. Lett.* **62**, 292 (1989).

24. F.F. Abraham, I.P. Batra and S. Ciraci, *Phys. Rev. Lett.* **60**, 1314 (1988).

25. U. Landman, W.D. Luedtke and M.W. Ribarsky, *J. Vac. Sci. Technol. A* **7**, 2829 (1989).

26. F.F. Abraham and I.P. Batra, *Surf. Sci.* **210**, L177 (1989).

27. M.C. Payne, N. Roberts, R.J. Needs, M. Needels and J.D. Joannopoulos, *Surf. Sci.* **211/212**, 1 (1989).

28. R. Car and M. Parrinello, *Phys. Rev. Lett.* **55**, 2471 (1985).

29. F. Ancilotto, W. Andreoni, A. Selloni, R. Car and M. Parrinello, *Phys. Rev. Lett.* **65**, 3148 (1990).

30. Inder P. Batra, N. Garcia, H. Rohrer, H. Salemink, E. Stoll and S. Ciraci, *Surf. Sci.* **181**, 126 (1987).

31. D. Tomanek and S.G. Louie, *Phys. Rev. B* **37**, 8327 (1988).

32. J. Tersoff and D.R. Hamann, *Phys. Rev. Lett.* **50**, 1998 (1983).

33. J. Tersoff and D.R. Hamann, *Phys. Rev. B* **31**, 805 (1985).

34. JJ.M. Soler, A.M. Baro, N. Garcia and H. Rohrer, *Phys. Rev. Lett.* **57**, 444 (1986).

35. H. Ohtani, R.J. Wilson, S. Chiang and C.M. Mate, *Phys. Rev. Lett.* **60**, 2398 (1988).

36. J. Wintterlin, J. Wiechers, H. Brune, T. Gritsch, H.Höfer and R.J. Behm, *Phys. Rev. Lett.* **62**, 59 (1989).

37. P. Hohenberg and W. Kohn, *Phys. Rev. B* **136**, 864 (1964).

38. W. Kohn and L.J. Sham, *Phys. Rev. A* **140**, 1133 (1965).

39. D.R. Hamann, M. Schlüter and C. Chiang, *Phys. Rev. Lett.* **43**, 1494 (1979).

40. I.P. Bata and S. Ciraci, *J. Vac. Sci. Techn. A* **6**, 313 (1988).

41. S. Ciraci, A. Baratoff and I.P. Batra, *Phys. Rev. B* **41**, 2763 (1990).

42. W. Zhong, G. Overney and D. Tomanek, *Europhys. Lett.* (1991).

43. D. Tomanek, W. Zhong and H. Thomas, *Europhys. Lett.* **15**, 887 (1991).
    D. Tomanek, p. 269 in *Scanning Tunneling Microscopy III*, Eds. R. Wiesendanger and H.-J. Güntherodt, Springer Berlin (1993).

44. W. Zhong and D. Tomanek, *Phys. Rev. Lett.* **64**, 3054 (1990).

45. A.L. Weisenhorn, P. Maivald, H.-J. Butt and P.K. Hansma, *Phys. Rev. B* **45**, 11226-11232 (1992).

46. F. Ohnesorge and G. Binnig, *Science* **260**, 1451 (1993).

47. F.O. Goodman and N. Garcia, *Phys. Rev. B* **43**, 4728 (1991).

48. G.M. McClelland, R. Erlandsson and S. Chiang, in *Review of Progress in Quantitative Non-Destructrive Evaluation*, edited by D.O. Thompson and D. E. Chimenti (Plenum, New York, 1987), Vol. 6B, p. 1307.

49. M. Nonnenmacher, J. Greschner, O. Wolter, and R. Kassing, *J. Vac. Sci. Technol. B* **9** 1358 (1991).

50. F.J. Giessibl, *Science* **267**, 68 (1995).

51. T.R. Albrecht, P. Grütter, D. Horne, and D. Rugar, J. Appl. Phys. **69**, 668 (1991)

52. P. Güthner, *J. Vac. Sci. Technol. B* **14**, 2428 (1996).

53. R. Lüthi et al., *Z. Phys. B* **100**, 165 (1996).

54. M. Guggisberg et al., to appear in *Phys. Rev. B* (1998).

55. R. Lüthi et al., *Surf. Rev. Lett.* **4**, 1025-1029 (1997).

56. R. Perez, M.C. Payne, I. Stich and K. Terakura, *Phys. Rev. Lett.* **78**, 678 (1997).

57. R. Erlandsson, L. Olsson and P. Martensson, *Phys. Rev. B* **54**, 8309 (1996).

58. Y. Sugawara, M. Ohta, H. Ueyama and S. Morita, *Science* **270**, 1646 (1995).

59. M. Bammerlin et al., *Probe Microscopy* **1**, 3 (1997). **1**, 1 (1997).

60. R.N. Barnett and U. Landman, *Phys. Rev. Lett.* **76**, 2302 (1996).

Chapter 4

Understanding of lateral forces

4.1 Geometrical effects: The role of topography

Lateral force maps of surfaces that are not atomically smooth are often dominated by the topography. We distinguish two cases:

1) The local slope of the topography, $\partial s/\partial x$ causes a lateral force $F_{topo} = \mu_{topo} \cdot F_N \cdot cos\delta \approx \mu_{topo} \cdot F_N$, where $F_N = F_L + F_A$ is the total normal force, F_L the externally applied loading force, F_A the attractive force, $\mu_{topo} = tan\delta = \partial s/\partial x$ and δ is indicated in Fig. 4.1. The approximation is valid for small slopes with $\delta << 1$. In principle, this kind of lateral force would even occur for frictionless sliding, originating simply from a component of the normal force. In contrast to frictional forces, the lateral force F_{topo} does not depend on scan direction and can be distinguished by examining the forward and backward scan. This effect is observed on homogeneous, rather smooth surfaces, where the radius of curvature of the probing tip is small compared with the surface roughness. Small hillocks of polycarbonate on a compact disc are shown in Fig. 4.1, where lateral forces independent of scan direction are found. The derivative image of the topography is in good agreement with the back- and forward scan. Similar examples are described in the literature[1].

2) The second topography effect is often dominating over the previously described topography effect and is related to local variations of the contact area. At steep slopes of the topography, where the radius of curvature becomes comparable with the curvature of the local topography, the contact area is changed and causes a direct increase or decrease of friction. In addition, the local topography can also lead to a change of attractive forces, such as van der Waals forces or capillary forces. As drawn schematically in Fig. 4.2, the van der Waals force on a summit of a hillock is weaker than in a valley, where the interaction volume is bigger. The larger attractive force has to be compensated by a larger repulsive contact force and leads to a larger contact zone and thus increases friction. Capillary forces play an analogous role: They are increased

in the valley, where the liquid film thickness is increased, too. The frictional forces depend on the total normal force by $F_F = \pm\mu \cdot (F_{load} + F_a)$ and on the scan direction. This observation was first made by Mate on carbon coatings[2]. An example is given in Fig. 4.2 showing a typical contrast reversal with scan direction. Interestingly, this observation is made on the same compact disc as in Fig. 4.1. Contrast reversal occurs at the steep edges of the bits of the compact disc.

In summary, the total lateral force is given by a scan direction dependent and a scan direction independent component:

$$F_{Lateral} = \pm\mu F_N + \partial s/\partial x F_N \qquad (4.1)$$

Both components are closely related to the local topography and are superimposed on most surfaces.

4.2 Step edges and Schwoebel barriers

Friction at step edges can be seen as a special case of the previously discussed topography effects. However, the classical treatment is not generally applicable. Other atomistic effects can become important at steps of small heights. Therefore, the phenomena that occur at step edges are discussed in this section in more detail.

On surfaces, which are measured under ambient conditions, large lateral forces of up to 0.1 to 50nN are measured at step edges. It is well known, that steps are often covered by contaminants (hydrocarbons, water...), causing capillary forces or interactions via physisorption to the tip. In order to exclude these contaminants, measurements have to be performed in well-defined environments. Binggeli et al.[3] performed FFM-measurements on potential controlled graphite in an electrolyte, where rather high lateral forces of 5 to 40nN were found at the steps. Marti et al.[4] found that the lateral forces at step edges strongly depend on the applied potential and thus can be minimized for a certain value. A simple explanation might be that the contaminants are removed under optimum conditions, leading to minimized values of friction. Another explanation is related to local changes of the Helmholtz layer.

The first investigation of steps on NaCl(001) in ultrahigh vacuum was performed by Meyer and Amer[5]. They found lateral forces of $4 \cdot 10^{-10}$N at the step edges, showing distinct contrast reversal with scanning direction. An energy of $\approx$50eV was found to be dissipated at the step edge. The authors suggested that the energy dissipation is related to a change of potential energy, when the tip climbs up the step. However, force microscopes are operated in

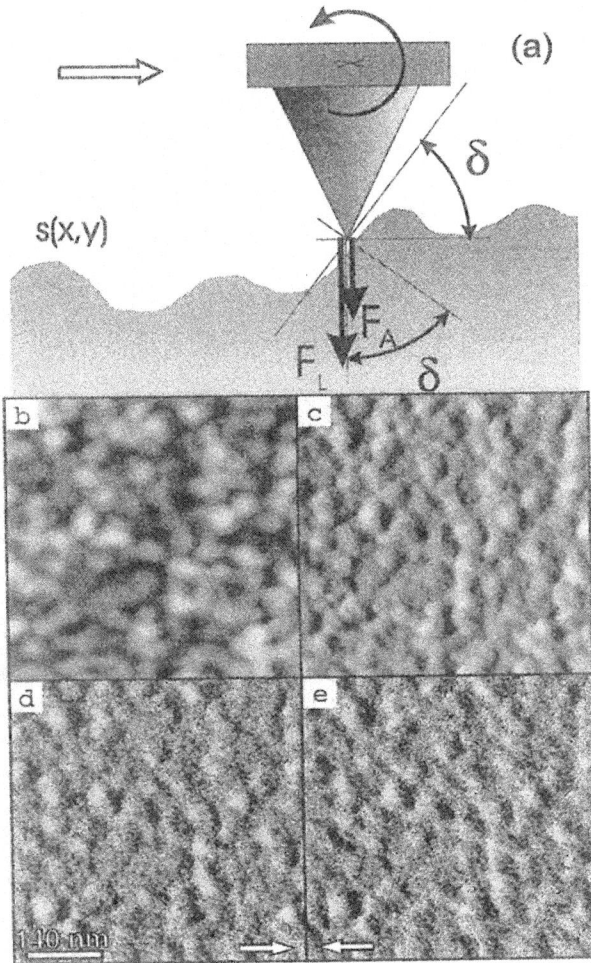

Figure 4.1: Influence of topography on lateral forces. (a) Schematic diagram of the relevant force vectors. Due to the local slope of topography, normal forces F_N cause a component of the lateral force. (b) Topography image of a compact disc (CD), consisting of small hillocks of polycarbonate. (c) Derivative of the topography. (d) Lateral force image acquired in the forward direction. (e) Lateral force image acquired in the backward direction. The contrast in (d) and (e) is independent of scan direction and is closely related to the topography image.

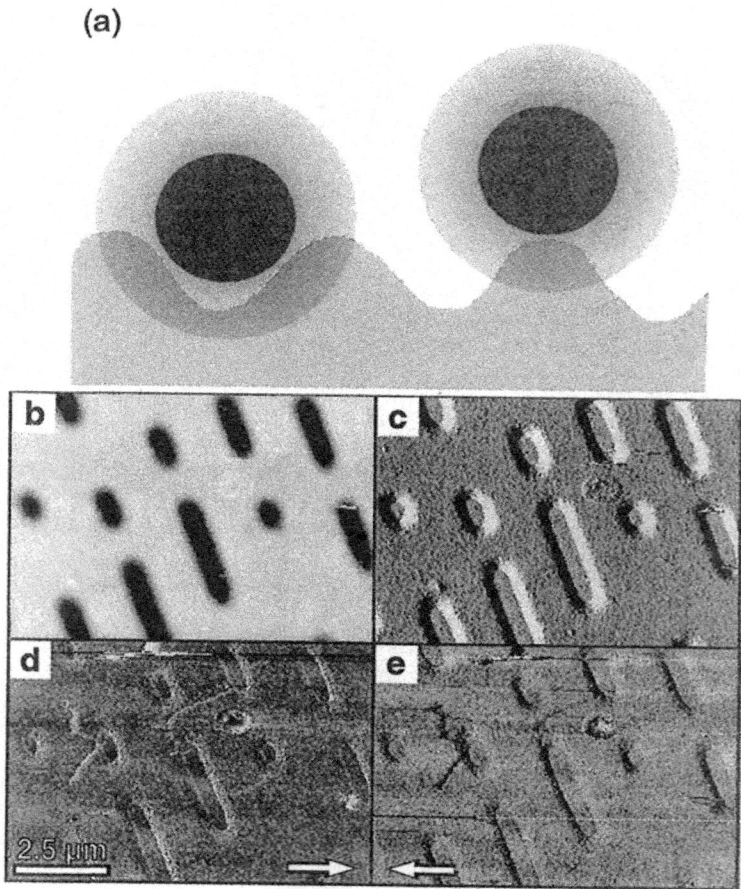

Figure 4.2: (a) Influence of topography on the attractive force F_a due to van der Waals forces. On a summit the attractive force is reduced because of the smaller volume, whereas in the valley the force is increased. The larger attractive force F_a leads to an increased effective normal force F_N, which causes an increased contact area and therefore causes larger frictional forces. Analogous effects are also caused by forces such as capillary forces. (b) Topography image of a compact disc (CD) on a larger scale compared to Fig. 1. The bits of the CD become visible. (c) Derivative of the topography. (d) Lateral forces in the forward direction. (e) Lateral forces in the backward direction. In contrast to the example in Fig. 1, the lateral forces at steep slopes do depend on the scan direction and therefore are related to frictional forces.

equiforce mode, which means that the normal force at the lower terrace and the normal force at the upper terrace is the same under the assumption that attractive forces are unchanged. Therefore, the potential energy is not likely to be changed by traversing from one terrace to the other.

Meyer et al.[6] demonstrated with non-contact force microscopy that steps of Si(111) are charged. Generally, the charging of step edges might play an important role, leading to rather strong coulombic attraction, increasing the total attractive force and thus the total normal force, which leads to an increase in friction. Cleavage steps on the surface of NaF(001) were investigated by Howald et al.[7] on the atomic scale under ultrahigh vacuum conditions. Using tips with high aspect-ratio, a transition region between the terraces as small as 1nm is found. These images can be interpreted as a convolution of tip shape with the step and give a direct measure of the contact area. Assuming a spherical tip shape, the width of the step region, S, is related to the tip radius, R, by

$$R = \frac{S^2 + H^2}{2H} \qquad (4.2)$$

where H is the step height. It is found that the transition region increases with step height and that equation 4.2 yields a different tip radius for a biatomic step, which is related to a deviation of the tip shape from the spherical geometry. These experiments are of importance because of two observations: 1) The transition region of the monatomic step is only 1nm wide, which means that the diameter of the contact area is of the same size and that the local radius of curvature is about R=2nm (H=0.28nm). 2) The lateral force map in the transition region appears similar to the atomic scale features on flat terraces, but is less ordered. We conclude that the contrast of friction at step edges is closely related to the atomic scale friction. The plucking mechanism, which will be discussed in more detail in the next section, might play an important role. Atoms at step edges have a reduced coordination number and are more weakly bound. From the point of view of the plucking mechanism this different bonding corresponds to a weaker spring and leads to a lower threshold for instabilities compared to the atoms within the terraces.

To summarize this first part, it has been found that processes at step edges are rather complex. Besides variations of contact area which are related to the tip shape, other processes have to be taken into account: Attractive forces, such as van der Waals, capillary and electrostatic forces can differ significantly at step edges and lead to different normal forces, which can cause different friction. Adsorption of contaminants at step edges affects the interaction. Different bonding of atoms at step sites can lead to different atomic scale

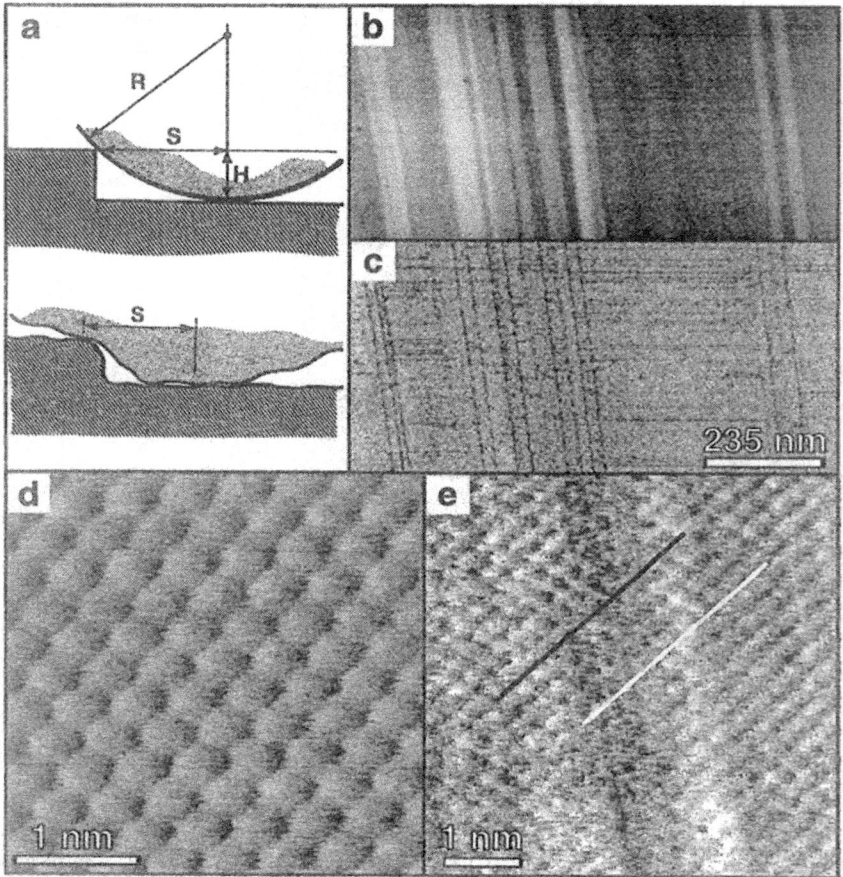

Figure 4.3: (a) Rigid model of the contrast formation at step edges related to the convolution with tip geometry. The simplest approximation of the tip is a sphere of radius R. From experiments at step edges of various heights, it is concluded that the tip geometry deviates from the spherical geometry. (b) Topography image of the NaF(001) surface measured under ultrahigh vacuum conditions. Cleavage steps are visible. (c) Lateral force image of the same area as in (b). (d) Lateral force images showing atomic-scale features of NaF(001) with a spacing of 3.1 ± 0.3Å, corresponding to the spacing between equally charged ions. (e) Lateral force image at a step edge. The transition region between the upper and lower terrace is only 1nm wide which gives a good estimate of the contact area between probing tip and sample. The bright and dark lines indicate that the ions are shifted which is in agreement with simple stacking models of crystals having the NaCl-structure.

friction. In the subsequent part, we will focus on the atomistic mechanisms at step edges, which will be related to Schwoebel barriers.

Step sites play an important role in thin film growth. The activation barriers for diffusion across step edges are significantly higher than the barriers on the terraces. The motion of adatoms during growth conditions is affected by these barriers, leading to 3d-island growth. A possible way to study these Schwoebel-barriers[8] is to measure the lateral force of an atomic-scale probing tip that crosses these step sites. The probing tip plays the role of the adatom: the higher the barrier the higher the force to pass the step region.

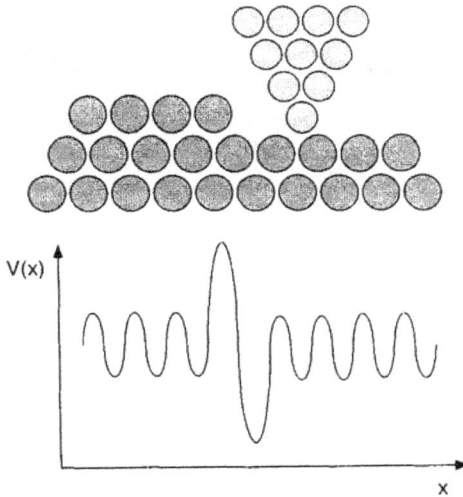

Figure 4.4: Due to the different coordination number, the potential at step edges is modified. Rather large barriers, known as Schwoebel barriers, can affect the motion of adatoms.

In this section, the site-specific loading dependence of NaCl(001) surfaces relative to silicon oxide tips are investigated by FFM. The surfaces are cleaved in ultrahigh vacuum, heated for 30min at 400°C and then transferred into the dry nitrogen chamber. Curved steps with a density of about $20 steps/\mu m^2$ are observable in the imaging mode (Fig. 4.5). Crossing the step down- and upwards leads to increased friction, which is a good indication that the lateral force is not a simple topography effect, but is actually related to the Schwoebel-barrier. Ramping the normal force from 0nN to 140nN and measuring friction loops on the NaCl(001) surface, the data for the 2d-histogram are acquired. As shown in Fig. 4.6, the (F_F,F_N)-pairs pile up in distinct regions, corresponding

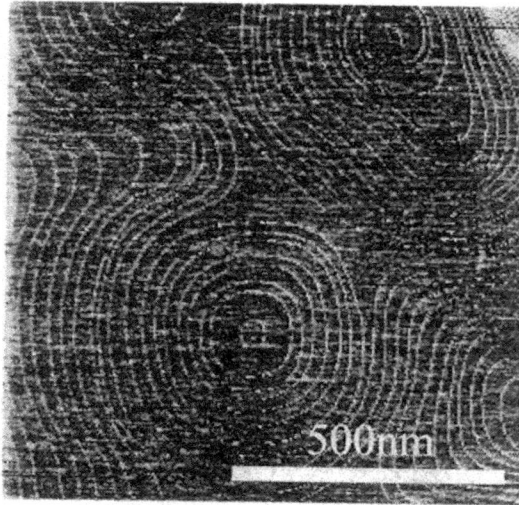

Figure 4.5: Friction force map on NaCl(001). Increased friction is observed on the step sites

to the case of strong correlation. An analysis of the data shows that the upper part of the highly occupied region corresponds to the step regions, whereas the the lower part (lower friction) corresponds to the terrace regions. Remarkably, both branches have a similar curve shape, which is another indication that the Schwoebel-barrier is the origin of the contrast. Digitizing the minima and maxima of these regions, we get the functional dependences $F_F^i(F_N)$ for the terrace regions (curve 1 in Fig. 4.7) and the step sites (curve 2 in Fig. 4.7). Obviously, both curves demonstrate strong deviations from the linear dependence. Qualitatively, a 2/3-power law becomes plausible. In order to yield more quantitative results, the probing tip radius and the elastic properties have to be further characterized. With high resolution electron microscopy measurements (field emission source), the radius of curvature of the probing tip has been determined to be $R = 60$nm. Using common values for the elastic properties of NaCl(001) and probing tip, a contact radius of 2.4nm is calculated for a normal force of 10nN, which corresponds to about 5 ion-ion distances on NaCl(001).

The first friction force microscopy measurements have shown a linear relationship between frictional forces and normal forces. A small friction coefficient $\mu=0.01$ was found in the case of a tungsten tip sliding on graphite. Mate et al.[20] suggested that with normal forces of 10^{-4}-10^{-6}N the contact is built up by many nm-sized asperities. The number of asperities increases with normal force leading to the observed linear dependence. The coefficient of friction is about an order of magnitude lower than in macroscopic experiments, which might be related to the lower forces and the smaller scale, where the formation of dislocation loops is less favourable and elastic deformation prevails.

At loads below 10^{-7}N, friction force microscopy definitively should enter the regime of the single asperity. Howald et al.[7] could demonstrate that the contact diameter of a silicon oxide tip on NaF(001) is about 1nm, by resolving the step region on the atomic scale. In the single asperity regime, the extended adhesion model predicts that the frictional force is proportional to the real area of contact. Assuming a Hertzian deformation of the contact, the frictional force is given by:

$$F_F = \tau \cdot A_R = \tau_0 \cdot \pi((R/E^*) \cdot (F_N - F_P))^{2/3} \qquad (4.3)$$

where $1/E^* = (1 - \nu_1^2)/E_1 + (1 - \nu_2^2)/E_2$ and ν_i and E_i are the poisson ratios and Youngs modulii of probing tip and sample and F_P the pull-off force. Thus, a $F_N^{2/3}$ power law is predicted under the assumption of constant shear stress τ. A more generalized shear stress can also depend on the normal pressure: $\tau = \tau_0 + \alpha \cdot p$. Then, the frictional force of a single asperity is given by :

$$F_F = \tau \cdot A_R = \tau_0 \cdot \pi((R/E^*) \cdot (F_N - F_P))^{2/3} + \alpha F_N \qquad (4.4)$$

Depending on the size of α, the frictional force of a single asperity contact can have a linear dependence (large α), a 2/3 power law (small α) or mixed behaviour. We believe that it is important to emphasize that a single asperity does not lead automatically to a 2/3 power law. Recently, two papers[10,11] have shown that in the case of a Si_3N_4-tip, probably covered by a silicon oxide layer, sliding on mica, a 2/3-power law is observed. Putman et al.[10] observed this 2/3 dependence only under humid conditions whereas a linear dependence is observed in dry conditions. They interpret this behaviour to be related to the formation of a water film on the probing tip leading to a smoother surface which is acting as a single asperity. In dry conditions, the probing tip is more rough and is believed to behave more like a multi-asperity contact.

Hertzian deformation is a rather simple approximation. Even though, we include an adhesive force with an offset of the normal force axis (pull of force F_P) the model is only of qualitative nature. An improved model has been given by Johnson, Kendall and Roberts[12], the so called JKR-theory, which includes adhesive forces accurately. Within the extended adhesion model, the lateral force is then given by:

$$F_F = \tau_0 \cdot \pi (R/E^*)^{2/3} \cdot ((F_N + F_P) + 2 \cdot F_P + \sqrt{4 \cdot F_P (F_N + F_P) + (2 \cdot F_P)^2})^{2/3}$$

(4.5)

Note, that the pull-off force is not treated as fit parameter, but is determined experimentally. The validity of the model can be verified by comparing with the predicted value $F_P = 3\pi R\gamma$, where γ is the surface energy. Similarly, than in the case of the Hertzian deformation, the model can be generalized by exchanging τ_0 by the pressure dependent shear strength $\tau = \tau_0 + \alpha \cdot P$.

Fits of the data were made for both the Hertzian deformation (equation 4.4) and the JKR-theory (equation 4.5). A shear strength τ_0=160±30MPa is found by the fit of curve 1 with equation 4.4 (cf. Fig. 4.7a). Note, that the shear strength is the only fit parameter whereas R and F_P are determined experimentally. The fit could not be improved by incorporating a linear term. Therefore, in the following discussion, α is neglected. Using the JKR-theory, better fits are achieved as can be seen in Fig. 4.7b. The shear strength is reduced to a value of 96MPa on the terrace. The general trend, that the shear strength is reduced with the JKR-theory compared with the Hertz-theory, makes sense because the contact area is underestimated with the Hertz-theory.

The data that were acquired at the step edges (curve 2) are more complex to be analyzed. Taking only the maximum frictional forces into account (upper margin of the highly occupied region), we assume that the contact area at step edge is maximum for this curve. The fit of this curve by equation 4.4 and 4.5 yields increased shear strengths at the step sites of 224MPa (Hertz) and 142MPa (JKR), which is in agreement with the increased barriers at the step sites. However, this analysis underestimates the barriers, because the contact area at the step edges is reduced. The extended adhesion model simply assumes, that the friction at the step edge can be subdivided into a part that comes from the shear on the terrace site and a part that originates from a narrow stripe where increased shear stress is to be expected.

$$F_F = \tau_{terrace} A_{terrace} + \tau_{step} A_{step}$$

(4.6)

Using the Hertz theory, fitting of curve 2 with equation 4.6 yields a shear strength at the terrace $\tau_{terrace}$=186MPa and a shear strength at the step edge τ_{step}=1.8GPa. Thus, the model gives approximatively consistent values for the shear strength on the terraces for both fits. The shear stress on the step site is about 10 times larger than on the terrace.

The same procedure has been applied with the JKR-theory where a shear strength of 145MPa on the terrace and 478MPa at the step edge have been found, resulting in a ratio $\tau_{step}/\tau_{terrace}=3.3$. Again, we find considerably reduced shear strengths with the JKR-theory. All the shear strengths are summarized in the table 1.

From these site-specific shear strength values, we can get a estimate of the energy barriers to cross the terrace or the step site. In first approximation, the barriers are proportional to the shear strength: $E_a = \tau \cdot b^3$ where $b = 3.98\text{Å}$ is the distance between equally charged ions. Thus, the estimate of the barrier on the terraces is: 0.07eV and on the steps: 0.7eV using the extended Hertz-theory whereas the barriers are determined to be 0.06eV on the terrace and 0.2eV on the step site using the JKR-theory (cf. Table 2).

TABLE 1. Shear strengths on NaCl(001) in MPa

site	Hertz	JKR	ext. Hertz	ext. JKR
$\tau_0(terrace)$	160	96	186	145
$\tau_0(step)$	224	142	1840	478
$\tau_0(step)/\tau_0(terrace)$	1.4	1.5	10.0	3.3

TABLE 2. Barrier heights in meV

site	Hertz	JKR	ext. Hertz	ext. JKR
V(terrace)	63	38	73	57
V(step)	88	56	724	188
V(step)/V(terrace)	1.4	1.5	10.0	3.3

To our knowledge, there are no other experimental values for surface diffusion barriers on NaCl(001). Therefore, we only can estimate the barrier heights by comparing to typical electrostatic energies of adatoms on the different sites. The electrostatic energy for an ionic crystal is given by: $W_{el.st.} = \frac{\alpha \cdot e^2}{d}$, where α is the Madelung constant and d the interatomic distance. Now, the Madelung constant depends strongly on the coordination number: Adatom on a terrace site: $\alpha_{terr}=0.0662$, adatom on a step site: $\alpha_{step}=0.1806$ and adatom on a kink site: $\alpha_{kink}=0.8736$. The ratios are then: $\frac{\alpha_{step}}{\alpha_{terr}}=2.73$ and $\frac{\alpha_{kink}}{\alpha_{terr}}=13.2$. Assuming that Madelung constants are proportional to the diffusion barriers and thus are proportional to the shear strengths: we conclude that the observed ratio of $\tau_{step}/\tau_{terrace}=2\text{-}9$ is roughly consistent with the ratio $\frac{\alpha_{step}}{\alpha_{terr}}$. The shear strengths differ significantly between the models. At present, we

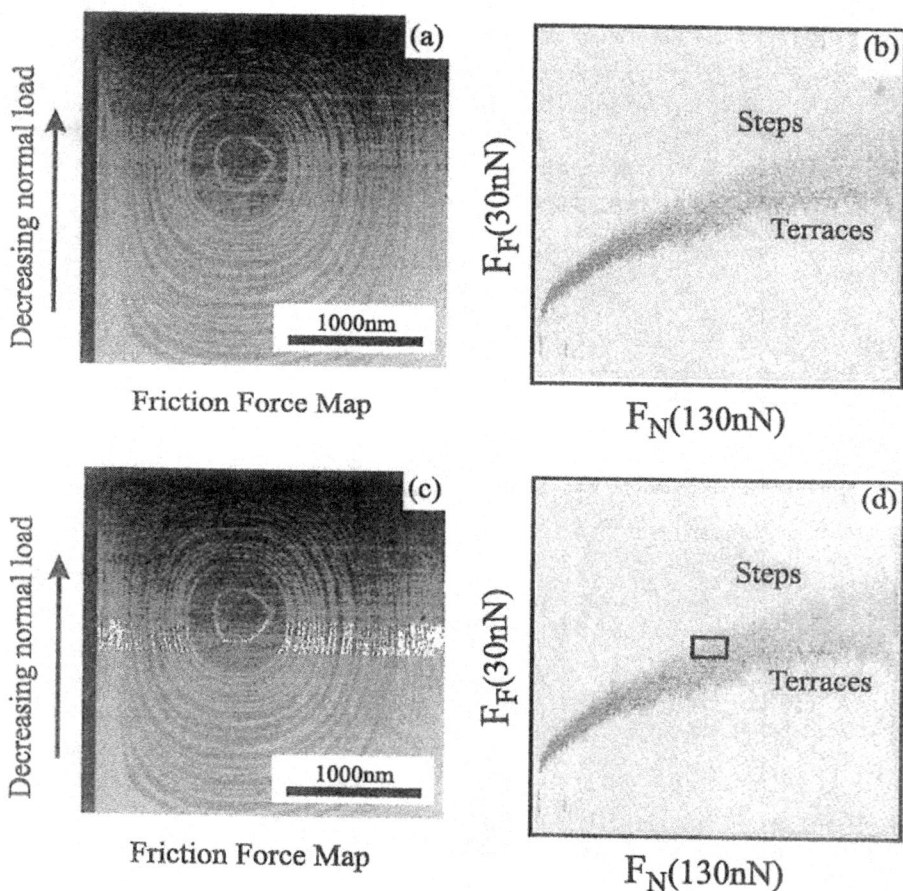

Figure 4.6: (a) Friction force map on NaCl(001). The loading is decreased from 120nN to -10nN (jump off point) during imaging. (b) 2d-histogram of (a). Each (F_F, F_N)-pair in (a) gives a contribution to the 2d-histogram. The data pile up in distinct regions. (c) and (d) are the same data as (a) and (b), but one region in (d) is selected (solid frame) and the corresponding region in the friction force map (c) is highlighted, clearly showing that the upper part of the highly occupied region in (d) originates mainly from step sites. Accordingly, it can be demonstrated that the lower part originates from terrace sites.

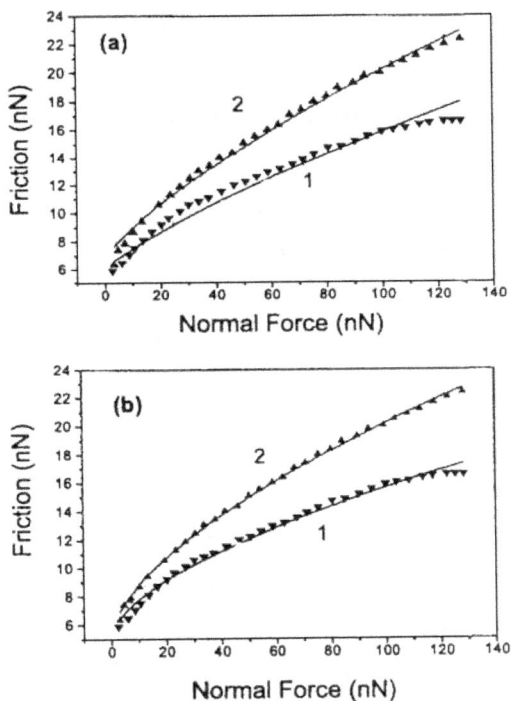

Figure 4.7: Data from Fig. 4.6b are digitized. Only maximum and minimum values are considered, corresponding to the maximum friction on step edges and the minimum friction on terraces. (a) Curve 1 (triangles downwards) is fitted with equation 4.4 (Hertz) and curve 2 (triangles upwards) is fitted with equation 4.6 using Hertz-theory. (b) Curve 1 (triangles downwards) is fitted with equation 4.5 (JKR) and curve 2 (triangles upwards) is fitted with equation 4.6 using JKR-theory.

suggest that the JKR-theory is more appropriate, because of the accurate inclusion of adhesive forces. More fundamental theories, including atomic-scale processes should be developed and applied in order to get firm conclusions. Concerning the experiments, some restrictions have to be made: Taking into account that our measurement of the shear strength on the step site is an averaged value over 5 to 20 ion-ion distances, it becomes reasonable that kinks affect the measurements. The curved shape of the steps makes plausible that there is a high density of kinks that influence the shear strength measurement on step sites. Future experiments are planned where samples with lower density of kink sites (cleaved surfaces with straight steps) are investigated. In this case the shear strength on the step site should be considerably lower. Other explanations could be related to a more complex charging of the step sites, that could change the barriers.

4.3 Atomic-scale friction: Tomlinson's mechanism

4.3.1 Introduction

As we learned in the previous section, in an FFM experiment, the topography of a surface gives rise to lateral forces which can be interpreted as friction. However, these forces are conservative and do not give rise to the typical dissipative character of friction, since the interatomic forces relevant to produce the effects discussed in the section IV 1 are conservative. Scanning in one direction produces the same force curve as scanning in the reverse direction.

The following question arises: How can conservative potentials between the atoms produce nonvanishing friction even in the wearless case, where the system refers to the same state as before the sliding process. A possible solution to this problem has been given by G.A. Tomlinson in 1929[13], which is nowadays known as the *Tomlinson mechanism*. The main emphasis of Tomlinson was to take into account the role of mechanical adiabaticity in the dry friction process:

"To explain friction it is necessary to suppose the existence of some irreversible stage in the passage of one atom past another, in which heat energy is developped at the expense of external work"

Historically it is interesting to note, that the famous physicist Prandtl proposed the Tomlinson model already 1928 in[14]. His figure for Tomlinson's model is given in Fig. prandtl.

Based on this idea he developped a possible wearless friction mechanism (see Fig. 4.9):

Given two surfaces sliding against each other: We consider the motion of

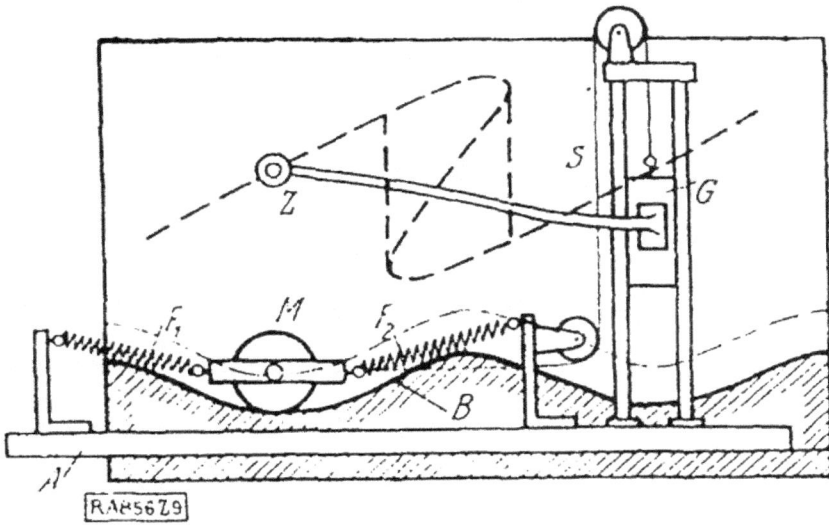

Figure 4.8: Already Prandtl described the Tomlinson model as a mechanical model for wearless friction.

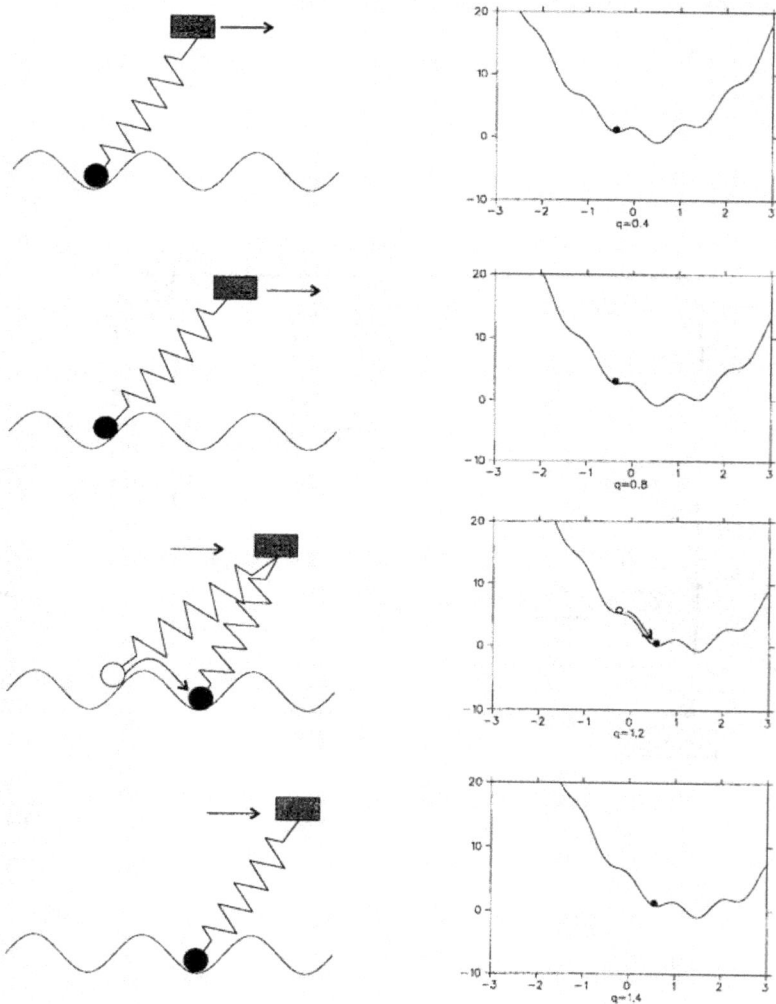

Figure 4.9: The steps of Tomlinson's model as snapshot (left) and the local total energy (right) as a functon of the particle position. Note, that the potential is a function of the support position q and therefore it varies.

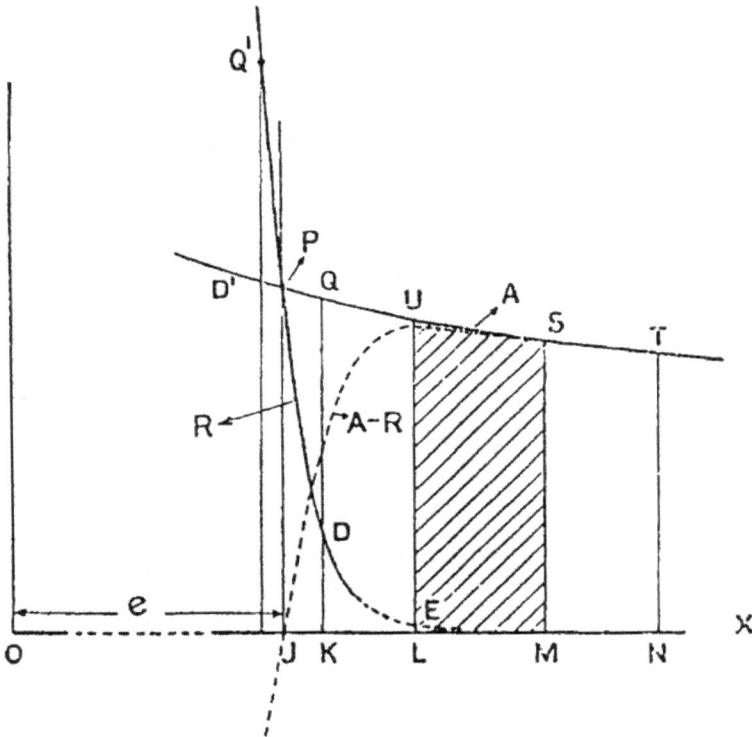

Figure 4.10: Figure of the force vs. distance behaviour in the Tomlinson model, taken from Tomlinson's work[13] (1929). Curve A denotes the attractive force, whereas R denotes the repulsive force. The dashed curve is the sum of repulsive and attractive force.

surface atom A, belonging to the upper surface, when atom B belonging to the lower surface is passing by. As long as B is far away, A is in a stable equilibrium due to the interaction forces with the bulk atoms. (Fig. 4.9a)

As B moves along, A follows the motion of B adiabatically (Fig. 4.9b). For a critical position of B, the equilibrium position of atom A becomes unstable. A non-adiabatic jump of A back to the nearest stable position occurs. (Fig. 4.9c)

The original figure of Tomlinson's work 1929 is shown in Fig. 4.10. The graph represents the forces which act on the three atoms B, C and D. Important are the three curves denoted A , R and $A - R$ (difference of A and R), which represent the repulsive(R) and the attractive(A) force respectively

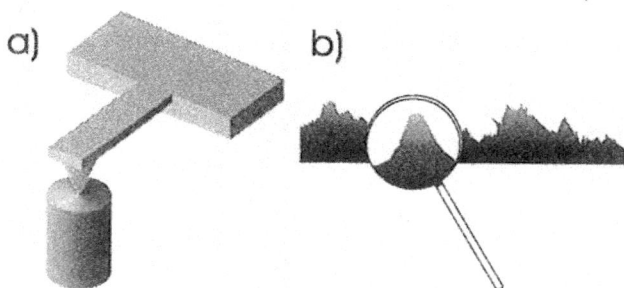

Figure 4.11: Tomlinson's model can be used to model an FFM but also a microscopic asperity.

between atom B and atom C. In principle, it is a simple force vs. distance diagram, but not easy to interprete since the vertical axis represents attractive forces for the curves A and $A - R$, but repulsive forces for the curve R. When the curve $A - R$ is mirrored at the horizontal axis, a Lennard-Jones-like potential is revealed.

4.4   A modern analysis of Tomlinson's mechanism

In driven systems one distinguish between control variables which are directly controlled by the outer world, and system variables which constitute the answer of the system to changes of the control variables. After a change of the control variables, the system variables follow them corresponding to a relation of the form:

$$systemvariables = f(\{controlvariables\}) \tag{4.7}$$

There is no problem to control the system, when the above function f is single-valued. In such a case, the history of the system is not important. Otherwise, when f has several branches, the history of the system becomes important to obtain the complete information on the system variables. This problem of invertibility of the relation f will turn out to be a very fundamental criterion for the occurence of hysteresis, energy dissipation and kinetic friction.

   In friction studies it is crucial to distinguish between different length scales. However, due to the strong physical analogy between a fully elastic asperity sliding against other asperities and the tip of an FFM[a] sliding over the corrugation of an atomically flat surface, Tomlinson's model may be valid in both cases and has been successfully used to model both systems. In the following sections, we will always use the FFM as an example, since the atomic corrugation is very regular and therefore easy to model with a sinusoidal potential.[b]

4.4.1   One-dimensional Tomlinson model

The following analysis describes in detail the friction for the case of one particle in the framework of the one-dimensional Tomlinson model. This approach has been used by Mc Clelland et al.[16], Zhong and Tomanek et al.[17,18], Colchero and Marti et al.[19] and others to model the FFM.

We consider the tip of a FFM in a one-dimensional periodic potential, which represents the interaction with an atomically flat surface. We will concentrate on the quasistatic limit, of vanishing relative velocity of the two bodies. The total energy is given by the sum of the potential at the position x of the tip on the surface and the elastic energy which is stored in the cantilever.

$$E_{tot} = V(x) + \frac{c}{2}(x - x_0)^2, \tag{4.8}$$

where c is the spring constant of the cantilever and x_0 is the position of the support of the microscope. In our case, it is the position of the tip on the atomic surface, which is the system variable, whereas x_0 is the control variable. In a quasi-static process, the system variables arrange themselves always in a stable equilibrium. Mathematically, this condition means that the derivative of the energy with respect to the system variable is zero and that the second derivative is positive:

$$\frac{dE}{dx} = \frac{dV}{dx} + c(x - x_0) = 0 \tag{4.9}$$

$$\frac{d^2E}{dx^2} > 0 \tag{4.10}$$

The equilibrium condition can be rewritten in the form

$$x_0 = \frac{1}{c}\frac{dV}{dx} + x \tag{4.11}$$

This equation consists of a linear and a periodic term. Therefore, for small values of the spring constant c it is not invertible (see Fig. 4.12). We define

---

[a]The FFM is discussed in more detail in chapter II
[b]Those readers who are just interested in the case of macroscopic asperities, may be reffered to the papers by Caroli et al.[15].

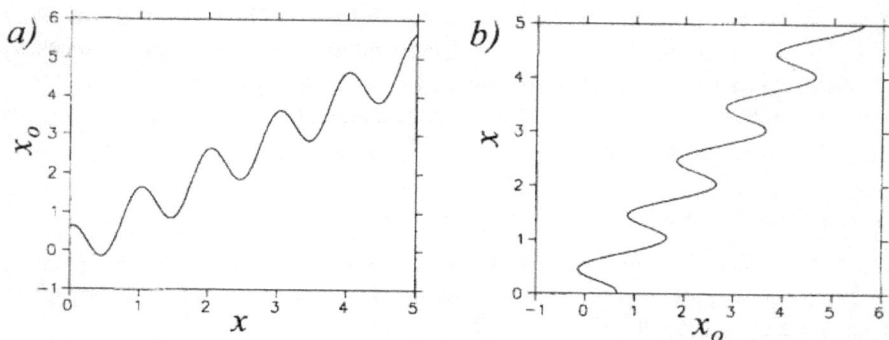

Figure 4.12: The mapping $\vec{r}(\vec{r}_0)$ contains folds for $c < c_{crit}$

the critical spring constant c_{crit} to be the smallest value of c above which the mapping (4.11) is invertible.

An interesting behaviour of the system occurs, when a stable equilibrium becomes unstable after an infinite change of x_0. Such a processes occurs when the second derivative changes its sign. Hence, these instabilities occur when:

$$\frac{\partial^2 E_{tot}}{dx^2} \;=\; V''(x) + c = 0 \tag{4.12}$$

$$\Updownarrow$$

$$V''(x) \;=\; -c \tag{4.13}$$

The solution can be illustrated graphically in a force-distance plot ($F_L = -dV/dx$) where $F' = c$, and is shown in Fig. 4.13 for a sinusoidal potential $V(x) = \sin(2\pi x)$.[c] It is possible, that even more than two solutions coexist, depending on the shape of the potential.

For the understanding of the processes it helps to take a look at the energy as a function of tip- and support-position, respectively. The relation $E(x, x_0)$ can be reduced to a function $E(x)$ using the equilibrium condition:

$$E'(x, x_0) = V'(x) + c(x - x_0) \;=\; 0 \tag{4.14}$$

$$\Updownarrow$$

$$x - x_0 \;=\; -\frac{1}{c} V'(x) \tag{4.15}$$

$$\Downarrow$$

---

[c]The same analysis can be used to understand the process of "snap to contact" (see chap. III).

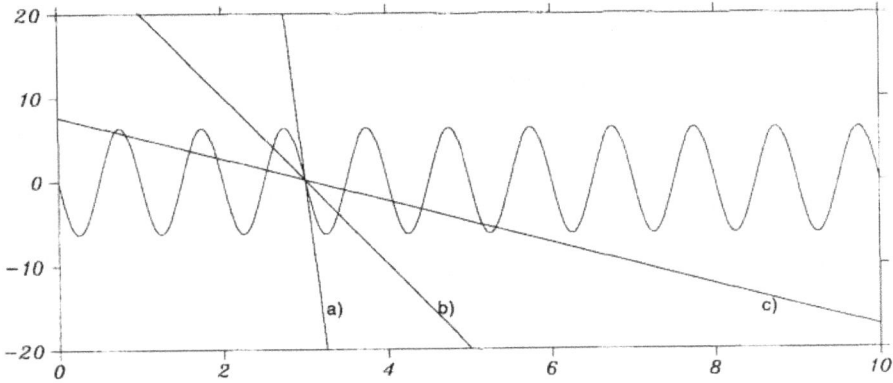

Figure 4.13: The graphic solution of eq. 4.9. An instability occurs when the force gradient is equal to the spring constant c.

$$E(x) = V(x) + \frac{c}{2}(x - x_0)^2 \quad = \quad V(x) + \frac{1}{2c}\left(\frac{\partial V}{\partial x}\right)^2 \qquad (4.16)$$

The graph of this dependence $E(x)$ shows clearly the behaviour of the system. In Fig. 4.14a the relation is shown graphically for a potential $V(x) = \sin 2\pi x$. In this case (4.16) becomes:

$$E(x) = \sin 2\pi x + \frac{4\pi^2}{2c}\cos^2 2\pi x \qquad (4.17)$$

The tip position is stable in the energy valleys, where the curvature is positive.

A relation $E(x_0)$ cannot be defined since relation (4.9) generally cannot be solved for x. However, a parametric function $x \mapsto (x_0(x), E(x))$ using the tip position x as parameter can be defined unambigously, and is shown in Fig. 4.14b for a sinusoidal potential $V(x) = sin(2\pi x)$.

4.4.2 Two-dimensional Tomlinson model

The Tomlinson model can be extended into two dimensions. The equations look similar to those of the one-dimensional case. The one-dimensional analysis can simply be extended to two dimensions:

$$E_{tot} = V(\vec{r}) + \frac{c}{2}(\vec{r} - \vec{r}_0)^2 \qquad (4.18)$$

where the symbol $\vec{r}$ corresponds to x in one dimension.

This is the most general model, that can be used under the following assumptions:

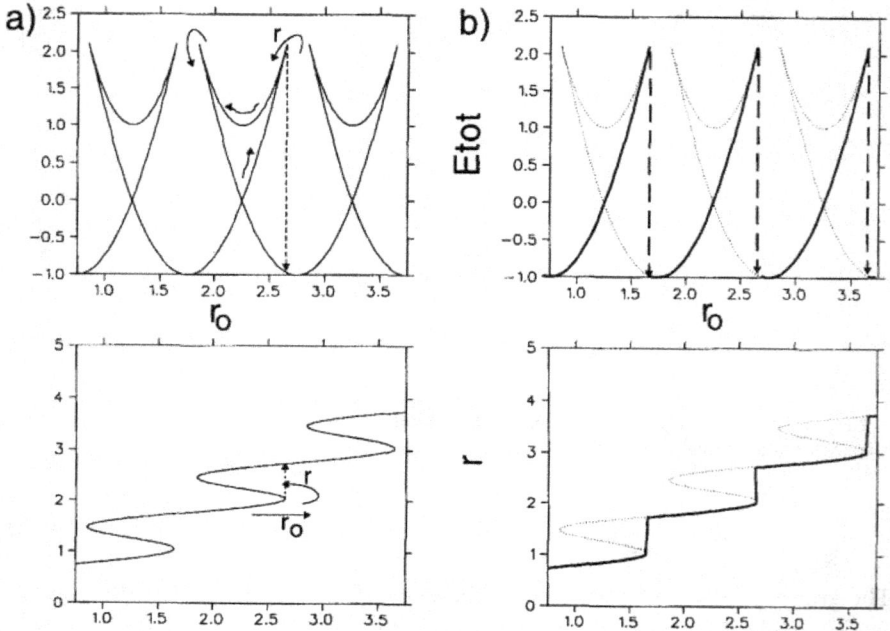

Figure 4.14: a) The relation $E(x_0)$ compared to the relation $x(x_0)$. Jumps occur at the positions $dr_0(r)/dr = 0$. The unstable continuation of $r_0(r)$ is shown. b) The energy contour in a forward scan is shown as a thick line.

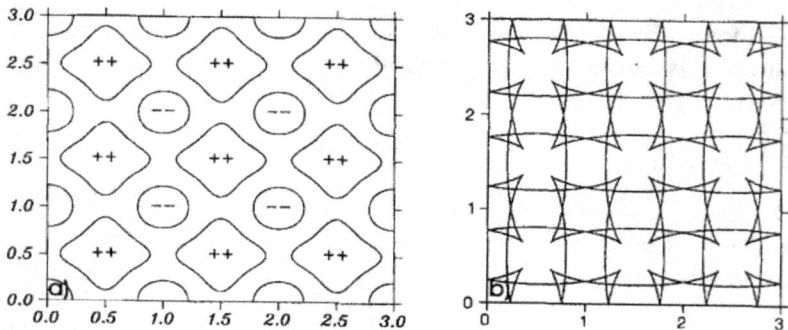

Figure 4.15: a) The regions in the tip plane are labelled according to the signs of the Eigenvalues of the Hessian matrix. b) The critical curve, corresponding to the regions in a)

1 The driving velocity is quasistatic, $\dot{\vec{r}_0} \mapsto 0$.

2 the atoms follow the motion of the tip adiabatically, no instabilities occur between the single atoms in the tip and in the surface.

3 The torsion of the cantilever remains in the harmonic regime.

We have not made any assumption about the potential. For explicit calculations, the only assumption which has been made, is that the potential is periodic due to a perfect crystal lattice, and hence it can be expanded into a Fourier series.

$$V(x, y) = \sum_{l=0}^{\infty} (a_l \cos(k_x x) + b_l \sin(k_x x))(c_l \cos(k_y y) + d_l \sin(k_y y)) \quad (4.19)$$

with Fourier components a_l, b_l, c_l, d_l and lattice constants $a_x = 2\pi/k_x$ and $a_y = 2\pi/k_x$ respectively.

The equilibrium condition becomes

$$\nabla_r E_{tot} = \nabla_r V + c(\vec{r} - \vec{r}_0) = 0 \quad (4.20)$$

The stability condition refers to the signs of the eigenvalues of the Hessian $H = (\partial_{\alpha\beta})$, $\alpha, \beta = x, y$[d].

$$H = \begin{pmatrix} \partial_{xx} V + c & \partial_{xy} V \\ \partial_{yx} V & \partial_{yy} V + c \end{pmatrix} \quad (4.21)$$

The position $\vec{r}$ is stable when both eigenvalues are positive. In the harmonic ansatz for the elastic energy of the cantilever, the Hessian does not depend on $\vec{r}_0$ but only on $\vec{r}$. We therefore can label every point in the $\vec{r}$−plane unambiguously corresponding to the signs of the eigenvalues. This is illustrated for a periodic potential of the form

$$V(x, y) = \cos kx + \cos ky + \gamma \cos kx \cos ky \quad (4.22)$$

and for various spring constants c in Fig. 4.15a.

The equilibrium condition can be solved for $\vec{r}_0$:

$$\vec{r}_0 = \frac{1}{c} \nabla_r V + \vec{r} \quad (4.23)$$

Hence, the mapping $\vec{r} \mapsto \vec{r}_0$ again is unambiguous as in the one-dimensional case. However, for small values of the spring constant c the mapping $\vec{r} \mapsto \vec{r}_0$

---

[d]The expression $\partial_{\alpha\beta} = \frac{\partial^2}{\partial x_\alpha \partial x_\beta}$.

is not invertible anymore, and the inverse mapping $\vec{r}_0 \mapsto \vec{r}$ contains folds, as shown in Fig. 4.12.

The parametric relation can be written in the form:

$$(x, y) \mapsto (x_0(x, y), y_0(x, y), E(x, y)) \qquad (4.24)$$

and is shown in Fig. 4.16 for a spring constant below the critical value c_{crit} and for a spring constant above c_{crit} for the potential given in (4.22). For values below c_{crit} one can see in the figure the critical curve, which is constituted by the edges of the fold (fig 4.16b).

The critical curve

It is interesting to calculate those points in the plane of support positions, where irreversible jumps occur. The structure of this so-called critical curve gives a good understanding of the shape of experimental friction force images, anisotropy of the friction coefficient and other features of friction on the atomic scale. For values of c below $c_{T,crit}$, the mapping (4.24) contains folds as shown in Fig. 4.16b. Whenever the tip is dragged across the edge of this fold, an irreversible jump occurs and the energy difference between the points denoted A and B in fig 4.16b, dissipates.

We derive a mathematical criterion for the positions of these edges, and therefore a criterion for the occurrence of an instability. The jump occurs, when the function $\vec{r}_0(\vec{r})$ has a local extremum, and for larger (smaller) values of $\vec{r}_0$ the corresponding energy branch does not have a continuation.

$$\nabla_{\vec{r}} \vec{r}_0(\vec{r}) = 0. \qquad (4.25)$$

Using (4.20), the above equation is equivalent to the condition:

$$\det(H) = 0. \qquad (4.26)$$

There exists a less abstract way to understand the role of the critical curve, which leads to the same results as obtained above: We make use of the unambigous labelling of the tip positions corresponding to the signs of the local Hessian (4.21) shown in Fig. 4.15a. The tip follows the support adiabatically as long as it remains in a $(++)$-region. When the tip is dragged to the border of the region, it suddenly jumps into the next $(++)$-region. Hence, the critical points in the plane of tip positions are the borders of the $(++)$-regions, where the eigenvalues $\lambda_{1,2}$ of the Hessian have the following form:

$$\lambda_1 = 0 \qquad\qquad \lambda_2 > 0 \qquad (4.27)$$

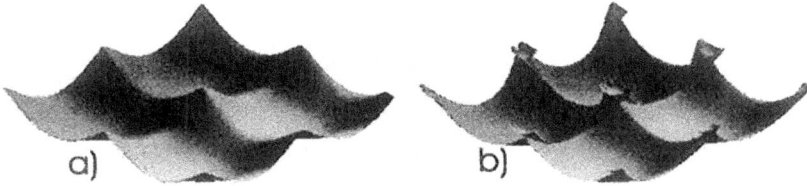

Figure 4.16: The parametric function $(x, y) \mapsto (x_0(x,y), y_0(x,y), E(x,y))$ shown as a surface in a three-dimensional space for a spring constant above the critical value (a) and below the critical value (b).

$$\Updownarrow$$

$$\det(H) = \lambda_1 \lambda_2 = 0 \qquad tr(H) = \lambda_1 + \lambda_2 > 0 \qquad (4.28)$$

which is the same condition we obtained using the shape of the folds. By making use of eq. (4.23) these points can be mapped into the plane of support positions, which can be controlled by the voltage applied to the piezo tubes of the microscope.

The critical curve in the plane of support positions corresponding to the systems shown in Fig. 4.15a are shown in Fig. 4.15b.

4.4.3 Instabilities and the superlubric phase

In the analysis of Tomlinson's mechanism we found that the important question for the occurrence of friction is whether instabilities occur in the system or not, since the dissipative character of friction in the quasistatic limit and the corresponding hysteresis effects are due to these instabilities.

This mathematical analysis makes a prediction which is somewhat counter-intuitive. If the ratio between spring constant c and potential strength v becomes smaller than a critical value, no instabilities occur at all, and hence no dissipation and no resulting friction force is detected. This does not imply that the lateral force $F_{frict}(\vec{r}_0) = 0$ is zerobut that the mean friction $\langle F_{frict} \rangle$ vanishes. The friction loop has no hysteresis and no sawtooth behaviour.

Experimental confirmation of this "superlubric state" has not been established so far, since is not easy to change the spring constant of the cantilever and signal-to-noise ratio which limit the experiments. On the other hand it is not astonishing that we have no hysteresis when the normal load is so small, that the distance between tip and surface is in the range of μm, where from an atomistic point of view the tip and the surface are still "in contact" since the range of the interatomic forces is infinite. Hence, there must be a critical normal load for every given cantilever and surface, where no friction occurs.

4.5   Comparison of atomic-scale stick slip with the Tomlinson plucking mechanism

In the last section we discussed, what should be seen in an FFM experiment in the limit of Tomlinson's model. In this section the experimental results with atomic-scale stick slip are discussed, which were obtained in the last decade by means of FFM (for more details of the FFM technique, see chapter II and VIII). The first friction force image on a nanometer scale has been obtained by Mate et al.[20] in 1986. They observed that friction force scans show a sawtooth-like behaviour with the periodicity of the crystalline substrate. They also observed a hysteresis after reversing the scan direction. From the area that is enclosed by the friction force loop, one can directly calculate the energy that is dissipated. In this pioneering work, the authors observed a linear loading dependence and determined a small friction coefficient of about 0.01 for the case of a tungsten tip on graphite. The cantilever spring constants were also changed from 155N/m to 2500N/m. In the latter case, a significant reduction of friction was observed. The results were interpreted in the context of the Tomlinson mechanism, also called plucking mechanism, where instabilities of the spring lead to the characteristic stick-slip behaviour. Two restrictions have to be made:

1. Rather high forces of up to $5 \cdot 10^{-5}$N were applied. Under these conditions, the formation of graphite flakes is likely. Therefore, the observed frictional forces are related to the shearing of rather large area of a graphite flake on top of the graphite surface[21].

2. The measurements were performed in ambient conditions where contaminants may exist at the interface between probing tip and sample.

The adsorption and desorption of these contaminants by the action of the probing tip can lead to atomic-scale features of the lateral forces. Therefore, the observed friction might not be purely wearless. In the field of surface force apparatus (SFA) shear forces are found to show characteristic oscillations as a function of the thickness of the liquid layer between the curved mica sheets. Actually, every maximum of the shear force is related to the expulsion of one molecular layer[22]. In other words, liquids confined between solid surfaces can exhibit a behaviour that is not described by simple viscous forces but resembles a solid-like behaviour. In an analogous way, confined molecules at the interface of the FFM-tip and the sample might lead to shear forces that vary on an atomic scale.

4.5.1 Atomic-scale stick slip under ultrahigh vacuum conditions

In order to exclude the influence of contaminants, the susequent experiments were conducted in ultrahigh vacuum. The aim was to establish wearless friction on an atomic scale on clean surfaces. The first observation of atomic-scale stick slip under UHV-conditions was reported by Germann et al.[23]. These results will be discussed in more detail in chapter VIII. In the following, we will focus on results on ionic crystals from our group. As shown in Fig. 4.3 atomic-scale features on NaF(001) can be observed in the friction force map under ultrahigh vacuum conditions, where the crystal has been cleaved in-situ. The spacing between the protrusions is 4.0Å which is in good agreement with the spacing between equally charged ions. Thus, atomic-scale friction is observed in the absence of contaminants. On the NaF(001) crystal no flakes are formed, because the easy shear plane is the (110)-plane. Therefore, the shearing between the incommensurable surfaces of the amorphous silicon oxide and the NaF(001)-surface is observed. From measurements at step edges the size of the contact diameter is estimated to be about 1 nm[7]. As a function of the applied normal force, a weak dependence is observed. A friction coefficient of 0.01-0.03 is determined, which is comparable to graphite. In conclusion, atomic-scale stick-slip exists also on non-layered materials in the absence of contaminants.

In the following various efforts have been taken to increase the resolution of the friction images. However, individual defects have not been observed so far. Apparantly, all the existing data may represent multi-atom contacts, which average over the surface, so called lattice imaging. More details about the resolution limits are given in chapter II. Two explanations may be valid:

1. The friction contributions of the atoms in contact are superimposed constructively (even for incommensurate tip structure), which may be related to a rearrangement of tip atoms at finite temperature. Recently, Shluger et al. have presented molecular dynamics calculations on ionic surfaces, where a self-lubrication effect was observed. Atoms from the sample are transferred to the tip and form rather commensurate tip structure[24].

2. Alternatively, one tip atom may dominate and the other atoms add up incoherently. Then, the question about the non-observation of individual defects arises. Shluger et al.[25] pointed out that even a monoatomic tip may cause rather large stresses in the sample, which then can lead to the motion of the defects, such as vacancies, which are much more mobile than the periodically arranged atoms that are stabilized by their neighbours. So, one may imagine that the probing tip moves defects in front of the contact zone, like delphins that swim in front of an ocean cruiser.

In Fig. 4.17, an experimental FFM image on KBr(001), obtained in UHV with a Si probing tip, is compared with an theoretical image, that has been calculated with the 2d-Tomlinson model[26,41]. Qualitatively, the data fits excellent with the theory. Even the spikes of the critical curve can be seen in the force scans. In close analogy of the macroscopic stick-slip phenomenon (discussed in chapter II), the sawtooth behaviour of the force scans is often referred to as the *atomic-scale stick-slip* phenomenon or nano stick-slip.[e]

The slope of the sticking part in Fig. 4.17 c) is about k^x_{eff}=7N/m. According to the analysis, which was explained in chapter II, the lateral contact stiffness can be determined:

$$k^x_{contact} = (\frac{1}{k^x_{eff}} - \frac{1}{k^x_{tip}} - \frac{1}{c_x})^{-1} \qquad (4.29)$$

where the torsional spring is c_x=35.5N/m and the lateral stiffness of the probing tip $k^x_{tip} \approx 84$N/m is estimated from the geometrical dimensions of the tip. Then, a value of $k^x_{contact}$=18N/m is found. Furthermore, the contact radius a can be estimated:

$$a = \frac{k^x_{contact}}{8G^*} \qquad (4.30)$$

where the effective shear modulus is given by

$$G^* = (\frac{2 - \nu_1^2}{G_1} + \frac{2 - \nu_2^2}{G_2})^{-1} \qquad (4.31)$$

and G_1 and G_2 are the shear moduli of the sample and probing tip and ν_1 and ν_2 are the Poisson ratios[19,27]. The Youngs modulus E_1, shear modulus G_1 and

---

[e]One must be careful. Although the two phenomena result in similar force curves and hysteresis loops, they have different origins. The periodicity of the sawtooth curve is nanoscopically given by the lattice constant of the surface, whereas macroscopically it depends on the elasticity (i.e. spring constant) of the slider and other parameters, such as the mass of the slider and the velocity.

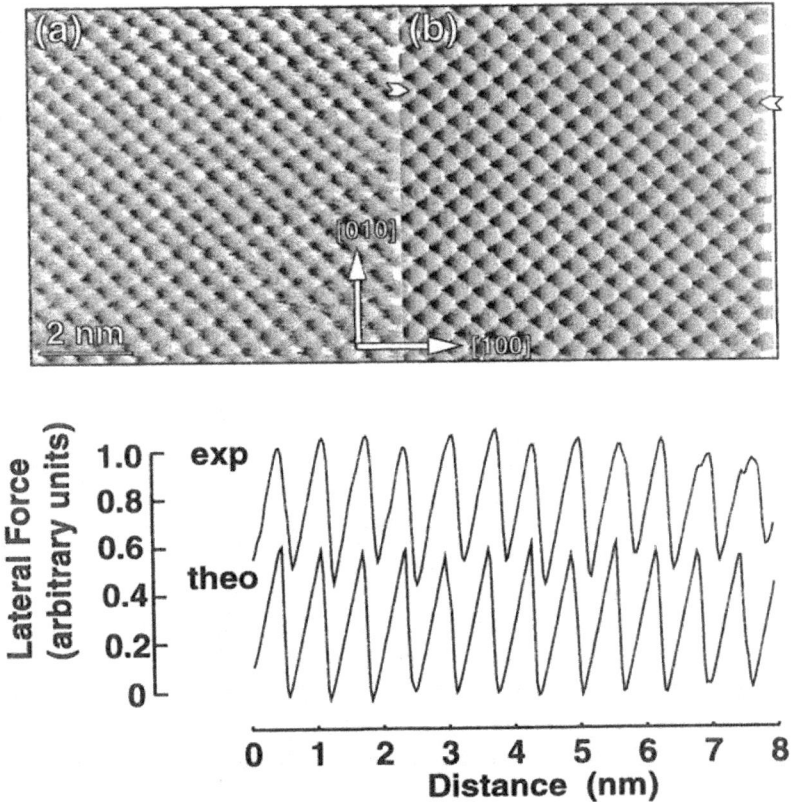

Figure 4.17: (a) Measured (b) theoretical lateral force map. The experiment was performed in UHV with a Si-tip. As shown in the profiles, the typical atomic-scale stick-slip is revealed in both images. The slope of the lateral force curve during the sticking part is about 7N/m. Taking into account the torsional spring constant of the cantilever of 32N/m, one can estimate the lateral contact stiffness to be about 10N/m. Thus, the softest spring is due to the sample/tip elasticity. From[26].

Poisson ratio ν_1 of KBr can be calculated in the following way:

$$E_1 = \frac{(c_{11} - c_{12} + 3c_{44})(c_{11} + 2c_{12})}{2c_{11} + 3c_{12} + c_{44}} \qquad (4.32)$$

$$G_1 = \frac{c_{11} - c_{12} + 3c_{44}}{5} \qquad (4.33)$$

$$K_1 = \frac{c_{11} + 2c_{12}}{3} \qquad (4.34)$$

$$\nu_1 = \frac{3K_1 - 2G_1}{2(3K_1 + G_1)} \qquad (4.35)$$

with $c_{11}=3.8 \cdot 10^{10} \mathrm{N/m^2}$, $c_{44}=0.64 \cdot 10^{10} \mathrm{N/m^2}$, $c_{12}=0.6 \cdot 10^{10} \mathrm{N/m^2}$[228].

According to these equations, the shear modulus of KBr $G_1=1.0 \cdot 10^{10} \mathrm{N/m^2}$ and the Poisson ratio of KBr is $\nu_1=0.25$. In combination with the shear modulus of silicon $G_2=6.8 \cdot 10^{10} \mathrm{N/m^2}$ and Poisson ratio $\nu_2=0.22$, the effective shear modulus $G^*=0.45 \cdot 10^{10} \mathrm{N/m^2}$. Then, a contact radius of $a=4.2$Å is found, which indicates that the contact may be formed with one single atom.

In order to test the consistency of the elasticity model, the Hertzian deformation in normal direction can also be used to estimate the contact radius:

$$a = \frac{3RF_N}{4E^*} \qquad (4.36)$$

where $R=15$nm was determined at step edges of KBr(001) and the effective Youngs modulus is given by

$$E^* = \left(\frac{1 - \nu_1^2}{E_1} + \frac{1 - \nu_2^2}{E_2} \right) \qquad (4.37)$$

with $E_1=2.55 \cdot 10^{10} \mathrm{N/m^2}$, $\nu_1=0.245$, $E_2=1.66 \cdot 10^{10} \mathrm{N/m^2}$, $\nu_2=0.217$. A value of $E^*=2.37 \cdot 10^{10} \mathrm{N/m^2}$ is calculated. Finally, the contact radius for 1nN normal force (including adhesive forces) is approximately 8Å. This is larger than the value, which was determined with the lateral contact stiffness. Equation 4.30 is independent of the elasticity model, whereas equation 4.36 is based on the Hertzian theory, which is again comparable to the lattice constant of 4.7Å. JKR or Maugis-Dugdale may appear as more suitable, but are not discussed here for briefness. More generally, we may have also arrived at the limit of continuum elasticity theory, because atomic dimensions are found. Nevertheless, there were recent MD-calculations which point out that elasticity theory may be valid with reasonable accuracy down to the nanometer scale[29].

Finally, we may determine the shear strength τ from the knowledge of the contact radius and the frictional force F_F:

$$\tau = \frac{64(G^*)^2 F_F}{\pi (k_{contact}^x)^2} \qquad (4.38)$$

With an average friction force F_F=1.5nN, a shear strength τ=0.19·10^{10}N/m^2 is found, which is 20% of the the shear modulus of KBr, G_1.

To summarize, we have observed that 2d-stick slip exists under UHV-conditions, but is limited to rather small normal forces. The contact radius, which can be estimated with lateral contact stiffness measurements, is in the range of atomic dimensions. Thus, it might be possible that single atom contacts are observable. High shear stresses are observed under these conditions. Due to these high lateral forces, which are applied to extremely small contacts (possibly single atom contacts), it becomes plausible, that only periodically arranged atoms, which are stabilized by their neighbours, are imaged, whereas vacancies and other defects are less stable and are not observed. Similar predictions were made by Shluger et al.[25] and Landman et al.[30].

4.5.2 Zig-zag walk

A careful analysis of 2d-stick slip has been made by Fujisawa et al.[31,32]. The samples were precisely orientated relative to the scan direction. The periodic stick-slip was observed in both direction (normal deflection and torsion). Scanning along and across the cantilever, characteristic changes were observed. First, the authors observed that the frictional force is not exactly in the direction of scanning, but gives contributions in both directions. This, observation is in contradiction to our macroscopic understanding of friction, where friction acts always in direction of motion. Second, Fujisawa et al. observe that the tip jumps always to the nearest sticking point. In terms of the previously discussed 2d-Tomlinson model, we may say that the tip jumps from one (++)-region to the closest (++)-region. It is known from calculations[34] that this behaviour is observed for the strongly damped case. Some of the data of Fujisawa et al. are shown in Fig. 4.18. Either a simple stick slip motion is observed, where the sticking points are in-line with the scan direction (cf. Fig. 4.18b) or a zig-zag motion is observed where the tip moves both in direction of scanning and perpendicular to the scanning direction (cf. Fig. 4.18a). The motion of the tip on the MoS$_2$-sample is schematically drawn in Fig. 4.18 c.

Other examples of the Fujisawa et al. are cited in the literature list[31], e.g., Si$_3$N$_4$-tip on mica or Si$_3$N$_4$-tip on NaF(001). The second case is remarkable, because the observation of 2d-stick slip on the non-layered material of NaF(001), where the formation of flakes is excluded, was limited to loads below 14nN. This in contrast to the results on layered materials, where loads up to 10^{-5}N were applied. Thus, the case of 2d-stick slip on NaF(001) is more likely to be related to a single-atom contact or "few-atoms" contact, where

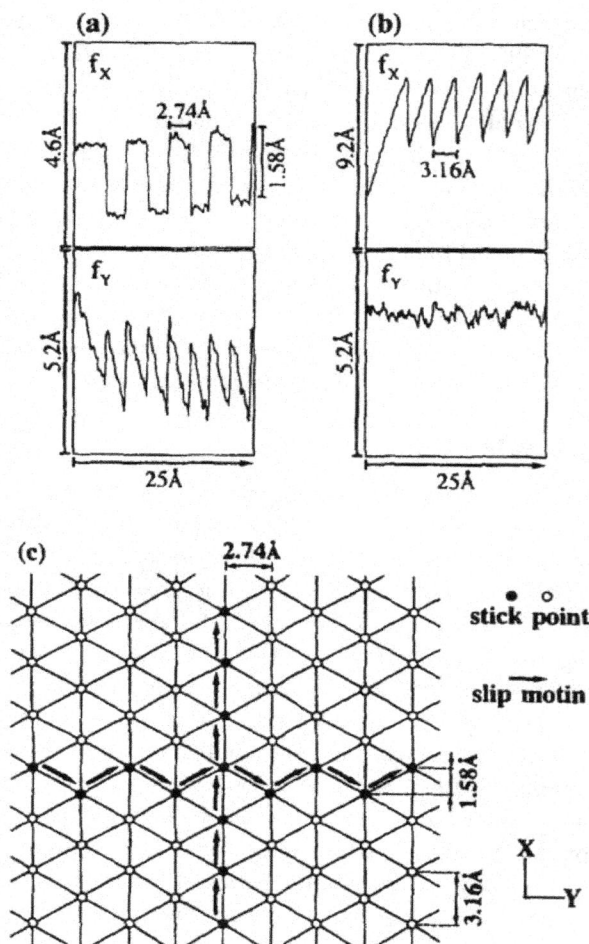

Figure 4.18: Profiles of FFM-measurements on MoS_2. (a) "Zig-zag walk" of the tip along Y-direction. Force changes are measured in both directions. Top: Square-wave behaviour with a periodicity of 5.4±0.6Å. Bottom: Sawtooth-like behaviour with a periodicity of 2.7±0.6Å. (b) Normal stick-slip in X-direction with a periodicity of 3.1±0.3Å, corresponding to the straight walk, where the scan direction is in-line with the positions of the sticking points. Top: Sawtooth-like behaviour. Bottom: no variations are detected. (c) Schematic diagram of the motion of the tip on the sample. The zig-zag walk in Y-direction corresponds to (a) and the straight walk corresponds to (b). From[32].

forces of the order of 1nN may be expected. Hölscher et al.[35] have performed dynamical calculations for the case of NaF(001), where the motion of the tip was followed as a function of time. They observed similar zig-zag walks and straigth walks as in the experiment of Fujisawa et al.[32] and Morita et al.[33]. They observed that the tip remains most of time in "areas of stability" regions and then jumps to the next region.

The results on graphite from Mate et al.[20] and later from Fujisawa et al.[31], were analysed theoretically from Sasaki and Tsukada[36], which confirmed the assumption of 2d-stick slip in the frame-work of the 2d-Tomlinson model. Recently, Hölscher et al. presented dynamical calculations, where the zig-zag walk was time-resolved. A "position probability density" showed that the tip stays most of the time in the hollow site of the carbon hexagon. These calculations explain the experimentally observed 3-fold symmetry. Ruan and Bhushan have postulated "every other atom" resolution by considering the lateral forces that act on a single atom ontop of the graphite surface, where also the difference between the inequivalent carbon atoms were taken into account (A-atom has neighbour in the lower layer, B-atom does not)[37].

A nice experimental confirmation of the zig-zag walk was presented by Kawawkatsu et al.[39]. The authors used two laser beams with two quadrant detectors. One beam was positioned at the end of the cantilever and close to the middle of the cantilever. With this set-up, they could separate normal bending from buckling. Finally, they could calculate the real motion of the tip on mica. In agreement with the 2d-Tomlinson model, the tip was found stay most of the time in sticking regions.

4.6 Friction between atomically flat surfaces

4.6.1 Introduction

The simple case of a single-asperity contact can be generalized to extended atomically flat surfaces in contact by making use of the perfect periodicity of a crystal lattice, neglecting any dislocation effects[41]. Also for the case of an FFM, this approximation should be better than the single-atom tip, since it is not possible to produce really mono-atomic tips. The above assumption (2), that every atom in the tip follows a global variable (in our case $\vec{r}$) adiabatically, generally does not hold. For extended periodic surfaces in contact the role of commensurability becomes important.

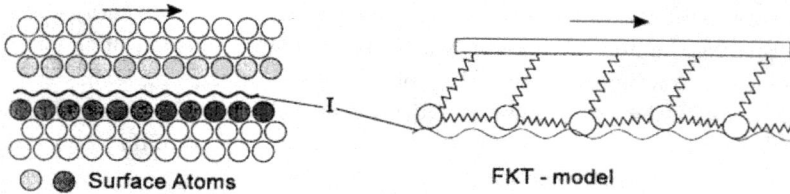

Figure 4.19: The lower surface is described through its surface potential (I) acting on the upper surface.

4.6.2 Commensurability in one dimension

Since for large contacts only a small percentage of the atoms is at the border, and additionally border effects have a short range, we can approximate the problem by studying the friction process between two infinite surfaces. It is well known from the one-dimensional case and intuitively clear, that friction between corrugated surfaces with lattice constants a_l (lower body) and a_u (upper body) depends strongly on their commensurability. This problem is strongly correlated to the problem of "rationality" of a fraction a_u/a_l, and a well-defined measure of "rationality" has not been developed yet. However, it is known that any number can be represented unambiguously as a continued fraction.

$$z = a_0 + \cfrac{1}{a_1 + \cfrac{1}{a_2 + \cfrac{1}{\ddots}}} \tag{4.39}$$

with a_i positive integers. The length of the continued fraction of rational numbers is finite, whereas for irrational numbers it is infinite. The irrational numbers can be approximated by their continued fraction up to a certain order, $z_N = (a_1, ..., a_N)$. The sequence z_N for $N \to \infty$ converges to the irrational number z. The continued fraction which converges most slowly is the one where all $a_i = 1$. This represents the golden mean $(\sqrt{5} - 1)/2$.

4.6.3 The one-dimensional Frenkel-Kontorova-Tomlinson model

In the FKT model the surfaces in contact are modelled as follows: The upper surface consists of surface particles with lattice constant a_u, which are harmonically coupled to their nearest neighbours. Additionally, every particle is coupled harmonically to a support which represents the bulk of the upper body. Friction is the total lateral force acting on the support. The lower surface is modelled by a periodic potential with lattice constant a_l. For computational reasons one often uses periodic boundary conditions. The size of the periodicity cell of the composite system in one dimension is $L = Na_u = Ma_l$, where N is the number of particles in the upper surface and M is the number of potential valleys. Since N and M are integer numbers, the ratio a_l/a_u must be rational.

Weiss und Elmer considered the friction dependence on commensurability in the one-dimensional FKT-model, shown in Fig. 4.19 by approaching the ratio between the lattice constants a_l/a_u subsequently to the golden mean, following the continued fractions, which are equal to the ratio of two successive Fibonacci numbers. They find that static friction decreases with decreasing commensurability, where static friction is defined as "smallest driving force above which no stable stationary state exists"! In Fig. 4.20 the static friction is shown as a function of the potential strength b. Clearly static friction vanishes for $b \to 0$. It is the important result of their work, that for incommensurate lattice constants static friction remains zero even for a finite potential strength and a critical potential strength b_c^S can be defined where static friction begins to increase.

4.6.4 Two-dimensional infinite surfaces

The effetct of commensurability can be treated in two dimensions, studying two given surfaces with fixed lattice constants rotated through an arbitrary misfit angle ϑ. The commensurability of the system is now represented as an angle.

4.6.5 Two-dimensional commensurate structures

We consider two infinitely extended bodies of the same material with identical surfaces, which are misaligned with respect to each other by an arbitrary angle ϑ. We decribe this system in terms of a two-dimensional FKT model: the upper surface, represented by a harmonic two-dimensional lattice, is dragged quasistatically across the lower surface, represented by an external potential with the same symmetry and periodicity as the upper surface (see Fig. 4.21). We restrict ourselves to quadratic symmetry. For simplicity, the lattice constant is chosen as unit of length. The lattice basis vectors of the upper and lower surface are denoted by $\vec{e}_{1,2}$ and $\vec{e}'_{1,2}$, respectively.

We expect important differences in friction behaviour between the case that the two surfaces form a commensurate structure with a finite periodicity

---

*The definition given of static friction may sound a little academic, but it is a good measure in terms of nonlinear dynamics.

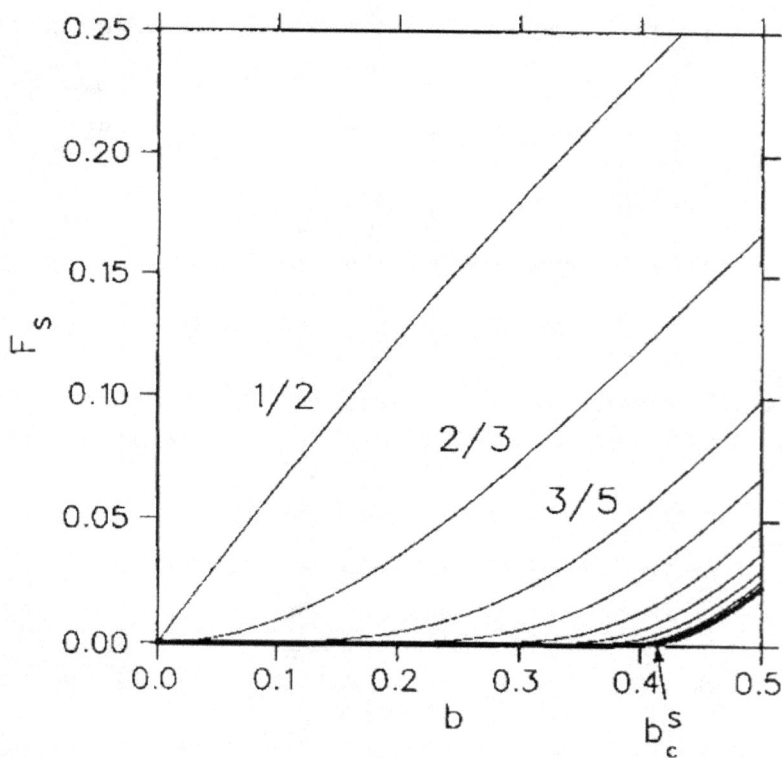

Figure 4.20: The commensurability dependence of friction

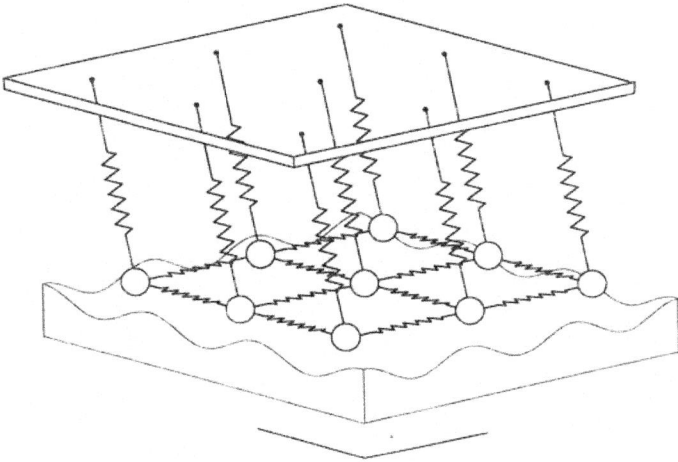

Figure 4.21: The two-dimensional FKT model.

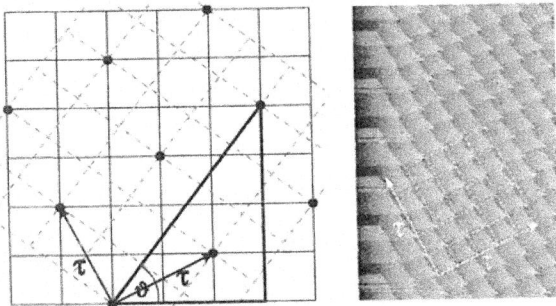

Figure 4.22: The allowed angles

cell of the composite system, and the incommensurate case. For computational reasons, incommensurate structures have to be approximated by suitable sequences of commensurate structures. Misfit angles ϑ giving rise to commensurate structures are of the form

$$\cos \vartheta = \frac{a}{N}, \quad \sin \vartheta = \frac{b}{N}, \quad a, b, N \text{ integer}, \tag{4.40}$$

(see Fig. 4.22). Clearly a, b and N constitute a pythagorean triplet. Since every pythagorean triplet can be generated from two integers p, q $(p > q)$ by $a = p^2 - q^2$, $b = 2pq$, $N = p^2 + q^2$, the permitted angles are given by

$$\tan \frac{\vartheta}{2} = \frac{q}{p}. \tag{4.41}$$

The elementary pythagorean triplets (a, b, N have no common divisor) are obtained from pairs p, q which are relative prime and of opposite parity. The ratios q/p form a dense subset of all rational numbers between 0 and 1, i. e., the corresponding angles lie dense on the ϑ-axis.

The smallest periodicity cell of the composite system has lattice constant $\sqrt{N}$ and contains N particles (see Fig. 4.22a). It is spanned by the vectors

$$\vec{\tau}_1 = \sqrt{N} \, (\vec{e}_1 \cos \frac{\vartheta}{2} + \vec{e}_2 \sin \frac{\vartheta}{2}), \quad \vec{\tau}_2 = \sqrt{N} \, (-\vec{e}_1 \sin \frac{\vartheta}{2} + \vec{e}_2 \cos \frac{\vartheta}{2}). \tag{4.42}$$

4.6.6   The two-dimensional FKT model

The position of the l-th particle in the periodicity cell is denoted by $\vec{r}_l = \vec{R}_l + \vec{\rho}_l$ where $\vec{\rho}_l$ is the displacement from its unperturbed equilibrium position $\vec{R}_l$, and the relative shift of the two bodies is denoted by $\vec{R}$. The configurations which determine the friction behaviour do not break the translational symmetry of the energy, i. e., they satisfy periodic boundary conditions $\vec{\rho}_l = \vec{\rho}(\vec{R}_l) = \vec{\rho}(\vec{R}_l + \vec{\tau}_{1,2})$. The total energy per periodicity cell of the model is

$$E_{tot} = \sum_{l=1}^{N} V(\vec{r}_l) + \frac{1}{2} \sum_{l,r=1}^{N} \vec{\rho}_l \cdot \mathsf{D}(\vec{R}_{lr}) \cdot \vec{\rho}_r \tag{4.43}$$

with $V(\vec{r}_l + \vec{e}_1') = V(\vec{r}_l + \vec{e}_2') = V(\vec{r}_l)$. The non-vanishing elements of the dynamical matrix $\mathsf{D}(\vec{R}_{lr})$ are

$$\mathsf{D}(0) \;=\; (c_T + 2c_{FK} + 2c_B) \, \mathbb{1} \tag{4.44}$$

$$\mathsf{D}(\pm \vec{e}_1) \;=\; -(c_{FK} \, \vec{e}_1\vec{e}_1 + c_B \, \vec{e}_2\vec{e}_2) \tag{4.45}$$

$$\mathsf{D}(\pm \vec{e}_2) \;=\; -(c_B \, \vec{e}_1\vec{e}_1 + c_{FK} \, \vec{e}_2\vec{e}_2). \tag{4.46}$$

Here, c_T denotes the (Tomlinson) spring constant between a particle and the bulk of the upper body, c_{FK} denotes the (Frenkel-Kontorova) spring constant for the bond stretching between adjoining particles, and in addition, we have introduced a spring constant c_A describing a stiffness of the bond angles.

The equilibrium values of the displacements $\vec{\rho}_l$ for a given shift $\vec{R}$ are determined by minimizing the energy,

$$\frac{\partial V}{\partial \vec{r}_l} + \sum_{r=1}^{N} \mathsf{D}(\vec{R}_{lr}) \cdot \vec{\rho}_r = 0. \tag{4.47}$$

We assume a sufficiently strong damping such that during quasistatic sliding the system moves with the local minimum, and at an instability, it follows the path of steepest descent.

The friction force is defined as the component of the lateral force acting on the bulk of the upper body along the pulling direction $\vec{e}_p$. For the friction force per particle one obtains in the limit of vanishing sliding velocity

$$F_{fric}(\vec{R}) = \frac{c_T}{N} \sum_{l=1}^{N} \vec{\rho}_l \cdot \vec{e}_p. \tag{4.48}$$

4.6.7 Symmetry of the force scan image

The positions $\vec{R}_l + \vec{\tau}_{1,2}/N$ are symmetry-equivalent to the positions $\vec{R}_{p_{1,2}(l)}$ where p_1 and p_2 are certain cyclic permutations of the particles in the periodicity cell. Therefore, in addition to the translation invariance $(\vec{\tau}_1, \vec{\tau}_2)$ due to the periodic boundary conditions, the total energy remains invariant under translations $(\vec{\tau}_1/N, \vec{\tau}_2/N)$, combined with permutations p_1, p_2 of the displacements $\vec{\rho}_l$ within the periodicity cell,

$$\vec{R}_l \mapsto \vec{R}_l + \frac{1}{N}\vec{\tau}_1, \qquad \vec{\rho}_l \mapsto \vec{\rho}_{p_1(l)} \tag{4.49}$$

and

$$\vec{R}_l \mapsto \vec{R}_l + \frac{1}{N}\vec{\tau}_2, \qquad \vec{\rho}_l \mapsto \vec{\rho}_{p_2(l)}. \tag{4.50}$$

Due to the damping, after some jumps the system follows a periodic orbit in configuration space with periods $\vec{\tau}_1$ and $\vec{\tau}_2$, respectively.

In a force scan only the sum over all displacements is recorded, and as soon as the system has reached the periodic orbit, the scan has the same translation symmetry $(\vec{\tau}_1/N, \vec{\tau}_2/N)$ as the total energy (see Fig. 4.22b). Thus, the force scan can be derived from an effective potential $\bar{V}(\vec{R})$ with translational symmetry $(\vec{\tau}_1/N, \vec{\tau}_2/N)$, which can be calculated analytically in the stiff-spring limit $c_{FK}, c_B \to \infty$.

The average friction force per particle is obtained by integrating over one unit cell Ω of the force scan,

$$\langle F_{fric} \rangle = \frac{1}{\Omega} \iint_{(\Omega)} F_{fric}(\vec{R}) \, d^2 R. \tag{4.51}$$

4.6.8   Friction in the FKT model

For sufficiently hard springs the mean friction force vanishes. There exists a well-defined spring constant $c_{T,crit}$ as a function of c_{FK} and c_B, such that no instabilities occur and friction becomes zero for $c_T > c_{T,crit}$. in the case of a separable potential

$$V(\vec{r}) = v \left[\cos(2\pi \vec{r} \cdot \vec{e}_1') + \cos(2\pi \vec{r} \cdot \vec{e}_2') \right] \tag{4.52}$$

it takes its maximum value $c_{T,crit} = 4\pi^2 v$ which is independent of the misfit angle in the Tomlinson limit $c_{FK}, c_B \to 0$, and decreases initially with slope -4. In the stiff-spring limit $c_{FK}, c_B \to \infty$, $c_{T,crit}$ approaches zero. This behaviour is specific for the potential given in (4.52). If the potential contains higher harmonics, then $c_{T,crit}$ remains positive in the stiff spring limit. Numerical results for various misfit angles are shown in Fig. 4.23. Corresponding calculations for the one-dimensional case reproduce the curves for $F_{frict} = 0$ calculated by Weiss and Elmer[40].

Dependence on the pulling direction

From the attributes of one-particle scans and by making use of the effective potential $\bar{V}(\vec{R})$ it becomes plausible that the friction force for a given misfit angle as a function of the pulling direction has a minimum for a pulling angle $\varphi = \vartheta/2$ and a maximum for $\varphi = \vartheta/2 + \pi/4$. Fig. 4.24a shows results for the dependence of the friction force on the pulling angle obtained by numerical simulations.

Dependence on the misfit angle

Since the misfit angles giving rise to commensurate structures form a dense subset, one may expect that the dependence of the friction force on the misfit angle is everywhere discontinuous, being large for structures with small periodicity cells and tending to zero for incommensurate structures. Numerical simulations show that in fact the dependence on the misfit angle ϑ is not continuous, and that friction tends to become smaller for increasing area N of the periodicity cell. As shown in Fig. 4.24b, this dependence becomes even more distinct if friction values are compared at misfit angles corresponding to ratios a/b which have continued fraction representations coinciding in the leading orders. A similar behaviour has been found in the one-dimensional case by Weiss and Elmer[40], who have shown that friction goes to zero for commensurability ratios approaching the golden mean.

Significance for the experiment

Since a force scan of the present model is equivalent to a force scan of a single particle in the effective potential $\tilde{V}(\vec{R})$, it cannot be decided from a single force scan whether it originates from a single-atom tip or from a large-area tip. In order to establish that one has in fact a single-atom tip, it is necessary to perform force scans for different orientations between tip and sample, which must show the same symmetry.

Domains, slip lines and misfit centers

In order to understand the friction mechanism, it is important to identify the coordinates which undergo instabilities and release the energy stored in the springs during the sliding process in irreversible jumps.

The structure changes occurring during the sliding process are relatively easy to follow for small misfit angles. The equilibrium structure consists of domains in which the particles of the upper surface try to fit into the potential valleys of the lower surface, such that there is exactly one particle in each valley. The misfit gives rise to the occurrence of slip lines which form a quadratic lattice with sides $(\vec{\tau}_1 \pm \vec{\tau}_2)/2$ (Fig. 4.25).

At the intersections of the slip lines, the displacements of the particles form vortex-like structures, half of them associated with an extra particle, the other half with a vacancy at its center (Fig. 4.25). These *disclinations* play a role similar to the discommensurations (kinks and antikinks) occurring in the one-dimensional case[40].

During the sliding process, this domain pattern moves across the surface approximately *perpendicular* to the sliding direction, and it is the configuration of the misfit centers which for $c_T < c_{T,crit}$ undergoes irreversible jumps.

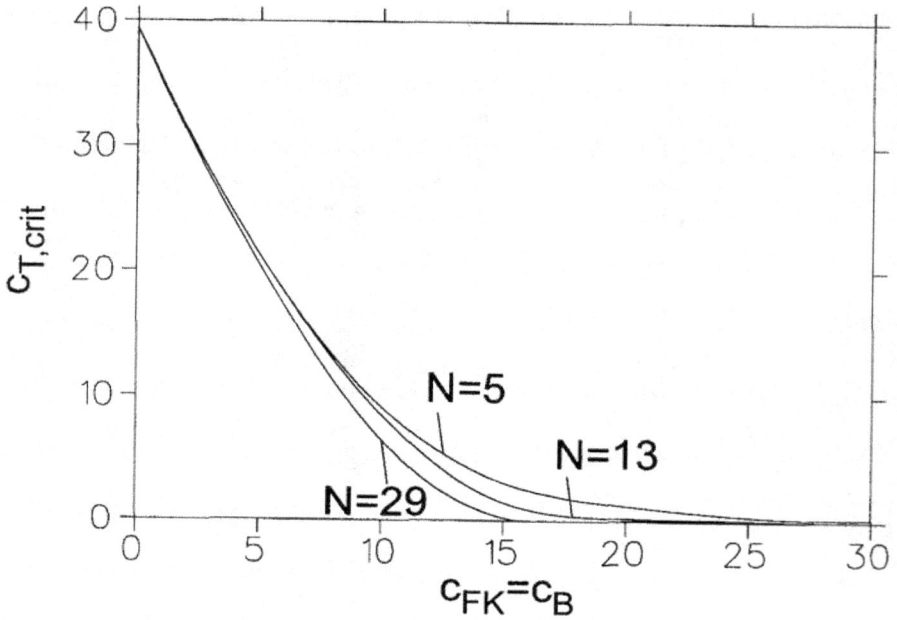

Figure 4.23: The critical value $c_{T,crit}$ of the Tomlinson spring constant where instabilities disappear as a function of $c_{FK} = c_B$ for a potential given in 4.22. (From[41])

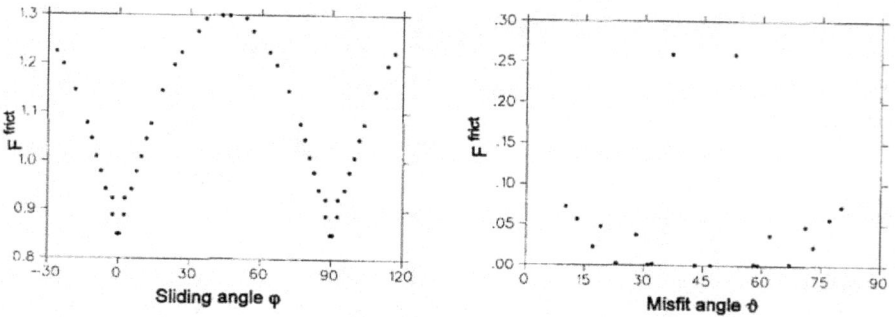

Figure 4.24: Anisotropy of atomic friction. a) as a function of sliding direction and fixed misfit angle and b) as a function of misfit angle and sliding direction corresponding to the maximum friction. (From[41])

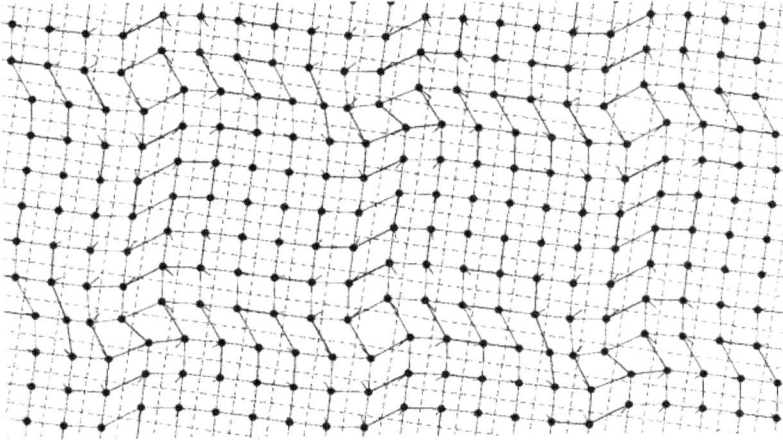

Figure 4.25: Disclination formation during a friction process. The pattern moves perpendicular to the sliding direction. (From[41])

4.6.9 Finite area of contact

For friction between a small cluster and an extended surface, periodic boundary conditions may not hold, that is, a tip containing 10–100 atoms. One may describe the tip as a cluster of atoms in the outer potential, which are coupled to another by an interatomic force $f(d\vec{r})$, which may be of the Lennard-Jones type.

We restrict ourselves to the case of wearless friction, where every atom remains more or less at its given position, which means, that it remains very near to the equilibrium position of the interatomic potential, and we therefore can expand the atom-atom interaction into a Taylor series and neglect terms higher than second order. Only nearest-neighbour interaction has to be considered in such a case, since the interatomic potential rapidly goes to zero at these length scales.

For simplicity we assume that the surface of the cluster has quadratic symmetry, same as the potential itself. but may be rotated by a misfit angle ϑ. The computed force scan images look a bit strange for small tip surfaces. For larger surfaces it becomes clearer because border effects reduce by a factor

$1/N$, where N^2 is the number of atoms of the tip surface.[g] It is therefore reasonable to neglect these border effects and to study infinitely large systems.

4.7 Molecular dynamics simulations: Quantitative results

In the last decade computer power has increased rapidly, and therefore new powerful numerical techniques have been developed. The most fascinating one used in surface physics is molecular dynamics (MD).

In a MD Simulation, the Newtonian equations of motion are solved numerically for a system containing up to ten thousands of particles. To be able to consider temperature effects, various methods have been developed to make simulations even at finite temperatures.

The potentials used in MD to determine the interaction forces are often sophisticated model potentials, precise enough to get even quantitative results comparable to experimental data. The MD technique makes it possible to reconstruct experiments with the big advantage that the processes can be visualized as a movie in real space.

The quality of the simulation strongly depends on the validity of the chosen potential. For simple Lennard-Jones potentials qualitative results can be obtained[44], but in order to get more realistic results, the chemical nature of the materials has to be considered.

For example, the Stilling-Weber potential, a realistic interatomic interaction potential, has been used by Landman et al.[30] to simulate the scanning process of an AFM tip on a Si (111) surface in both the constant height and the constant force mode.

In the constant force mode for a relatively strong force, where the tip is in contact with the substrate, stick-slip motion has been observed. In Fig. 4.26d a plot of friction force vs. distance is given. In figs. 4.26a-c, real space snapshots of the system are shown with the trajectories of all the particles in the system. The motion of the tip atoms in contact with the substrate is seen to lie in the scanning direction.

The tip atoms closest to the substrate attempt to remain in a favorable bonding environment as the tip-holder assembly proceeds to scan. When the forces on these atoms due to the other tip atoms exceed the forces from the substrate, they move rapidly to minimize the F_x force, by breaking their current bonds to the surface and forming new bonds in a region shifted by one unit

---

[g]The border grows $\propto N$, the area grows $\propto N^2$, hence the percentage of atoms in the border decreases with $1/N$.

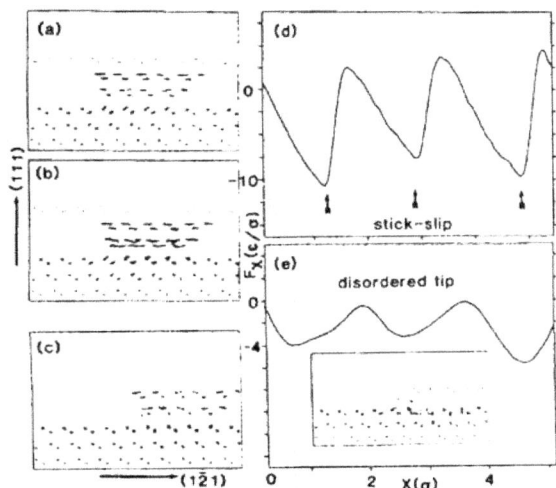

Figure 4.26: The stick slip process in the silicon system using molecular dynamics calculations. From Landman et al.[30]

cell along the scan direction. The corresponding friction coefficient $\mu = 0.77$ is in the range of typical values obtained under UHV conditions.

It is important to note, that in these simulations it is not the tip undergoing an instability, but the tip atoms in contact themselves, since the cantilever is stiff compared to the elasticity of the chemical bonds.[h]

Molecular dynamical simulations have prooved their applicability also to friction physics of lubricated systems. Bonner et al.[45] considered the fricton process between the tip of an AFM and a self-assembled monolayer (SAM) of alkane-thiolates chemisorbed on a gold substrate by means of MD. The experimental results of Liu et al.[47] for an approach-retraction cycle are reproduced on the computer during simulation. A continuous increase of the number of gauche defects is observed during an increase of the normal force. After a shift of the tip at constant height, the typical stick-slip behaviour of the force scan is observed. In[46] Mc Clelland and Glosli present a MD simulation of a friction process between two close-packed alkane films with identical atomic lattices, no rotation, schematically shown in Fig. 4.27a. Each alkane chain contains 6 carbon atoms, which are presented with their attached hydrogene atoms as one single particle. These atom groups interact with each other through Lennard-

---

[h]In the work discussed, an infinitely stiff cantilever was assumed

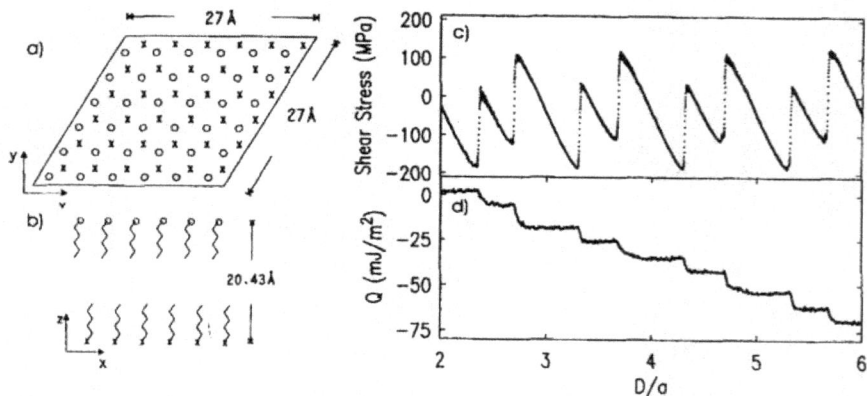

Figure 4.27: MD Simulation from Mc Clelland and Glosli. (See text for details). From[46].

Jones potentials. Additionally, there are realistic potentials within the chain wich allow bending and twisting, but no stretching. Simulations are performed with nearly no extrinsic load, so that the effective load is due to adhesive forces between the surfaces. During simulation, the shear stress, i.e. friction force per unit area, is recorded as a function of the sliding distance. The result for a small sliding speed of 2.3 m/s at a low temperature of 20 K shows typical sawtooth behaviour and is given in Fig. 4.27b. In Fig. 4.27c the cumulative heat flow into the heat reservoir of the MD system is plotted. The heat flow is non-zero only for positions corresponding to a slip (compare parts c) and d) of Fig. 4.27) of the relative motion of the surfaces.

1. A.J. den Boef, The influence of lateral forces in scanning force microscopy, *Rev. Sci. Instrum.* **62**, 88-92 (1991).
   E. Meyer, R. Overney, D. Brodbeck, L. Howald, R. Lüthi, J. Frommer and H.-J. Güntherodt in *Fundamentals of Friction*, p. 427, Eds. Singer I. and Pollock, H. Vol. 220, Kluwer Academic Publisher (1992).
   S. Fujisawa, Y. Sugawara and S. Morita, *Microbeam Analysis* **2** (1993).
   S. Grafström, M. Neitzert, T. Hagen, J. Ackermann, R. Neumann, O. Probst and M. Wörtge, The role of topography and friction for the image contrast in lateral force microscopy, *Nanotechnology* **4**, 143-151 (1993).
2. C.M. Mate, Nanotribology studies of carbon surfaces by force microscopy, *Wear*, **168**, 17-20 (1993).

C.M. Mate, Nanotribology of lubricated and unlubricated carbon over-coats on magnetic disks studied by friction force microscopy, *Surface and Coatings Technology*, **62**, 373-379 (1993).

3. M. Binggeli, R. Christoph, H.-E. Hintermann, J. Colchero and O. Marti, Friction force measurements on potential controlled graphite in an electrolyte environment, *Nanotechnology*, **4**, 59-63 (1993).

4. O. Marti, (private communication).

5. G. Meyer and N. Amer, Simultaneous measurement of lateral and normal forces with an optical-beam-deflection atomic force microscope, *Appl. Phys. Lett.* **57** 2089-2090 (1990).

6. E. Meyer, L. Howald, R. Lüthi, H. Haefke, M. Rüetschi, T. Bonner, R. Overney, J. Frommer, R. Hofer and H.-J. Güntherodt, Force microscopy on the surface of Si(111), *J. Vac. Sci. Techn. B* **12**, 2060-2063 (1994).

7. L. Howald, H. Haefke, R. Lüthi, E. Meyer, G. Gerth, H. Rudin, and H.-J. Güntherodt, Ultrahigh vacuum scanning force microscopy: Atomic resolution at monatomic cleavage steps, *Phys. Rev. B* **49**, 5651-5656 (1993).
L. Howald, R. Lüthi, E. Meyer, G. Gerth, H. Haefke, R. Overney and H.-J. Güntherodt, Friction force microscopy on clean surfaces, *J. Vac. Sci. Techn. B* **12**, 2227-2230 (1994).

8. R.L. Schwoebel and and E.J. Shipsey, *J. of Appl. Phys.* **37**, 3682, (1966).

9. E. Meyer, R. Lüthi, L. Howald, W. Gutmannsbauer, H. Haefke and H.-J. Güntherodt, Friction force microscopy on well defined surfaces, *Nanotechnology* **7**, 340-344 (1996).

10. C.A.J. Putmann, M. Igarshi and R. Kaneko, Experimental observation of single-asperity friction at the atomic scale, *Appl. Phys. Lett.* **66**, 3221 (1995).

11. M. Hu, X.-d. Xiao, D.F. Ogletree, and M. Salmeron, Atomic scale friction and wear of mica, *Surf. Sci.* **327**, 358-370 (1995).

12. K.L. Johnson, K. Kendall and A.D. Roberts, *Proc. R. Soc. Lond.* **A324**, 301 (1971).

13. G.A. Tomlinson, *Philosoph. Mag. Ser.* **7**, 905 (1929).

14. L. Prandtl, (1913). See e.g. Ein Gedankenmodell zur kinetischen Theorie der festen Koerper, *Z. angew. Math. Mechanik* **8**, 85-106 (1928).

15. C. Caroli and P. Nozieres, in *Physics of Sliding Friction,* B.N.J. Persson and E. Tosatti (Eds.), p. 27-49, Kluwer Academic Publishers (1996).
16. G.M. McClelland and J.N. Glosli, Friction at the atomic scale, in *Fundamentals of Friction: Macroscopic and Microscopic Processes* edited by I.L. Singer and H.M. Pollock, p. 405-425, NATO ASI Series E: Applied Sciences, Vol. 220, Kluwer Academic Publishers (1992).
17. W. Zhong and D. Tomanek, *Phys. Rev. Lett.* **64**, 3054 (1990).
18. D. Tomanek, W. Zhong and H. Thomas, *Europhys. Lett.* **15**, 887 (1991).
    D. Tomanek, p. 269 in *Scanning Tunneling Microscopy III*, Eds. R. Wiesendanger and H.-J. Güntherodt, Springer Berlin (1993).
19. J. Colchero, O. Marti and J. Mlynek, p. 345 in *Forces in Scanning Probe Methods*, Eds. H.-J. Güntherodt, D. Anselmetti and E. Meyer, NATO ASI Series E, Vol. 286, Kluwer Academic Publishers (1995).
20. C.M. Mate, G.M. McClelland, R. Erlandsson, and S. Chiang, Atomic-Scale Friction of a Tungsten Tip on a Graphite Surface, *Phys. Rev. Lett.* **59**, 1942-1945 (1987).
21. J.B. Pethica, Comment on interatomic forces in scanning tunneling microscopy: Giant corrugations of the graphite surface, *Phys. Rev. Lett.* **57**, 3235 (1986).
22. J.N. Israelachvili, *Intermolecular and Surface Forces*, Academic Press, London (1985).
23. G.J. Germann, S.R. Cohen, G. Neubauer, G.M. McClelland and H. Seki, Atomic scale friction of a diamond on diamond(100) and (111) surfaces, *J. Appl. Phys.*, **73**, 163-167 (1993).
    G.J. Germann, G.M. McClelland, Y. Mitsuda, M. Buck and H. Seki, Diamond force microscope tips fabricated by chemical vapor deposition, *Rev. Sci. Instrum.*, **63**, 4053-4055 (1992).
24. A.I. Livshits and A.L. Shluger, Self-lubrication in scanning-force-microscope image formation on ionic surfaces, *Phys. Rev B* **56**, 12482-12489 (1997).
    A.I. Livshits and A.L. Sluger, Role of tip contamination in scanning force microscopy imaging of ionic surfaces, *Faraday Discuss.* **106**, 425-442 (1997).
25. A.L. Shluger, R.T. Williams and A.L. Rohl, *Surf. Sci.* **343**, 273 (1995).
26. R. Lüthi, E. Meyer, M. Bammerlin, L. Howald, H. Haefke, T. Lehmann, C. Loppacher, H.-J. Güntherodt, T. Gyalog and H. Thomas.

Friction on the atomic scale: An ultrahigh vacuum atomic force microscopy study on ionic crystals, *J. Vac. Sci. Technol.* B **14**, 1280-1284 (1996).

27. K.L. Johnson, *Contact Mechanics*, Cambridge University Press, Cambridge, United Kingdom, (1985).

28. S. Hearman, Elastic Constants of Anisotropic Materials, *Adv. in Physics*, **5**, 323-382 (1956).

29. J. Belak and I.F. Stowers, The Indentation and scraping of a metal surface: A molecular dynamics study, p. 511-520, in *Fundamentals of Friction: Macroscopic and Microscopic Processes*, I.L. Singer and H.M. Pollock (eds.), NATO ASI Series E: Applied Sciences, Vol. 220, Kluwer Academic Publishers (1992).

30. U. Landman, W.D. Luedtke and M.W. Ribarsky, *J. Vac. Sci. Technol.* A **7**, 2829 (1989).

31. S. Fujisawa, Y. Sugawara, S. Ito, S. Mishima, T. Okada and S. Morita, The two-dimensional stick-slip phenomeno with atomic resolution, *Nanotechnology* **4**, 138-142 (1993).
S. Fujisawa, E. Kishi, Y. Sugawara and S. Morita, Two-dimensionally discrete friction on the NaF(100) surface with the lattice periodicity, *Nanotechnology* **5**, 8-11 (1994).
S. Fujisawa, E. Kishi, Y. Sugawara and S. Morita, Lateral force curve for atomic force/lateral force microscope calibration, *Appl. Phys. Lett.* **66**, 526-528 (1994).
S. Fujisawa, E. Kishi, Y. Sugawara and S. Morita, Fluctuation in Two-Dimensional Stick-Slip Phenomenon Observed with Two-Dimensional Frictional Force Microscopy, *Jpn. J. Appl. Phys. Lett.* **33**, 3752-3755 (1994).
S. Fujisawa, Y. Sugawara and S. Morita, *Microbeam Analysis* **2** (1993).
S. Morita, S. Fujisawa and Y. Sugawara, *Surf. Sci. Rep.* **23**, 3 (1996).
S. Fujisawa, E. Kishi, Y. Sugawara and S. Morita, Atomic-scale friction observed with a two-dimensional friction force microscope, *Phys. Rev.* B **51**, 7849-7857 (1995).
S. Fujisawa, M. Ohta, T. Konishi, Y. Sugawara and S. Morita, Difference between the forces measured by an optical lever deflection and an optical interferometer in atomic force microscope, *Rev. Sci. Instrum.* **65**, 644-647 (1994).

32. S. Fujisawa, E. Kishi, Y. Sugawara and S. Morita, Atomic-scale friction observed with a two-dimensional friction force microscope, *Phys. Rev.* B **51**, 7849-7857 (1995).

33. S. Morita, S. Fujisawa and Y. Sugawara, *Surf. Sci. Rep.* **23**, 3 (1996).

34. T. Gyalog, M. Bammerlin, R. Lüthi, E. Meyer and H. Thomas, Mechanism of Atomic Friction, *Europhysics Letters* **31**, 5-6 (1995).

35. H. Hölscher, U.D. Schwarz and R. Wiesendanger, Modelling of the scan process in lateral force microscopy, *Surf. Sci.* **375**, 395-402 (1997).

    H. Hölscher, U.D. Schwarz and R. Wiesendanger, Simulation of a scanned tip n a NaF(001) surface in friction force microscopy, *Europhysics Letters* **36**, 19-24 (1996).

    H. Hölscher, U.D. Schwarz, O. Zwörner and R. Wiesendanger, Stick-slip movement of a scanned tip on a graphite surface in scanning force microscopy, *Z. Phys. B* **17**, (1997).

    U.D. Schwarz, H. Bluhm, H. Hölscher, W. Allers and R. Wiesendanger, p. 389, in *The Physics of Sliding Friction*, Eds. B.N.J. Persson and E. Tosatti, NATO ASI Series E, Vol. 311, Kluwer Academic Publisher (1996).

36. N. Sasaki, M. Kobayashi and M. Tsukada, *Phys. Rev. B* **54**, 2138 (1996).

    N. Sasaki, M. Kobayashi and M. Tsukada, *Surf. Sci.* **357/358**, 92 (1996).

37. J. Ruan and B. Bhushan, *J. Appl. Phys.* **76**, 5022 (1994).

38. J. Kerssemakers and J.Th.M. De Hosson, Influence of spring stiffness and anisotropy on stick-slip atomic force microscopy imaging, *J. Appl. Phys.* **80**, 623-632 (1996).

39. H. Kawakatsu and T. Saito, p. 55, in *Micro/Nanotribology and its Applications*, Ed. B. Bhushan, NATO ASI Series E, Vol. 330, Kluwer Academic Publishers (1997).

    H. Kawakatsu and T. Saito, *J. Vac. Sci. Technol. B*, **14** 872 (1996).

40. M. Weiss and F. J. Elmer, Phys. Rev. B **53**, 7539 (1996)

41. T. Gyalog and H. Thomas, Friction Between Atomically Flat Surfaces, *Europhysics Letters* **37**, 195 (1997).

42. J.B. Sokoloff, *Surf. Sci.* **144**, 267 (1984).

43. J.A. Harrison, C.T. White, R.J. Colton and D.W. Brenner, *Phys. Rev. B* **46**, 9700 (1992).

44. M. Cieplak, E.D. Smith and M.O. Robbins, *Science* **265**, 1209 (1994).

45. T. Bonner and A. Baratoff, "Molecular dynamics study of scanning force microscopy on self-assembled monolayers", Surf. Science **377-379** 1082 (1997)

46. I. L. Singer and H. M. Pollock (eds.) Fundamentals of Friction: Macroscopic and Microscopic Processes, Kluwer Academic 1992, Dordrecht NL

47. G.-Y. Liu and M. B. Salmeron, Langmuir **10** 367 (1994)

Chapter 5

Dissipation mechanisms

5.1 Introduction

In order to understand the dissipation problem in a friction process, it is crucial to take a look at the basics of the phenomenon of dissipation in a more general way. In such a dissipation process, energy is transferred from the macroscopic degree of freedom in an irreversible manner into a set of microscopic degrees of freedom which act as a heat bath. This process defines a time arrow. The task to understand a specific dissipation mechanism explicitly is to find out about the heat bath itself and its coupling to the macroscopic degree of freedom.

Some general remarks about dissipation:

- Dissipation does only exist in a closed and not in a open system.

- Dissipation is the positive net transfer of energy crossing the system border.

- Very generally dissipation can be understood as the transfer of energy from a lower dimensionality (less degrees of freedom) to a higher dimensionality. The positive irreversible net transfer of energy is strongly related to the probability that energy flows in or out of the system. The probability, P_{21}, that energy flows from the higher dimensionality to the lower dimensionality is smaller than the probability, P_{12}, for energy flow in the opposite direction. (This sentence explains Entropy). For friction we usually consider the transfer of energy from 2D into in a high dimensional system. Without talking about phonons (or single molecular vibrations) dissipation can occur also in a macroscopic model with probabilities. For instance, the cantilever in a force microscope, which can vibrate in all kinds of different modes.

The phenomenon of dissipation is still poorly understood. So far, two dissipation channels are believed to be important in a friction process: The so-called phononic friction due to the emission of acoustic waves spreading out through the solid or the liquid and the electronic friction due to electron-hole pair excitation in conducting materials. In liquids additionally viscous friction is important.

5.2   Friction behaviour in the limit $v \mapsto 0$

Dissipation channels can be classified according to their behaviour in the quasistatic limit. For zero velocity only dissipation channels that undergo mechanical instabilities may be open. For finite velocity additional dissipation channels may become important, i.e. van der Waals friction, but also various different mechanisms. ([3]) Another mechanisms, which will not be discussed here in detail are Schallamach waves (dislocation instead of sliding)[4].

5.3   Phononic friction

Phononic friction is the most important dissipation mechanism between sliding solids. In a mechanistic picture of friction, e.g. in Tomlinson's model, the emission of phonons is obvious and the coupling mechanism from the macroscopic degree of freedom to the heat bath, consisting of the vibrational modes of the solid, is trivially given mechanically since the vibration modes of the surfaces naturally are directly coupled to the irreversible jumps of certain collective coordinates. However in a multi-asperity contact or even between atomically flat surfaces the coupling through the internal degrees of freedom of the interface between the bodies in contact might be very complicated and the coupling between the relative distance of the macroscopic bodies into their vibrational modes might be very complicated.

Even a quantitative description at least in the harmonic approximation has been developed[8,7] to compute the phononic friction kernel $\eta_{ph}(t)$, which describes the energy transfer into the normal modes of a harmonic crystal.

5.4   Electronic friction

In conductors there can be electronic friction due to electron-hole pair excitations in addition to the phononic friction. From a theoretical point of view it is rather complicated to calculate the electronic contributions correctly, since neither the heat bath nor its coupling to macroscopic degrees of freedom are naturally given. However, electronic friction is observed experimentally by

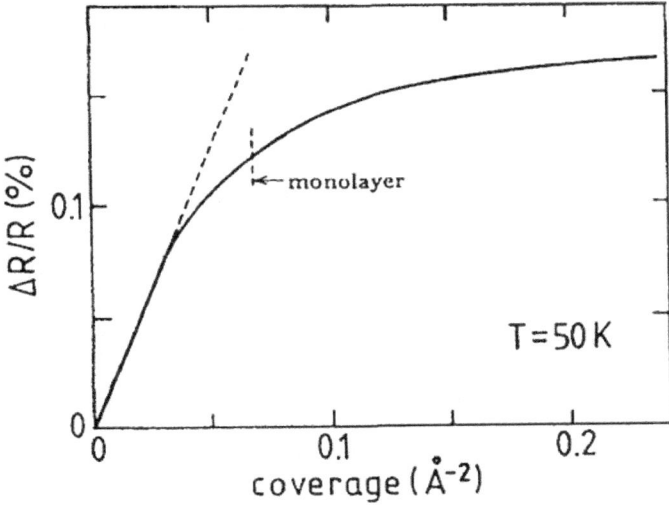

Figure 5.1: Measurement of electronic friction due to adsorption of Xe-atoms. From Persson et al.[5]

making use of the change of effective resistance of the metals in contact. In order to understand this point, we consider a thin conducting film and adsorbed on this film there are small particles of mass M in rest. When a current of density J flows within the film, the conduction electrons are in relative motion to the adsorbed particles, with a relative velocity $v = J/ne$, where n is the number of conduction electrons per unit volume within the film. Persson[9] showed that whenever a certain kind of electronic friction $F_{frict,el} = M\eta_{el}v$ exists, it reduces the current within the film due to energy transfer and therefore gives rise to an additional specific resistance $\Delta\rho$ of the film, which is proportional to the coverage $n_a = N/A$ of the adsorbed particles and independent of the current:

$$\Delta\rho = \eta_{el}\frac{Mn_a}{(ne)^2 d} \qquad (5.1)$$

This electronic friction of adparticles has been measured using the QCM, described in chapter II. The resulting additional resistance $\Delta\rho$ as a function of the coverage is shown in Fig. 5.1. Recently, Krim and coworkers could demonstrate that friction on lead is reduced below the superconducting transition temperature, which indicates that electronic friction is of importance[18].

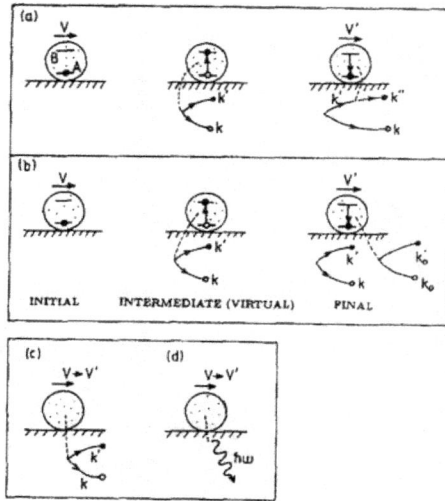

Figure 5.2: The main processes of electronic friction. (a) Second order process resulting in one electron-hole pair. (b) Second order process resulting in two electron-hole pairs. (c) First order process, where the effective one-particle. potential of the adsorbate excites an electron-hole pair. (d) Excitation of a phonon. From[5]

Theories of electronic friction have been developed for several limiting cases of the adsorption bond, namely covalent, ionic and van der Waals bonding. Most relevant is the case of van der Waals bonding (physisorption), where the adsorbate-substrate interaction consists of the long-ranged attractive van der Waals interaction plus a short ranged repulsion (Pauli repulsion) resulting from the overlap of the electron clouds of the adsorbate and the substrate.

Semi-infinite jellium model calculations[6] show that the processes described in Fig. 5.2 are mostly important for electronic friction, mainly first and second order processes.

5.5   Van der Waals friction

Levitov[19] considered another kind of dissipative forces due to the Van der Waals interaction between solids in relative motion, the so called Van der Waals friction. Two dielectric bodies (body A and B) in relative motion (velocity v) are considered. If a virtual photon, emitted in body A with frequency ω_0 and momentum k_0 is absorbed in body B an energy gain due to a change of frequency ($\omega^* = \omega_0 - k_0 * v$) can be computed. The related friction force is proportional to the relative velocity of the bodies, and vanishes in the limit $v \mapsto 0$ because this small effect is not due to an instability but due to the finite velocity of light.

5.6 Comparison

The dissipation channels discussed above are connected to a resulting shear strength, which is the quantity, that can be measured. However, the shear strengths produced in typical setups differ by orders of magnitude. In table 5.1 the dissipation channels and the related shear strengths are shown. It is assumed, that in FFM the main contribution to friction is due to phonons and that the results from the QCM technique have a large contribution from electronic excitations.

It can be seen that, like one would estimate, the phononic friction is by orders of magnitude larger than the electronic friction. Van der Waals friction is so small, that it is not observable with the present experimental techniques.

	Method	Shear Strength
Phononic Friction	Friction Force Microscopy[1]	$10^9 \, Pa$
	Island moving (FFM)[2]	$10^5 \, Pa$
Electronic Friction	Quartz Crystal Microbalance[16]	$10^{-1} \, Pa$
Van der Waals Friction	Theoretical estimation[19]	$10^{-9} \, Pa$

Table 5.1: Comparison of the shear strength due to the various friction mechanisms

1. E. Meyer, R. Overney, D. Brodbeck, L. Howald, R. Lüthi, J. Frommer, and H.-J. Güntherodt, *Phys. Rev. Lett.* **69**, 1777 (1992)
2. R. Lüthi, E. Meyer, H. Haefke, L Howald, W. Gutmannsbauer and H.-J. Güntherodt, *Science* **266**, (1994)
3. T. Strunz and F. J. Elmer in[5]
4. B. Briscoe in *Fundamentals of Friction*, edited by I.L. Singer and H.M. Pollock, Series E: Applied Sciences, Vol. 220, Kluwer Academic Publishers (1992).
5. B. N. J. Persson and E. Tossatti (eds.), *Physics of Sliding Friction*, Kluwer Academic Publishers (1996)
6. C. Holzapfel, F. Stubenrauch, D. Schumacher and A. Otto, *Thin solid films* **188**, 7 (1990)
7. R. Zwanzig, *J. of stat. Physics* **9**, 215 (1973)

8. P. Hänggi, P. Talkner, and M. Borkovec, Reaction-rate theory, *Rev. Mod. Phys.* **62**, 251 (1990)

9. B. N. J. Persson, *Phys. Rev.* **B 44**, 3277 (1991) see also: B. N. J. Persson in *Physics of Sliding Friction* (NATO ASI Series E, Vol. 311, Eds. B. N. J. Persson and E. Tossatti) (1995)

10. J. B. Sokoloff, *Phys. Rev.* **B 48**, 9134 (1993)

11. J. Krim and A. Widom, *Phys. Rev.* **B 38**, 12184 (1988)

12. E. T. Watts, J. Krim and A. Widom, *Phys. Rev.* **B 41**, 3466 (1990)

13. J. Krim, E. T. Watts and J. Digel, *J. Vac. Sci. Technol.* **A 8**, 3417 (1990)

14. J. Krim, D. H. Solina and R. Chiarello, *Phys. Rev. Lett.* **66**, 3417 (1991)

15. J. Krim and R. Chiarello, *J. Vac. Sci. Technol.* **A 9**, 2566 (1991)

16. J. Krim and C. Daly in *Physics of Sliding Friction* (NATO ASI Series E, Vol. 311, Eds. B. N. J. Persson and E. Tossatti) (1995)

17. G. Wedler, H. Reichenberger and H. Wenzel, *Z. Naturforsch.* **26a**, 1452 (1971)

18. A. Dayo, W. Alnasrallah, and J. Krim *Phys. Rev. Lett.* **80**, 1690 (1998)

19. L. S. Levitov, *Europhys. Lett.* **8**, 499 (1989)

Chapter 6

Nanorheology and nanoconfinement

6.1 Introduction

The field of tribology combines various disciplines in natural sciences. Physics is involved with classical mechanics, fluid mechanics, thermodynamics, electrodynamics, and the classical field of acoustics. Chemistry ties with tribology in fields such as oxidation processes, chemical bond formation at interfaces, and surfactant chemistry of lubricants. And in considering joints and implants in human bodies, microbiology is involved in the field of cell regeneration and formation. All these fields are partially compounded in engineering and are found in contact mechanics, transport phenomena, chemical and bioengineering.

Materials sciences is one of the key players in modern tribology. New materials are in demand for instance for internal combustion (IC) engines which are run at significantly higher temperatures to increase fuel efficiency, coatings in the electronic industry to provide higher accuracy and faster performance, and new additives and base material for faster production of casted films. A class of materials that is in high demand in modern technology, polymers, raises significant tribological issues. Polymers are generally very viscoelastic materials with characteristic mechanical properties of both a liquid and a solid. The mechanical properties of polymers are described by a combination of elasticity and hydrodynamic theories. The science which deals with the mechanical properties of polymers is called *Rheology*.

In general, rheology handles phenomena that occur during deformation and flow of fluid, colloidal or solid systems under the effect of externally acting forces. Rheology plays a major role in the understanding of lateral forces that take place during a relative sliding process of two bodies in contact. Lateral forces (the dissipative part is called friction force) cause solid systems to deform and fluid lubricants to flow.

This chapter is intended to be an introduction to the field of nanoconfinement of ultrathin liquid and polymeric films. It discusses rheology on the submicrometer scale where confinement influences significantly material properties and processes. The chapter is subdivided into essentially three parts. The first section provides an introduction into the basics of rheology and fluid dynamics with a particular emphasis in thin films. In the following two sections nanorheological and shear behavior of confined simple and complex liquids are discussed. The last section summarizes selected results achieved by scanning force microscopy in the field of nano-rheology and nano-contact-mechanics.

6.2 Continuum mechanics

6.2.1 Elastic moduli and free energy relations

The elastic concept of continuum mechanics starts with Hooke's law which linearly relates the stress (σ_{ij} or τ_{ij}) and the strain (ε_{ij} or γ_{ij}) of an ideal elastic body. In the case of an uniaxial elongation/compression in x-direction or simple shear in y-direction of an isotropic material, Hooke's law has the following simple form:

$$\sigma_{xx} = E\varepsilon_{xx} \qquad (6.1)$$
$$\tau_{xy} = G\gamma_{xy}$$

with the modulus of elasticity (Young's modulus) E and the shear modulus G. The left and right indices of the stress and the shear denote the normal vector of the shear plane and the direction of the force, respectively. The dimensions of the stress is a force per unit area (i.e., a pressure) while the strain is dimensionless - a relative measure of the local displacement u. The strain is defined for symmetry reason as

$$\varepsilon_{ij} = \frac{1}{2}\left(\frac{\partial u_i}{\partial x_j} + \frac{\partial u_j}{\partial x_i}\right); i = 1, 2, 3. \qquad (6.2)$$

If ε_{ij} is independent of the location **r**, the deformation can be expressed as $u = \underline{\varepsilon} \cdot \mathbf{r}$.[a] The index i represents the normal vector of the plane in respect to which the deformation occurs. The index j determines the direction of the deformation.

In the case of a hydrostatic compression, the non-zero-components of the stress tensor correspond to a constant pressure p, i.e. $\sigma_{ij} = -p\delta_{ij}$.[b] The specific work (i.e., the work per volume element) resulting from internal strain is

$$\delta w = -\sigma_{ij} \cdot \delta\varepsilon_{ij} \qquad (6.3)$$

[a] A vector is denoted in bold face and a tensor is doubled underlined
[b] δ_{ij} is the Kronecker symbol

where $\delta\varepsilon_{ij}$ represents the changes in the strain tensor. Internal strain in a body that results from external forces ceases to exist if the external forces are removed and the body behaves fully elastic. If the body is plastically deformed, a partial strain remains. Deformations which are slow enough, i.e., the body is in thermal equilibrium at all times, are reversible. In the case of a reversible deformation, the differential of the internal energy per volume element is

$$de = Tds + \sigma_{ij}d\varepsilon_{ij} \tag{6.4}$$

with T as the absolute temperature, ds as the entropy per volume element, and Tds the heat involved during the deformation. Hence, the free energy $F = E - TS$ can expressed as

$$df = -sdT + \sigma_{ij}d\varepsilon_{ij} \tag{6.5}$$

where f is the free energy per volume element. Defining the enthalpy as

$$h = e - Ts - \sigma_{ij}\varepsilon_{ij} = f - \sigma_{ij}u_{ij} \tag{6.6}$$

the following thermodynamic relationships are found

$$\sigma_{ij} = \left(\frac{\partial e}{\partial \varepsilon_{ij}}\right)_{s=const} = \left(\frac{\partial f}{\partial \varepsilon_{ij}}\right)_{T=const} ; \tag{6.7}$$

$$\varepsilon_{ij} = -\left(\frac{\partial h}{\partial \sigma_{ij}}\right)_{T=const}$$

The free energy has to be expressed as a function of the shear tensor in order to apply the thermodynamic relationships. For very small deformation the general expression for the free energy of a deformed isotropic body is

$$f = f_o + \frac{\lambda}{2}\varepsilon_{ii}^2 + \mu\varepsilon_{ij}^2 \tag{6.8}$$

with the Lamé coefficient λ and μ.[c] Any deformation can be written as the sum of pure shear and homogeneous dilatation by

$$\varepsilon_{ij} = \left(\left(\varepsilon_{ij} - \frac{1}{3}\delta_{ij}\varepsilon_{kk}\right) + \frac{1}{3}\delta_{ij}\varepsilon_{kk}\right). \tag{6.9}$$

The free energy can then be written as follows:

$$f = f_o + \mu\left(\varepsilon_{ij} - \frac{1}{3}\delta_{ij}\varepsilon_{kk}\right)^2 + \frac{K}{2}\varepsilon_{kk}^2 \tag{6.10}$$

---

[c]Tensor algebra (Einstein relation): $\varepsilon_{ii}^2 = \varepsilon_{11}^2 + \varepsilon_{22}^2 + \varepsilon_{33}^2$ and $\varepsilon_{ii}^2 \equiv \sum_{i,k}\varepsilon_{ij}^2$.

where μ is the modulus of rigidity (also called shear modulus) and $K = \lambda + 2/3\mu$ is the modulus of compression. Both moduli should be larger than zero. Hence, the stress tensor for a small deformation of an isotropic solid body can be expressed as the sum of a dilatation and pure shear component as

$$\sigma_{ij} = \left(\frac{\partial f}{\partial \varepsilon_{ij}} \right)_{T=const} = K\varepsilon_{kk}\delta ij + 2\mu \left(\varepsilon_{ij} - \frac{1}{3}\delta_{ij}\varepsilon_{kk} \right) \qquad (6.11)$$

The reversed expression is

$$\varepsilon_{ij} = \frac{1}{9K}\delta_{ij}\sigma_{kk} + \frac{1}{2\mu} \left(\sigma_{ij} - \frac{1}{3}\delta_{ij}\sigma_{kk} \right) \qquad (6.12)$$

In the case of a uniaxial deformation, equation (6.12) reduces to

$$\varepsilon_{33} = \frac{1}{3} \left(\frac{1}{3K} + \frac{1}{\mu} \right) \sigma_{33} \qquad (6.13)$$

which leads with the equation (6.2) to the fundamental definition of the Young's modulus

$$E = \frac{\sigma_{33}}{\varepsilon_{33}} = \frac{9K\mu}{3K + \mu} = \mu\frac{3\lambda + 2\mu}{\lambda + \mu} \qquad (6.14)$$

The shear modulus μ (to avoid any conflict with the coefficient of friction also called G modulus) can be expressed as

$$\mu \equiv G = \frac{E}{2(1 + \nu)} \qquad (6.15)$$

where ν is the ratio between the deformation parallel to the applied force and the deformation perpendicular to the applied load - also known as Poisson's ratio.[d] The modulus of compression K is

$$K = \frac{E}{3(1 - 2\nu)} \qquad (6.16)$$

Finally the specific free energy can be written as

$$f = f_o + \frac{E}{2(1 + \nu)} \left(\varepsilon_{ij}^2 + \frac{\nu}{1 - 2\nu}\varepsilon_{kk}^2 \right) \qquad (6.17)$$

---

[d] $\nu_{32} = -\frac{\varepsilon_{22}}{\varepsilon_{33}} = \frac{1}{2}\frac{3K-2\mu}{3K+\mu} = \frac{\lambda}{2(\lambda+\mu)}$

6.2.2 Special cases of elasticity and methods

Thermodynamically ν is only restricted to $[-1, 1/2]$ which would allow the body to extend its cross-sectional area during longitudinal extension. To avoid that cross-sectional area extension it suggests restricting the regime of ν further to $[0, 1/2]$. The stress and the shear tensors are expressed by the Young's modulus and the Poisson's ratio for a small longitudinal deformation as

$$\sigma_{ij} = \frac{E}{1+\nu}\left(\varepsilon_{ij} + \frac{\nu}{1-2\nu}\varepsilon_{kk}\delta_{ij}\right)$$

and

$$\varepsilon_{ij} = \frac{1}{E}\left((1+\nu)\sigma_{ij} - \nu\sigma_{kk}\delta ij\right) \quad , \tag{6.18}$$

respectively. The volume change of bulk polymers during compression or tension is in most cases very small in comparison to the deformation. Hence, $\nu \cong 1/2$ for polymers which simplifies the Young's modulus - shear modulus relation to $E = 3G$.

In the case of a plane stress assumption (used, for example, in thin plates), where the stress components perpendicular to the plane of the film σ_{33}, σ_{13} and σ_{23} are neglected, the two-dimensional stress-strain relation can be obtained from equations (6.18) as

$$\begin{pmatrix} \varepsilon_{11} \\ \varepsilon_{22} \\ \varepsilon_{12} \end{pmatrix} = \begin{pmatrix} \frac{1}{E} & -\frac{\nu}{E} & 0 \\ \frac{-\nu}{E} & \frac{1}{E} & 0 \\ 0 & 0 & \frac{1+\nu}{E} \end{pmatrix} = \begin{pmatrix} \sigma_{11} \\ \sigma_{22} \\ \sigma_{12} \end{pmatrix} \tag{6.19}$$

$$\varepsilon_{33} = \frac{\nu}{1-\nu}\left(\varepsilon_{11} + \varepsilon_{22}\right) \tag{6.20}$$

$$\varepsilon_{13} = \varepsilon_{23} = 0 \tag{6.21}$$

An anisotropic material has 21 independent elastic constants. Once the orthotropic axes of symmetry are known, the number of elastic constants needed to fully characterize the material reduces to 9. Orthotropic materials are unidirectional composites, woods, and laminated metallic products. With the principal axes of orthotropy as the reference axes, the strain-stress relation can be written as:

$$
\begin{pmatrix} \varepsilon_{11} \\ \varepsilon_{22} \\ \varepsilon_{33} \\ \varepsilon_{12} \\ \varepsilon_{23} \\ \varepsilon_{13} \end{pmatrix} = \begin{pmatrix} \frac{1}{E_1} & -\frac{\nu_{12}}{E_1} & -\frac{\nu_{13}}{E_1} & 0 & 0 & 0 \\ \frac{-\nu_{12}}{E_1} & \frac{1}{E_1} & -\frac{\nu_{13}}{E_1} & 0 & 0 & 0 \\ \frac{-\nu_{13}}{E_1} & -\frac{\nu_{13}}{E_1} & \frac{1}{E_3} & 0 & 0 & 0 \\ 0 & 0 & 0 & \frac{1+\nu_{12}}{E_1} & 0 & 0 \\ 0 & 0 & 0 & 0 & \frac{1}{2G_{13}} & 0 \\ 0 & 0 & 0 & 0 & 0 & \frac{1}{2G_{13}} \end{pmatrix} = \begin{pmatrix} \sigma_{11} \\ \sigma_{22} \\ \sigma_{33} \\ \sigma_{12} \\ \sigma_{23} \\ \sigma_{13} \end{pmatrix}
$$

$$(6.22)$$

Five coefficients characterize the material, E_1, E_3, G_{13}, ν_{12} and ν_{13}.

There are various techniques that are used to determine elastic constants of material. The techniques can be grouped in roughly three groups: (i) static tension-compression or bending tests, (ii) vibration tests, and (iii) wave propagation tests. Static measurements are technically easy to conduct and provide information about the extensional moduli, the shear moduli and the Poisson's ratios. An example of the static method is the cantilever beam method, Figure 6.1. The cantilever beam, the substrate, is coated with a film, the sample. The thickness of the film t_f is assumed to be much smaller than substrate thickness t_s. Further, it is assumed that the distortion at the end of the beam δ is smaller than the sample film thickness. Based on this assumption the following relation is used to determine the stress σ_f in the film which acts along the cantilever beam:

$$
\sigma_f t_f = -\frac{E_s \delta}{3(1 - \nu_s)} \left(\frac{t_s}{L} \right)^2,
$$

$$(6.23)$$

where E_s and ν_s are the Young's modulus and Poisson's ratio of the substrate.

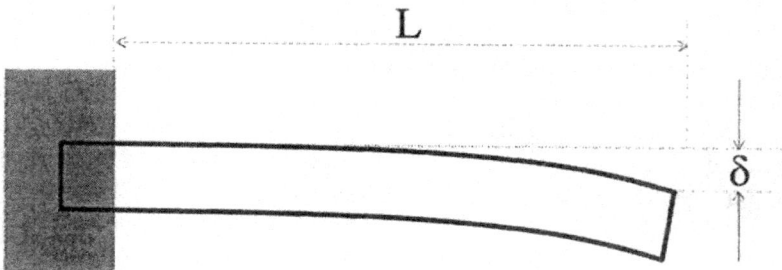

Figure 6.1: Cantilever beam method used to measure in-plane stresses of ultrathin films.

X-ray and electron diffraction methods are applied to determine in-plane stresses of ultrathin films supported by solid substrates using the plane stress assumption from above and assuming in-plane isotropic deformation, i.e., ε_{11} and ε_{22}. These methods work for crystalline material and are sensitive to variations in the lattice parameters. For a crystal, stress and strain are related by

$$\sigma_{kl} = \sum_{ij} c_{klij} \varepsilon_{ij} \qquad (6.24)$$

$$\varepsilon_{ij} = \sum_{ij} s_{ijkl} \sigma_{kl} \qquad (6.25)$$

with the elastic constants and coefficients (compliances) c_{klij} and s_{ijkl}, respectively. For films of isotropic elastic behavior, it follows from equation (6.20) that $\varepsilon_{11} = \varepsilon_{22} = \varepsilon_{33}(\nu - 1)/(2\nu)$.

The Poisson's ratios, ν_{12} and ν_{21}, of an orthotropic ultrathin polymeric material can be determined using vibrational holographic interferometry[1]. The tested polymer film resembles a vibrating membrane and its residual stresses are measured and analytically analyzed by the wave equation for a vibrating membrane. Principle stresses are measured for 2D constrained and 1D constrained films. Poisson's ratios are determined from

$$\nu_{12} = \frac{\sigma_{11}^{2D} - \sigma_{11}^{1D}}{\sigma_{11}^{2D}} \qquad (6.26)$$

$$\nu_{21} = \frac{\sigma_{22}^{2D} - \sigma_{22}^{1D}}{\sigma_{22}^{2D}} \qquad (6.27)$$

Other methods that are used to establish elastic constants of materials are, for instance, tensile testing, high pressure gas dilatometer, pressure-volume-temperature apparatus, torsion pendulum and scanning force microscopy. Before we, however, further discuss techniques another material property, besides elasticity, has to be introduced - viscosity.

6.2.3 Fundamental equations of fluid flow

The basic equation of simple flow is described one-dimensionally by Newton's law of viscosity,

$$\tau_{yx} = -\eta_o \frac{dv_x}{dy}, \qquad (6.28)$$

which relates proportionally the shear force per unit area, τ_{xy}, to the negative of the local velocity gradient with a constant viscosity value, η_o.[e] Liquids and

---

[e] x denotes the direction of momentum and y the direction of momentum transfer

gases that follow this law are called Newtonian fluids. They are incapable of storing mechanical energy. Experiments have shown that homogeneous non-polymeric liquids, such as for instance water, methane, toluene, and glycerol are well described by the Newtonian model. Many industrially important liquid materials, however, do not or only partially follow Newton's law of viscosity, and are hence called non-Newtonian liquids. The subject of non-Newtonian flow is a subdivision of the science of rheology. A more generalized form of equation (6.28) is provided by

$$\tau_{yx} = -\eta \frac{dv_x}{dy}, \tag{6.29}$$

where η is a function of either the shear or the velocity gradient.[f] Many empirical models have been developed over the past to describe experimental data. New terms were introduced such as *pseudoplastic, dilatant,* and *Bingham-plastic* which are illustrated in Figure 6.2. Pseudoplastic fluids include polymer solutions or melts, greases, starch suspensions, biological fluids, detergent slurries, paints, and dispersion media in various pharmaceuticals. In regions of pseudoplastic behavior the apparent viscosity η decreases with increasing rate of shear. An opposite behavior is observed for dilatant fluids, such as potassium silicate in water or solutions containing high concentrations of powder in water.

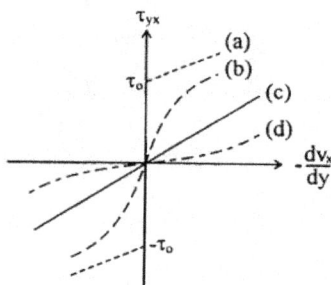

Figure 6.2: Shear diagram for Newtonian and non-Newtonian fluids: (a) Bingham-plastic; (b) pseudoplastic (also called shear thinning); (c) Newtonian; and (d) dilatant.

Bingham-plastic behavior are maybe the most intriguing of the non-Newtonian manifestation because it combines, in a very simple manner, elastic

---

[f] v_x denotes the velocity in x-direction

behavior with viscous Newtonian behavior. Below a critical absolute shear amount τ_o, a Bingham plastic is independent of the shear rate which describes fully elastic materials. Above τ_o, however, the material behaves like a Newtonian liquid. This model has been found useful for many fine suspensions and pastes, such as soap, paper pulp and chocolate.

So far only time-independent fluid behavior has been introduced and differentiated. There are additional types of non-Newtonian behavior under unsteady-state conditions: *thixotropic fluids, rheopectic fluids*, and *viscoelastic* fluids. Thixotropic and rheopectic fluids exhibit a reversible decrease and increase, respectively, in shear stress with time at a constant rate of shear. Examples of thixotropic fluids are some polymer solutions and paints, and rheopectic fluids certain sols. Viscoelastic fluids exhibit elastic recovery from flow deformations, which will be discussed in more detail below because of their importance for polymers.

Any fluid flow where the mass is conserved can be described by two equations, the *equation of continuity* (or mass conservation) and the *equation of motion*. The equation of continuity, which describes the rate of change of density for an observer which floats along with the fluid is

$$\frac{D\rho}{Dt} = -\rho(\nabla \cdot \mathbf{v}) = -\rho div \mathbf{v}, \tag{6.30}$$

in which the operator D/Dt is the substantial time derivative, ρ is the density of the fluid at a fixed point, and $\mathbf{v}$ is the velocity vector. The equation of motions is

$$\rho \frac{D\mathbf{v}}{Dt} = -\nabla p - [\nabla \cdot \underline{\tau}] + \rho \mathbf{g}, \tag{6.31}$$

which is a statement of Newton's second law in the form of mass times acceleration per unit volume, on the *right* of equation (6.31), and the sum of three forces (the pressure force, the viscous force and the gravitational force) per unit volume, on the *left*. The challenge in using these equations is to find, for the various stresses in the stress tensor $\underline{\tau}$, the appropriate expressions for velocity gradients and fluid properties.

Each element of the stress tensor should be a linear function of the velocity gradients. Viscous fluids can be considered isotropic in bulk fluid which leads to the following definition of the stress tensor for a viscous fluid:

$$\tau_{ij} = 2\eta \left(\frac{\partial v_i}{\partial x_j} - \frac{1}{3}\delta ij \frac{\partial v_i}{\partial x_i} \right) + \kappa \delta_{ij} \frac{\partial v_i}{\partial x_i} \tag{6.32}$$

Two viscosity values are introduced: η, the viscosity which is responsible for the momentum transport during flow, and κ, the "volume viscosity". The volume viscosity is a measure of the resistance to volume changes and usually set to zero for liquids. For constant density and constant viscosity, the equation of motion simplifies to:

$$\rho \frac{D\mathbf{v}}{Dt} = -\nabla p - \eta[\nabla^2 \mathbf{v}] + \rho \mathbf{g}, \tag{6.33}$$

This is the *Navier-Stokes equation* which describes the fluid flow of a Newtonian liquid.[9]

6.2.4   Unsteady flow and viscous boundary layers

Unsteady flow and viscous flow in boundary layers are of particular interest in tribology. Start-up and control problems are demanding the study of transient phenomena that occur when a fluid near a wall is suddenly set in motion. Under heavy load the liquid boundary layers are responsible for the viscous drag of lubricated systems.

Unsteady viscous flow near a solid flat surface, Figure 6.3, can be expressed with the following differential equation:

$$\rho \frac{\partial v_x}{\partial t} = \eta \frac{\partial^2 v_x}{\partial y^2} \tag{6.34}$$

that was derived from eqs. (6.30) and (6.33), assuming a constant density and viscosity, and a sudden surface motion in x-direction with a velocity v_{surf}. This is the classical one-dimensional diffusion equation. With the boundary conditions set to:

$$v_x = 0 \quad \text{at } t < 0, \; v_x = 0 \text{ for all } y,$$

$$v_x = v_{surf} \quad \text{at } y = 0 \text{ for } t > 0, \text{ and}$$

$$v_x = 0 \quad \text{at } y = \infty$$

equation (6.34) has the solution:

$$\frac{v_x}{v_{surf}} = 1 - \frac{2}{\sqrt{\pi}} \int_0^\xi e^{-\xi^2} d\xi; \quad \xi \equiv y \left(4\frac{\eta}{\rho} t \right)^{-\frac{1}{2}} \tag{6.35}$$

The integral is the well known *error integral*, abbreviated as $erf(\xi)$. Its asymptotic behavior, $erf(\infty) = 1$, allows one to define a characteristic thickness, called *boundary layer thickness* δ^* as the distance y for which v_x has

---

[9]In literature, the Navier-Stokes equation sometimes also refers to the equations of motion for a Newtonian liquid but with local variable density and viscosity.

dropped to $0.01 \times v_{surf}$. Hence, a boundary layer thickness of

$$\delta^* = 4\sqrt{\frac{\eta}{\rho}t^*} \qquad (6.36)$$

can be found. The time t^* represents a characteristic time of the momentum diffusion through the boundary layer. Low viscous and high viscous fluids will significantly affect the variable ξ, and hence the error signal and the boundary layer thickness.

$t < 0$ fluid at rest	$t = 0$ surface set in motion	$t > 0$ fluid in unsteady flow

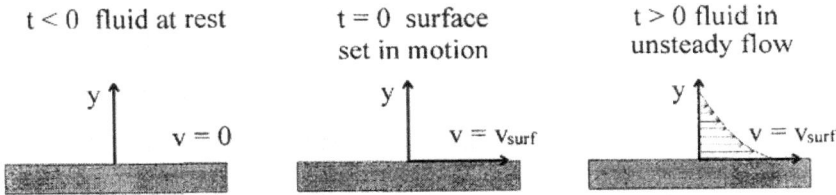

Figure 6.3: Unsteady viscous flow of a fluid near a solid flat surface at sudden motion.

It is reasonable to assume that the boundary layer is insignificantly growing for $t > t^*$ and finite body sizes. If two bodies, separated by a fluid film, are in relative motion, two sliding regimes can be differentiated, (a) a *turbulent sliding regime*; the thickness of the confined fluid film is larger than δ^*, and (b) a *laminar sliding regime*; the thickness of the confined fluid film is smaller than δ^*.

The relative sliding motion allows one to assume that the lower body is at rest while the opposite upper body is sliding at constant velocity v_{surf}. Any frictional losses in the turbulent sliding regime of a Newtonian liquid are due to velocity fluctuations which give rise to turbulent shear stresses. The friction force related to turbulent flow past a flat surface can be approximated as:

$$F_{fric}^{turb} = 0.072 \rho \frac{v_{turb}^2}{2} W L \left(\frac{L v_{turb} \rho}{\eta} \right)^{-\frac{1}{5}}, \qquad (6.37)$$

with the plate area WL and length L and the flow velocity v_{turb} of the turbulent fluid lubricant[2]. Many semiemperical theories have been developed over the years using various expressions for the turbulent momentum flux, also called *Reynolds stresses*, which will not be further discussed.

Within body distances smaller than the boundary layer thickness, momentum is transferred from the upper body to the lower body. Blasius' numerical solutions of flow near the leading edge of a flat plate provide a drag force per surface of

$$F_{fric}^{laminar} = 0.664\sqrt{\rho\eta L W^2 v_\infty^3} \qquad (6.38)$$

where v_∞ is the flow velocity far away from the solid boundary[2]. The laminar regime of lubricated sliding is, in literature, referred to hydrodynamic lubrication and is extensively studied with the *Reynolds equations* which we are discussing in the following paragraph.

6.2.5   Hydrodynamic lubrication

As long as a laminar flowing lubricant film can support the load between two sliding surfaces, the fluid can be considered to be in the hydrodynamic regime. The opposing surfaces must be conformal for hydrodynamic lubrication. The study of hydrodynamic lubrication is the study of a particular form of the Navier-Stokes equations, or more generally the equation of motion - the *Reynolds equations*. The Reynolds equations contain, as parameters, the viscosity, the density and the film thickness. These three parameters can vary locally and depend on temperature, pressure fields and the elastic behavior of the bearing surfaces. The following assumptions are made to reduce the equation of motion for Newtonian fluids to the Reynolds equation:

1. The height of the fluid film y is very small compared to the dimensions of the contact area,

2. the pressure is constant across the fluid film,

3. the flow is laminar, i.e., no turbulence occur,

4. the inertia of the fluid is small compared to the viscous shear (examples of inertia forces are fluid gravity and acceleration of the fluid),

5. the fluid velocity at the bearing surfaces is zero (no-slip condition), and

6. no external forces act on the film.

In general form, i.e., valid for compressible and incompressible Newtonian fluids, the Reynolds equation is

$$\frac{\partial}{\partial x}\left(\frac{\rho h^3}{\eta}\frac{\partial p}{\partial x}\right) + \frac{\partial}{\partial z}\left(\frac{\rho h^3}{\eta}\frac{\partial p}{\partial z}\right) = 6\left(U_1 - U_2\right)\frac{\partial\left(\rho h\right)}{\partial x} + 6\rho h\frac{\partial}{\partial x}\left(U_1 + U_2\right) + 12\rho V$$

$$(6.39)$$

with the film thickness, h, the sliding direction and its perpendicular surface direction, x and z, respectively, the velocity of the two bearing surfaces in x-direction, U_i, the radial component of the velocity, V, and the pressure, p, which is the mean of the diagonal shear elements, τ_{ii}. Only in the case of a very particular geometry, such as a journal bearing, there is a general analytical solution of the equation (6.39).

Between two moving bearings, the lubricant flows in x-direction because of pressure flow and shear flow, and in z-direction because of pressure flow only. Hence, shear stresses can be introduced for a Newtonian fluid as

$$\tau_x = -\frac{1}{2}\frac{\partial p}{\partial x}(2y - h) + \frac{\mu}{h}(U_2 - U_1), \tag{6.40}$$

$$\tau_z = -\frac{1}{2}\frac{\partial p}{\partial z}(2y - h), \tag{6.41}$$

The total drag, defined as

$$F = \int\int \tau dA, \tag{6.42}$$

exerted by the moving bearing surface is, at $y = 0$ or $y = h$,

$$F = \int_o^z\int_o^x \{\pm\frac{1}{2}\frac{\partial p}{\partial x}h + \frac{\mu}{h}(U_2 - U_1)\}dxdz, \tag{6.43}$$

since F_z is perpendicular to the moving direction. The challenge is now to derive from the Reynolds equation an analytical expression for the pressure distribution. Assuming steady loading, incompressible lubricants and a simple plane slider (which is a model for a one-dimensional thrust bearing), Fig. 6.4, the Reynolds equation reads

$$\frac{dp}{dx} = \frac{-6\eta U(h - h_o)}{h^3}, \tag{6.44}$$

where h_o is the height of maximum pressure. As illustrated in Figure 6.4, the film thickness, h, can be expressed as

$$h = \alpha x = \frac{h_2(a - 1)}{L}x. \tag{6.45}$$

Equations (6.41), (6.43) and (6.44) lead to expressions for the shear stress and the drag force

$$\tau = \eta U\frac{4h - 3h_0}{h^2} \tag{6.46}$$

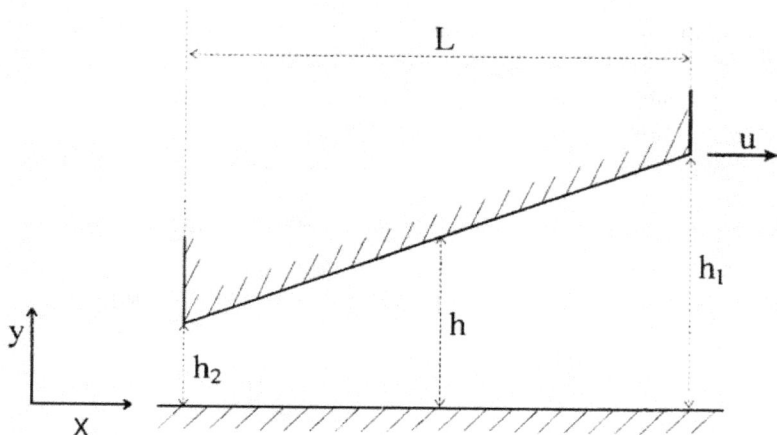

Figure 6.4: Plane slider - a one-dimensional model for thrust bearings.

$$F = \frac{N h_2 2(a^2 - 1) lna - 3(a - 1)^2}{L 3(a + 1) lna - 6(a - 1)} = \eta U W \frac{L}{h_2} C_f \qquad (6.47)$$

with the width of the bearing, W, the normal load exerted by the fluid, N,

$$N = W \int_{h_2}^{h_1} p \, dx = \frac{6\eta U W L^2}{h_2^2} \left(\frac{1}{a - 1} \right)^2 \left[lna - \frac{2(a - 1)}{a + 1} \right] = \eta U W \left(\frac{L}{h_2} \right)^2 C_p, \qquad (6.48)$$

the inlet and outlet height ratio, a,

$$a = \frac{h_1}{h_2} \qquad (6.49)$$

and the parameters of maximum pressure

$$h_o = \frac{2 h_1 h_2}{h_1 + h_2} = \frac{2a}{1 + a} h_2, \qquad (6.50)$$

$$p_o > \frac{2\eta U \alpha (a - 1)^2}{2 \alpha a}; \alpha = \frac{h_1 - h_2}{L} \qquad (6.51)$$

The dimensional coefficients C_p and C_f are measures for the load capacity and friction, respectively, of plane sliders. A maximum load capacity can be found by setting $dW/da = 0$ which yields an inlet and outlet height ratio of $a = 2.2$, and hence a maximum load of

$$N_{max} = 0.1602 \left(\eta UW\right) \left(\frac{L}{h_2}\right)^2 C_p \qquad (6.52)$$

and a resulting drag force of

$$F_{max} = 4.7\frac{h_2 N}{L} \qquad (6.53)$$

If the load capacity of a hydrodynamic bearing is exceeded, the lubricant wedge separating the bearing surfaces will eventually diminish in volume.

A friction coefficient, μ, can be introduced as the ratio between drag force and normal load

$$\mu = \frac{F}{N} = \frac{h_2}{L}\frac{C_f}{C_p} \qquad (6.54)$$

In Figure 6.5, the load capacity and the friction coefficient are plotted as a function of the height ratio. The friction coefficient reaches its minimum at $a = 2.55$.

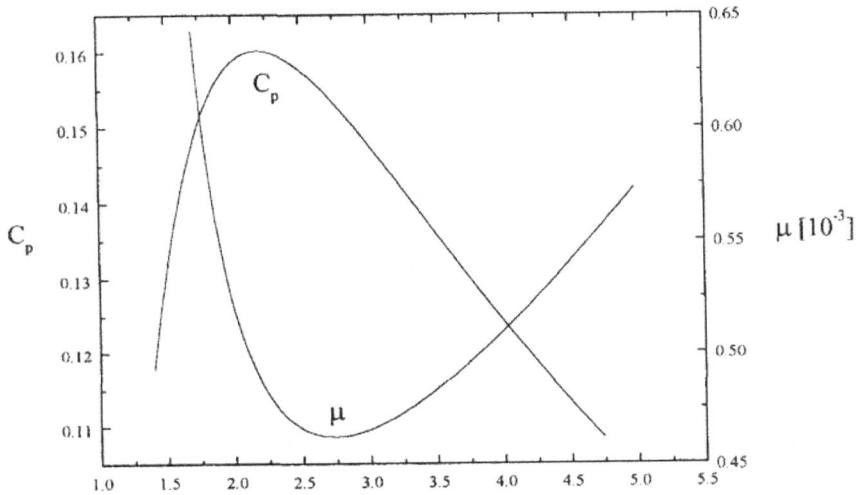

Figure 6.5: Load capacity and friction coefficient.

In the case the plane slider in Figure 6.5 slides parallel with distance h with very slow relative velocity U, the pressure distribution can be neglected and the drag force directly related to the momentum transfer, i.e.:

$$F = \frac{\eta}{h} ULW, \qquad (6.55)$$

The load, N, is constant provided by the weight of the slider. This is, of course, a very simplified solution of a parallel slider where entry effects, i.e., flow near the leading edge of a flat plate as described in equation (6.38), are entirely neglected. We will later see how this equation was used for liquid gap distances orders of magnitude smaller than the contact area when we discuss surface forces apparatus measurements.

Calculations with other shaped sliders showed the height ratio to be very important but find the effect of the shape of the lubricant insignificant for load capacity and drag force determinations. A film shape of

$$h = e^{\beta x}, \qquad (6.56)$$

which provides exact solutions to the Reynolds equation for constant viscosity is, therefore, justified. Because of the lengthy expressions of the exact solutions that can be found in any theoretical hydrodynamic book[3] only the one-dimensional solution is presented. Following the procedure above the load capacity and drag force is

$$N = \frac{3\eta UWL^2}{(ah_2 \ln a)^2} \left[\frac{a^2 - 1}{6} - \frac{a^2(a-1)\ln a}{a^3 - 1} \right], \qquad (6.57)$$

and

$$F = \frac{\eta UWL}{h_2} \frac{3}{2a \ln a} \left[\frac{(a-1)^4}{a^2 - 1} \right] \qquad (6.58)$$

respectively, with a load capacity maximum value at $a = 2.3$ of

$$N_{max} = 0.165 \eta UW \left(\frac{L}{h_2} \right)^2 \qquad (6.59)$$

6.2.6 Extended regimes of lubrication

As shown above, it is very challenging to attain solutions to the Reynolds equation even if the configuration is very simple and any lubricating artifacts are neglected. Features of lubricating artifacts are, for instance, striation with incomplete lubricant, surface roughness, or elastic deformations of the bearings. Most of the calculations above assume constant viscosity, which is however known to change with temperature and under high pressure. If the contacting surfaces are counterformal (i.e, non-conforming as assumed for hydrodynamic lubrication), local pressures in the contact zone will be very high - up to several Gigapascal. Examples of non-conforming contact regimes are shown in Figure 6.6.

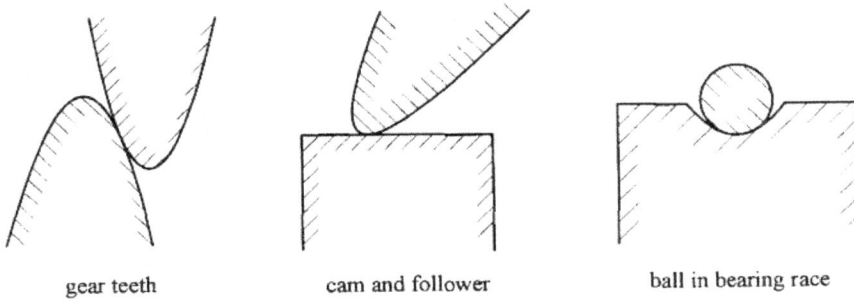

gear teeth   cam and follower   ball in bearing race

Figure 6.6: Devices of non-conforming contact regimes.

Very high local pressures in the film cause the viscosity of the lubricant to increase with the tendency to expand the film thickness over the predictions of the hydrodynamic theory. Also, the elastic deformation of the bearing surface has to be considered. High pressure lubrication is approached with an extended hydrodynamic theory - the theory of *elastohydrodynamic lubrication (EHL)*. Pressure spikes and sharp constriction of the film in the exit region of sliding bearings are poorly investigated.

The viscosity of oil, as one of the most common lubricants, shows a fairly close exponential relationship,

$$\eta = \eta_o \exp \alpha P, \tag{6.60}$$

with the hydrostatic pressure, P, and the constant parameters, η_o (viscosity at zero pressure), and α.[h] The pressure coefficient a is of the order of 10^{-8}Pa^{-1} for typical mineral oils. While in a hydrodynamically lubricated bearing any increase in the viscosity will be only a few percent, in the EHL regime the viscosity can increase by over 20,000 times, and that at atmospheric pressure. Under these circumstances the liquid can show a solid-like behavior.

Roughness, sliding velocity and pressure determine when full fluid lubrication begins to break down and lubrication enters new regimes. The *mixed regime of lubrication* is reached when in addition lubricant films, adhering to

---

[h]Emprical relationship: $\alpha \approx (0.6 + 0.965_{10} \lg \eta_o) \times 10^{-8}$, $[\eta_o] = [cPoise]$

the surface contours, should be considered. If there is no bulk liquid left and the lubricant is reduced to a ultrathin layer, a few molecular layers thick, *boundary lubrication* comes into effect.[i] In reality, high pressure lubrication is due to roughness of the bearing surfaces found in a mixed regime, assuming there is no severe wear. Mere boundary lubrication is rather academic of nature, however, can be the dominant factor in mixed lubrication, especially if the number of contact asperities is high. The likelihood of asperity contact is expressed as the ratio between the minimum film thickness of the bulk fluid and the r.m.s. roughness. The different regimes of lubrication are very nicely illustrated with the Stribeck curve, which is also known as the Reynolds-Sommerfeld curve, Fig. 6.7.

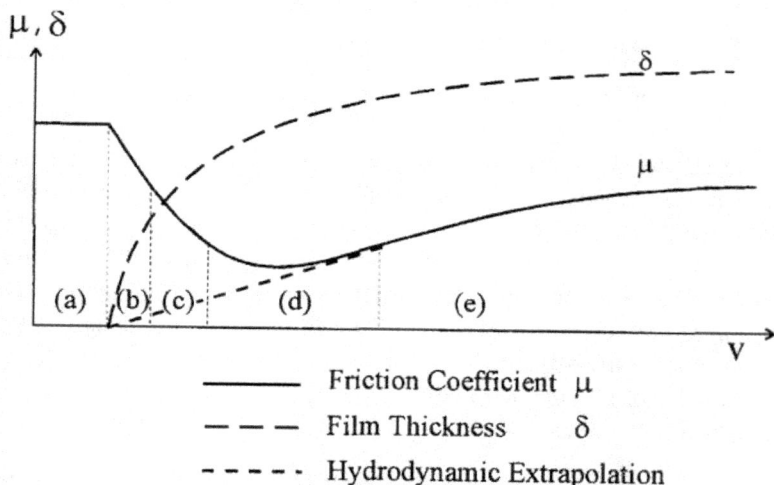

Figure 6.7: Stribeck Curve (schematic: (a) dry contact regime, (b) boundary lubrication, (c) mixed lubrication, (d) elasto-hdydrodynamic lubrication, (e) hydrodynamic lubrication. The friction force F is the product of the normal load and the friction coefficient.

The Stribeck curve relates the friction force ($F = \mu N$) to the hydrodynamic drag forces of equation (6.55), i.e.,

$$F = \frac{\eta_b S}{h} U \tag{6.61}$$

---

[i] Note that "boundary lubrication" should not be confused with "viscous boundary layer"

where S is the apparent surface area, and η_b is the liquid bulk viscosity. The regime in which hydrodynamic drag forces and friction forces correspond to each other is known as *Couette flow* regime. It is recognized from various experiments with surface forces apparatus (see below) that the bulk viscosity η_b in equation (6.61) has to be replaced for rough surfaces by the effective viscosity $\eta_{eff} \propto (U/h)^{-2/3}$[4,5,6,7]. Note that this famous power law has been observed for "simple liquids", i.e., liquids which are independent on the shear rate[8,9,10]. There is however a controversy of how universal this power law is. In a theoretical work of *Urbakh, Klafter* and co-workers it has been concluded that the exponent can vary between -2/3 and -1.0[9,11] in the shear thinning regime which has also been experimentally observed by *Israelachvili et al.*[12]. An exponent of -1 leads to a velocity independent friction force in equation (6.61) which may be contrasted with the behavior of a purely Newtonian liquid where the viscous friction force is proportional to the velocity. It is however important to note that the liquids studied by *Israelachvili* and co-workers were complex fluids, i.e., fluids in which the shear forces are velocity or frequency dependent as it will be discussed below.

6.2.7 Viscoelastic lubricants

As discussed above, fluids can show non-Newtonian behavior in which the viscosity is either a function of the shear stress or the shear rate. One special kind of a non-Newtonian fluid behavior is the viscoelastic behavior. A viscoelastic fluid exhibits both viscous flow and elastic restoring forces. All real liquids show viscoelastic behavior if stressed fast enough. The elastic response to stress is a fluid property such as the viscosity. As per equation (6.12) the stress in a solid body can be expressed as

$$\tau = G\gamma \tag{6.62}$$

where τ and γ replace σ_{ik} and $2\varepsilon_{ik}$, respectively, to express simple shear deformation.[j] Hence, the stress rate behavior of a solid is

$$\frac{d\tau}{dt} = G\frac{d\gamma}{dt} \tag{6.63}$$

Considering the case of low frequency in the stress applied, i.e. a frequency which is much lower than the inverse of the relaxation time t^* of the fluid ($\omega t^* \ll 1$), the fluid shows predominantly viscous behavior (i.e., Newtonian, no strain is built up). On the other hand, the fluid behaves more like an elastic

---

[j] $u_{ii} = 0$ and $\sigma_{ii} = 0$ for simple shear

body if the frequency of the applied stress is large ($\omega t^* \gg 1$). Based on this asymptotic stress behavior, the following extended differential equation,

$$\frac{1}{G}\frac{d\tau}{dt} + \frac{1}{\eta}\tau = \frac{d\gamma}{dt} \tag{6.64}$$

provides a reasonable description for viscoelastic fluids, where G is the shear modulus. The first and second term on the left side of the last expression describes the rate of elastic and flow deformation, respectively. Considering the rate of deformation $d\gamma/dt = 0$, after a fluid element has been rapidly deformed and constrained in its deformed shape, it yields that the stress exponentially decays with time. This is expressed by

$$\tau = \tau_0 exp\left(-\frac{G}{\eta}t\right) \tag{6.65}$$

The ratio $t^* = \eta/G$ is called the Maxwell relaxation time. Assuming again a bearing configuration as discussed with one of the one-dimensional sliders, as discussed above, the differential equation for viscous fluids can be rewritten as

$$\frac{d^2p}{dx^2} + \left(\frac{G}{\eta U}\right)\frac{dp}{dx} = \frac{12G}{U}\left(\frac{Uh-q}{h^3}\right) \tag{6.66}$$

with

$$q = Uh - \frac{h^3}{12}\left(\frac{1}{\eta}\frac{dp}{dx} + \frac{U}{G}\frac{d^2p}{dx^2}\right) \tag{6.67}$$

This expression corresponds to the Reynolds equation (6.39) if $1/G$ is set equal to zero and the Reynolds equation is reduced to one-dimension.

6.2.8   Linear viscoelasticity of solids

In the previous paragraphs, the basic ideas of elasticity theory of solids and viscous behavior of fluids have been discussed. Viscoelastic properties of fluids have been introduced in the special case of thin liquid films under high pressure. In addition to fluids, solids can also behave viscoelastically under mechanical stresses. Polymeric materials, in particular, can show viscoelastic behavior because of their molecular chain structure. Based on the Boltzmann's superposition principle, we consider a linear superposition of the present and past deformations by describing the stress as follows:

$$\tau(t) = \int_{-\infty}^{t} G(t-t')\frac{d\gamma}{dt'}dt'. \tag{6.68}$$

This equation is the corresponding integral equation to equation (6.63) with the relaxation function $G(t - t')$ as the replacement for the shear modulus. The inverse representation is

$$\gamma(t) = \int_{-\infty}^{t} J(t - t') \frac{d\tau}{dt'} dt',$$
(6.69)

where $J(t - t')$ describes the deformation as response to the stress. Fully viscous or elastic behavior is achieved by setting $\tau(-\infty) = 0$ or $\gamma(-\infty) = 0$, respectively. If viscous flow is neglected then the time-dependent functions retain only elastic relaxation components. Therefore, as $t \to \infty$, the stress and the deformation become proportional to one another, i.e.,

$$G_{\infty} = \frac{\tau_{\infty}}{\gamma_{\infty}}$$
(6.70)

The mechanical stress-strain behavior can then be described, with the help of relaxation times t^*, as

$$\frac{d\gamma}{dt} = \frac{1}{t_{\gamma}^*} (\gamma_{\infty} - \gamma),$$
(6.71)

and

$$\frac{d\tau}{dt} = \frac{1}{t_{\tau}^*} (\tau_{\infty} - \tau)$$
(6.72)

or, by substituting equation (6.70)

$$\tau + t_{\tau}^* \frac{d\tau}{dt} = G_{\infty} \left(\gamma + t_{\gamma}^* \frac{d\gamma}{dt} \right).$$
(6.73)

In the case of a one-time-disturbance in form a step function, equations (6.68) and (6.69) can be replaced by

$$\tau(t) = \gamma_o G(t), \text{and}$$
$$\gamma(t) = \tau_o J(t).$$
(6.74)

$J(t)$ and $G(t)$ resemble a creep function or a relaxation function, respectively, Fig. 6.8. In the case of a periodic disturbance, the strain can be rewritten in the form $\gamma = \gamma_o \sin \omega t$ which substituted in equation (6.68) leads to

$$\tau(t) = \gamma_o [G' \sin (\omega t) + G'' \sin (\omega t)]$$
(6.75)

with

$$G' = \omega \int_o^{\infty} G(s) \sin (\omega s) ds, \text{ and } \quad G'' = \omega \int_o^{\infty} G(s) \cos (\omega s) ds$$
(6.76)

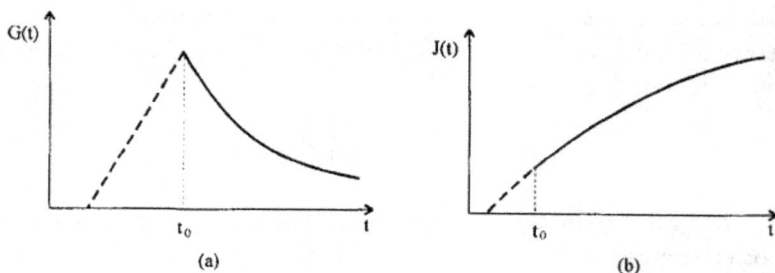

Figure 6.8: (a) Relaxation, (b) Creep.

Thus, the response of the system is constructed of an in-phase and out-of-phase component. The two harmonic functions can be replace by

$$G' = \left(\frac{\tau_o}{\gamma_o}\right) \cos \delta, \text{and} \quad G'' = \left(\frac{\tau_o}{\gamma_o}\right) \sin \delta, \tag{6.77}$$

with

$$\frac{G''}{G'} = \tan \delta = \frac{J''}{J'}, \tag{6.78}$$

which introduces the phase relation (loss tangents δ) between the disturbance and the response. The energy, W, that is dissipated during a viscoelastic deformation is

$$W = \int_0^T \tau d\gamma = \int_0^T \tau \frac{d\gamma}{dt} dt = \pi \gamma_o G'' = \pi \gamma_o G' \tan \delta \tag{6.79}$$

An oscillating load can also be mathematically expressed in a complex notation, so that the following identities and equations apply for

- complex stress

$$\tau(t) = \tau_o e^{i\omega t}, \tag{6.80}$$

- complex strain

$$\gamma(t) = \gamma_o e^{i(\omega t - \delta)}, \tag{6.81}$$

- complex viscosity

$$\eta^* = \frac{\tau^*}{\gamma^*}, \tag{6.82}$$

- complex relaxation function

$$G^*(t) = \frac{\tau^*}{\gamma^*} = \frac{1}{J^*(t)},$$ (6.83)

with the relation

$$G^* = G' + iG'', \text{and,} \quad J^* = J' + iJ''$$ (6.84)

to the in- and out-of-phase components G', G'', J' and J''. Many relationships follow from these equations. For instance,

$$J' = \frac{1/G'}{1 + \tan^2\delta},$$ (6.85)

$$J'' = \frac{1/G''}{1 + (\tan^2\delta)^{-1}},$$ (6.86)

$$\eta' = \frac{G''}{\omega},$$ (6.87)

$$\eta'' = \frac{G'}{\omega}$$ (6.88)

Finally, by substituting equations (6.74) into equation (6.73) with $G_{t\to\infty} \equiv G_{\omega\to0}$, the shear stress and modulus is

$$\tau(1 + i\omega t_\tau^*) = G_o\gamma \left(1 + i\omega t_\gamma^*\right), \text{ and, } \quad G_\infty = \frac{\tau(\omega \to \infty)}{\gamma(\omega \to \infty)} = G_o\frac{t_\gamma^*}{t_\tau^*}.$$ (6.89)

Since $G_\infty > G_o$, the characteristic time for the creep process is longer than the relaxation time.

6.2.9 Mechanical models

Based on the discussed linear viscoelastic theory of stress-strain and rate behaviors, the following summary for the possible behaviors of materials can be obtained:

(a) fully elastic behavior: $\tau = G\gamma$
(b) fully viscous behavior: $\frac{d\tau}{dt} = \eta\frac{d\gamma}{dt}$,or
(c) linear viscoelastic behavior: $\tau + t_\tau^*\frac{d\tau}{dt} = G_\infty\left(\gamma + t_\gamma^*\frac{d\gamma}{dt}\right)$

Simple phenomenological models have been developed by using mechanical springs and dampers to describe Hooke's elasticity and Newton's viscosity, respectively. Maxwell's and Kelvin-Voigt's models. in their simplest form. are sketched in Figure 6.9.

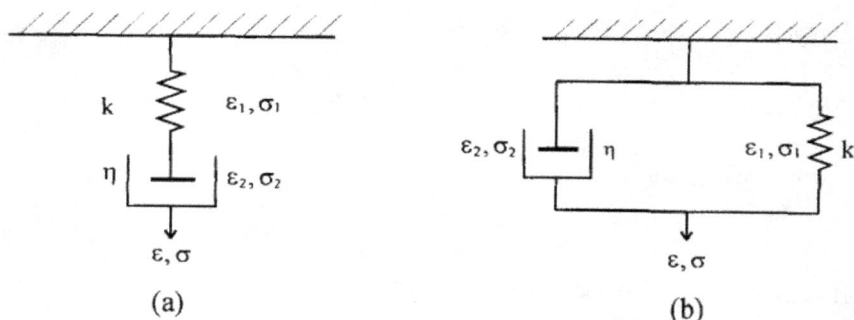

Figure 6.9: Viscoelastic models: (a) Maxwell model, (b) Voigt Model.

The models contain a single spring with a damping term either attached in series or in parallel. The significant difference between the two basic mechanical viscoelastic models is, that the system based on the Kelvin-Voigt model relaxes to a finite displacement in the event of a step-like disturbance, and the Maxwell model not. For $t \to \infty$, the Kelvin-Voigt model behaves fully elastic, and the Maxwell model fully viscous. Hence, the two models are only of limited use in describing the behavior of real systems. The Maxwell model cannot account for the time-dependent aspect of creep, and the Kelvin-Voigt model fails to explain stress relaxation.

The Kelvin-Voigt model corresponds to the following choices of variables:

- the strain γ is the observable variable,

- the stress is the strain's associated variable, and

- the stress is divided into an "elastic" part $\tau_e = G\gamma$ and an "inelastic" one $\tau_i = \eta d\gamma/dt$ so that the total stress is $\tau = G\gamma + \eta d\gamma/dt$.

The Maxwell model corresponds to the following choices of variables:

- the strain γ is always the observable variable,

- the stress is the strain's associated variable, and

- the strain is divided into an "elastic" part $d\gamma_e/dt = d\tau/dt*1/G$ and an "inelastic" one $d\gamma_i/dt = \tau/\eta$ so that the total stress is $d\gamma/dt = d\tau/dt*1/G + \tau/\eta$.

- Setting the stress equal to a periodical function (equation (6.74)) and using equation (6.89) and its counterpart for $\omega \to \infty$, equations of modulus and compliance can be derived which is summarized in Table 6.1.

Kelvin-Voigt	Maxwell
Diff. Equation	Diff. Equation
$\eta_0 \frac{d\gamma}{dt} + G_0 \gamma = \tau$	$\tau = G_0 \frac{d\gamma}{dt}$
$G(t) = G$	$G(t) = G e^{-t/t^*}$
$J(t) = J \left(1 - e^{-t/t^*}\right)$	$J(t) = J + t/\eta$
$G'(\omega) = G, G''(\omega) = \omega\eta$	$G'(\omega) = G \frac{(\omega t^*)^2}{1+(\omega t^*)^2}$
$\tan\delta = \omega t^*$, $t^* = \frac{\eta}{G}$	$G''(\omega) = G \frac{\omega t^*}{1+(\omega t^*)^2}$
$J^* = \frac{1}{G^*}$	
(relationship between complex compliance and complex modulus)	$\tan\delta = \frac{1}{\omega t^*}$, $t^* = \frac{\eta}{G}$
$J' = \frac{J}{1+(\omega t^*)^2}$, $J'' = J \frac{\omega t^*}{1+(\omega t^*)^2}$	$J'(\omega) = J = \frac{1}{G}$,
	$G''(\omega) = \frac{1}{\omega\eta}$

Table 6.1: Kelvin-Voigt and Maxwell models

6.3 Nanorheological and shear behavior of confined liquids

The rheological properties of liquids depend on *external parameters* such as pressure, shear rate, and temperature. If a homogeneous liquid is confined between two structureless surfaces of infinitesimal stiffness, direct correlations between the rheological properties and the external parameters can be drawn. These boundary conditions are usually made in the *hydrodynamic regime* of lubrication, and described for incompressible Newtonian fluids with the Reynolds equation (6.39). At higher pressures and thinner films more sophisticated theories are necessary to explain the complex shear force behavior which is reflected in the multidimensional Stribeck curve, Fig. 6.7. Assumptions of structureless and hard surfaces should be dropped and new parameters included; e.g., statistical parameters for roughness and Young's and shear modulus.

The complexity of the shear behavior of an ultrathin film demands the breakdown of the problem into very distinct regimes:

(a) How do rheological properties and shear properties of a homogeneous simple fluid change if it is normally confined between two "nearly" structureless solid surfaces?

(b) How does a rheological or structural complex but homogeneous liquid behave if it is normally confined between two "nearly" structureless solid surfaces?

(c) How does a surface adsorbed liquid or polymer melt or brush behave under shear?

(d) What is the impact of interfacial confinement on the material properties of a liquid or liquid-like polymer film?

(e) What is the effect of roughness on the rheological or structural properties of a thin liquid film? And,

(f) how are viscous properties of single phase liquids influenced by interfacial interactions? This section will deal with these problems by reviewing recent surface forces apparatus (SFA) and scanning force microscopy (SFM) studies and theoretical models and simulations.

This section will deal with these problems by reviewing recent surface forces apparatus (SFA) and scanning force microscopy (SFM) studies and theoretical models and simulations.

6.3.1   Dynamic surface forces apparatus studies on confined liquids

Liquids are in many practical applications confined to ultrathin films. The study of ultrathin confined liquids is relevant, for example, for lubrication, flow of liquids through porous media, flow of liquids in biological systems, formation of polymer composites, and thin film castings.

In the previous paragraphs, the viscosity of liquids has been discussed in terms of the Navier-Stokes equations of continuum hydrodynamics. This theory has been extended to viscoelasticity by introducing Boltzmann's linear superposition principle and simple mechanical models consisting of springs and damping terms.

Montfort and *Hadziioannou* applied corresponding phenomenological theories to their surface forces apparatus (SFA) experiments of ultrathin confined films of long chain molecules[10], Fig. 6.10.

Pure polymer liquids, perfluorinated polyether $(CF_2CF_2O)_m - (CF_2O)_n$, were confined between two mica surfaces. Static forces were measured as the surfaces approached each other very slowly, Fig. 6.11. Repulsive forces were found to extend out to separations greater than 10 times the radius of gyration, R_g, of the sample polymer in the bulk. The steep repulsive slope in the force at small separation distances is due to a hard wall effect. The separation can be decreased no further because of the softness of the SFA spring, or in

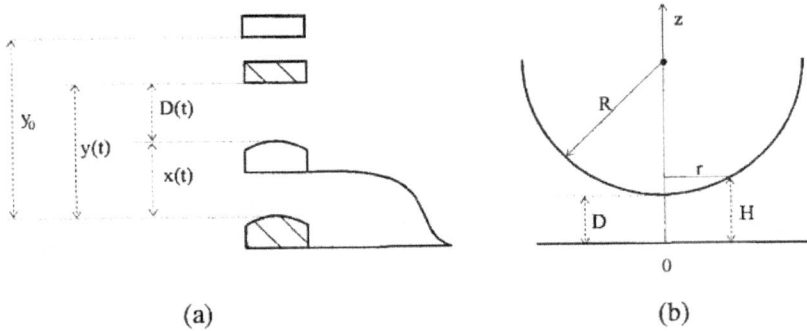

(a)   (b)

Figure 6.10: (a) SFA schematic. Upper surface moves with $y(t)$, and lower surface responds with $x(t)$. (b) The liquid thickness, D, is chosen to be much smaller than the effective radius, R, and hence $r \ll R$.

other words, the stiffness of the confined sample exceeds the spring constant of the double cantilever. The authors conclude, based on the high wettability of perfluorinated polyether to clean mica surfaces, that the formation of surface films on each side, plus unattached chains in between, are causing the measured long-range repulsive forces.

Montfort and *Hadziioannou* considered the following forces acting on the lower surface (with mass m) of their SFA, Fig. 6.10:

(a) the inertial force, $F_I = m\frac{d^2x}{dt^2}$,
(b) the restoring force of the spring, $F_R = -kx$,
(c) the surface forces, F_S, and
(d) the hydrodynamic forces, F_H.

The surface force can be assumed to corresponds with the static forces as discussed above as long as the perturbation of the distance, D, around the mean distance, $\overline{D}$, is small. Hence, the surface forces can be described, in a first order approximation, as:

$$F_S(D) = F_S(\overline{D}) + k_{eff} \times (D - \overline{D}), \qquad (6.90)$$

where the slope, k_{eff}, is determined from static force-distance measurements. The slope is a measure of the effective stiffness of the systems which includes the sample liquid and the SFA. The hydrodynamic force F_H of the confined liquid has been treated by *Montfort* and *Hadziioannou* as a first-order linear, viscoelastic fluid by combining the continuity equation and the equation of motion for incompressible liquids with the Maxwell model. In the crossed

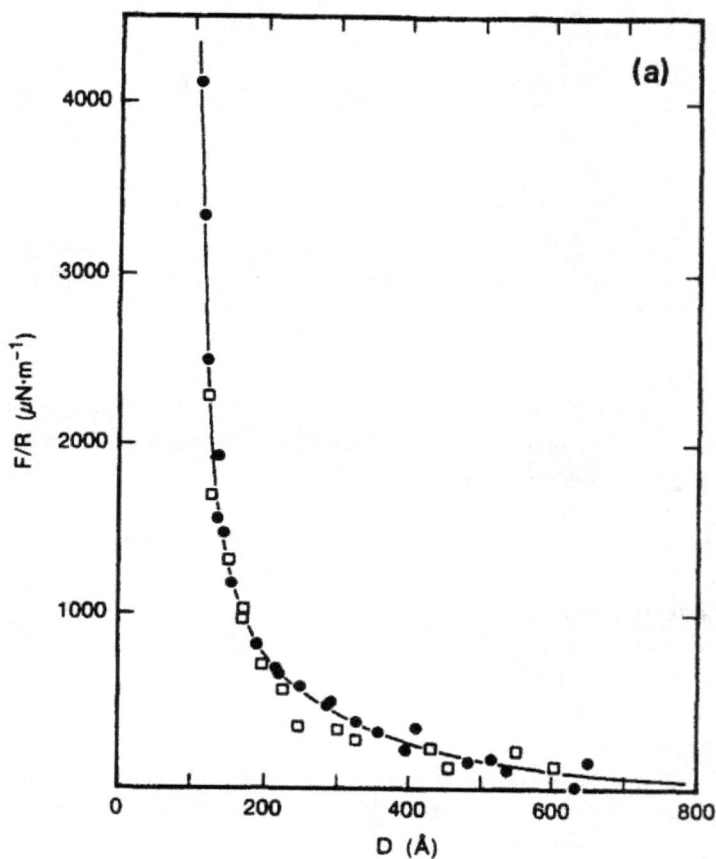

Figure 6.11: Logarithmic force/(radius of curvature), plot vs. distance, d, of perfluorinated polyether liquid at 25°C (● droplet between the surfaces) (□ immersed surfaces in the liquid). (From[13]).

cylinder configuration of the SFA experiment, the equations are:

$$\frac{1}{r}\frac{\partial}{\partial r}(rv_r) + \frac{\partial v_z}{\partial z} = 0 \text{(equation of continuity)}^k$$

and

$$\frac{dp}{dr} = \frac{\partial \tau_{rz}}{\partial z} \text{equation of motion}^l \qquad (6.91)$$

assuming that the local radii of curvature of the SFA are large in comparison with the distance between the surfaces. In this approximation, the flow is similar to that between parallel plates. The stress tensor τ for a viscous liquid without any "memory" can be expressed by

$$\tau(t) = \eta \frac{d\gamma}{dt}, \qquad (6.92)$$

which is also predicted by the Maxwell model for very slow motion. *Chan* and *Horn* obtained for the hydrodynamic force, F_H , based on the following relation for a plane-sphere geometry[14]

$$F_H = -\frac{6\pi R^2}{D}\eta\frac{dD}{dt}. \qquad (6.93)$$

This relation was extended by *Montfort* and *Hadziioannou* to viscoelastic liquids with Boltzmann's superposition principle which yield a hydrodynamic force of

$$F_H = -\frac{6\pi R^2}{D(t)}\int_{-\infty}^{t} G(t-t')\frac{dD}{dt'}dt'. \qquad (6.94)$$

Hence, the equation of motion of the lower SFA surface is

$$m\frac{d^2x}{dt^2} + kx - \frac{6\pi R^2}{D}\int_{-\infty}^{t} G(t-t')\frac{dD}{dt'}dt' + F_S(\overline{D}) + k_{eff} \times (D - \overline{D}) = 0. \quad (6.95)$$

This differential equation can be solved for a sinusoidal movement of amplitude A_{in} and frequency ω that is applied to the upper surface, Figure 6.10, by using the complex notation

$$\int_{-\infty}^{t} G(t-t')\frac{dD^*}{dt'}dt' = i\omega\eta^*(D^* - \overline{D}), \qquad (6.96)$$

with the complex quantities , where $D^* = \overline{D} + (A_{in} - A_{out})e^{i(\omega t + \Phi)}$ where A_{in} and A_{out} are the input and output modulation amplitudes at the upper and lower surface, respectively, and $\eta^* = \eta' + i\eta''$. The solution yields a viscous component of the complex viscosity,

$$\eta' = -\frac{\overline{D}}{6\pi R^2 \omega}(k - m\omega^2)A'\sin\Phi, \qquad (6.97)$$

and an elastic component,

$$\eta'' = -\frac{\overline{D}}{6\pi R^2 \omega}[(k - m\omega^2)(A'\cos\Phi - 1) + k_{eff}] = \frac{G'}{\omega}, \qquad (6.98)$$

where $A' = A_{in}/(A_{in} - A_{out})$ and Φ is the phase lag in regard to the input modulation."For purely viscous liquids the viscosity, η, is equal to the dynamic viscosity, η', which can be rewritten in the form

$$\eta = -\frac{\overline{D}}{6\pi R^2 \omega}(k - m\omega^2)\sqrt{(A')^2 - \left(\frac{k - k_{eff} - m\omega^2}{k - m\omega^2}\right)^2}. \qquad (6.99)$$

Israelachvili used a simplified form of this equation, i.e.,

$$\eta = -\frac{\overline{D}k}{6\pi R^2 \omega}\sqrt{(A')^2 - \left(1 - \frac{k_{eff}}{k}\right)^2}, \qquad (6.100)$$

for viscous liquids probed by SFA under low frequency conditions by neglecting the inertial terms of equation 6.99, i.e., by setting $k = k - m\omega^2$[215]. The complex response function, η^*, is connected to the relaxation function $G(t)$ through the superposition principle as follows:

$$\eta^* = \int_0^\infty G(t)e^{-i\omega t}dt, \qquad (6.101)$$

which leads, together with the Maxwell model for $G(t)$, to:

$$\eta^* = \frac{\eta_0}{1 + i\omega t_0}, \qquad (6.102)$$

with the relaxation time, $t_o = \eta_o/G_o$.

Montfort and Hadziioannou combined equations (6.97), (6.98) and (6.102) to obtain

$$\frac{1}{A'} = \frac{[\frac{\omega\eta_o}{\alpha(1+\omega^2 t_o^2)}]^{-1}}{\sqrt{1 + [\omega t_o + \alpha\left(1 - \frac{k_{eff}}{k}\right)\frac{1+\omega^2 t_o^2}{\omega\eta_o}]^2}}; \alpha = \frac{k\overline{D}}{6\pi R^2}, \qquad (6.103)$$

and studied the functional behavior of the amplitude of oscillations for a Maxwell fluid in the presence of surface forces, Fig 6.12. They found that in the case of repulsive forces ($k_{eff} < 0$), the variation in A' are comparable to the variations observed without surface forces[13].

"Note that $k\overline{x} + F_S(\overline{D}) = 0$ for one full period.

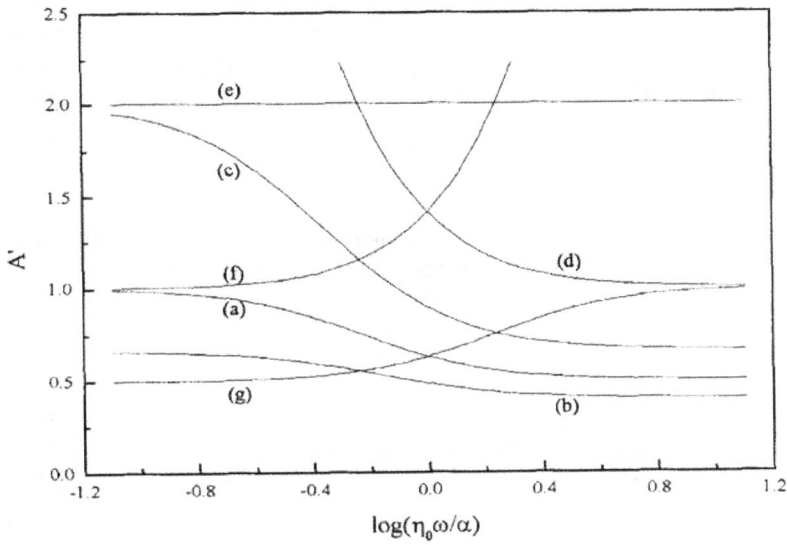

Figure 6.12: Amplitude of oscillations for a Maxwell fluid in the presence of surface forces: (a) $k_{eff} < 0$, (b) no surface forces: $k_{eff} = 0$, (c) $0 < k_{eff}/k < 1$, (d) $k_{eff}/k = 1$, (e) $k_{eff}/k = 1 + G_o/\alpha$, (f) $k_{eff}/k.1 + Go/\alpha$; for $G_o/\alpha = 1$.

With the asymptotic behavior

$$\lim_{\omega \to 0} \frac{1}{A'} = \left(\frac{1}{A'}\right)^{(0)} = | 1 - \frac{k_{eff}}{k} |^{-1}, \tag{6.104}$$

$$\lim_{\omega \to \infty} \frac{1}{A'} = \left(\frac{1}{A'}\right)^{(\infty)} = | \frac{G_o}{\alpha} + 1 - \frac{k_{eff}}{k} |^{-1}, \tag{6.105}$$

the effective spring constant, k_{eff}, and the elastic modulus, G_o, can be calculated. In the case of attractive surface forces, different classes should be distinguished, Fig. 6.13. It is important to note that depending on the strength of the attractive surface forces the response amplitude increases, decreases or remains constant with the frequency. The different qualitative behavior of the amplitude variation is found in the frequency dependence of the phase shift, Φ, which is

$$\tan \Phi = - \left[\omega t_o + \alpha \left(1 - \frac{k_{eff}}{k} \right) \frac{1 + \omega^2 t_o^2}{\omega \eta_o} \right]^{-1} \tag{6.106}$$

and presented in Figure 6.13. While for the amplitude the asymptotic behavior is interesting, it is the minimum in the phase shift, Φ_m expected for

$$\omega_m^2 t_o^2 = \frac{(\frac{1}{A'})^{(\infty)}}{(\frac{1}{A'})^{(0)}}, \tag{6.107}$$

where ω_m is the frequency to the minimum phase shift, which provides additional information about the Maxwell relaxation time t_o.

Montfort and *Hadziioannou* found equation (6.100) experimentally confirmed for a polymer liquid of perfluorinated polyether, until a mean separation distance of around 200 nm[13]. The measured viscosity of the confined film was reported to be 2.47 P, which is in correspondence to the bulk polymer. At a distance of 84 nm, the viscosity was found to increase to a value of 3.32 P. The authors claim that as the gap decreases to a size comparable to the dimensions of the pinned chains, the mobility of the chains decreases and, as a result, the viscosity increases. The first finding of *Montfort* and *Hadziioannou* was that the phenomenological theory developed above predicts well the viscoelastic behavior of liquids, e.g., semidilute polymer solutions up to a certain thickness. Their second important finding deals with ultrathin liquid films below a certain critical thickness where interfacial interactions and molecular dimensions become noticeable. Also other groups observed, at lesser separations, that hydrodynamic forces are larger than expected from the phenomenological calculations discussed above[16,17,18].

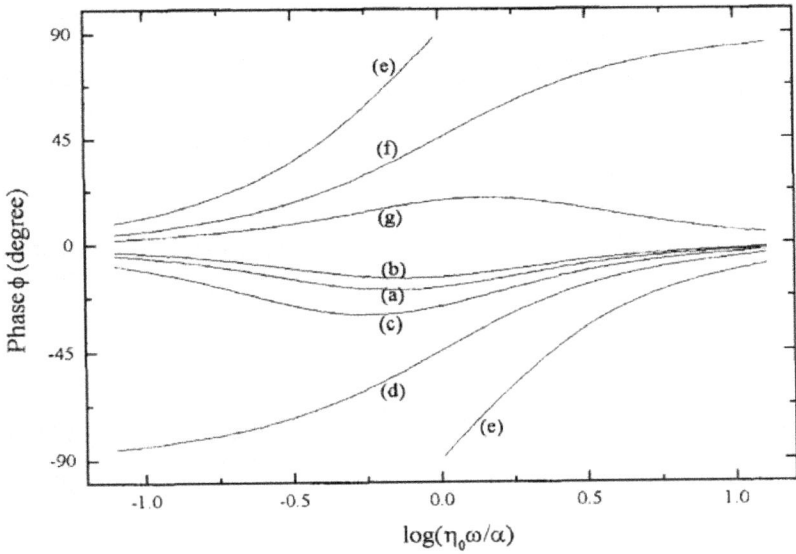

Figure 6.13: Phase shift for a Maxwell fluid in the presence of surface forces: (a) $k_{eff} < 0$, (b) no surface forces: $k_{eff} = 0$, (c) $0 < k_{eff}/k < 1$, (d) $k_{eff}/k = 1$, (e) $k_{eff}/k = 1 + G_o/\alpha$, (f) $k_{eff}/k.1 + Go/\alpha$; for $Go/\alpha = 1$.

The extended theories of lubrication, for instance the exponential relationship of the viscosity with hydrostatic pressure, equation (6.60), were found to predict viscosity values larger than experimentally encountered. Thus, an adsorbed polymer layer or a brush is not well described by a rigid film. Further theories were developed by *Milner*[19] and *Pincus*[20] who were modeling the complex flow of solvents between grafted polymer layers. Their theories are in analogy of flow through porous media.

6.3.2   Dynamic force microscopy study on liquids

Dynamic viscoelastic properties of ultrathin liquid films have also been studied with another technique, the scanning force microscope (SFM)[21,22]. *Friedenberg* and *Mate* applied the SFM to a low-molecular-weight polymer liquid, poly(dimethylsiloxane) (PDMS), by using spherical glass beads of 22 mm radius, R, at the end of a tungsten wire with a calculated spring constant of 40 N/m, k_L from a measured resonance frequency of 7.8 kHz, Fig. 6.14. Also *Friedenberg* and *Mate* used Maxwell's steady creep model consisting of a Newtonian damping term b_p for the polymer sample at small shear rates, by setting the samples elastic response k_P to zero, and connecting the cantilever spring constant in series, Fig. 6.14. The authors further assumed that the surface force between the glass bead and the polymer surface is dominated by the meniscus force, F,

$$F \cong -4\pi R \gamma_L \left(1 + \frac{\delta}{2r} \right), \tag{6.108}$$

where γ_L is the surface tension, $\delta > 0$ is the average penetration depth of the sphere in the polymer liquid, $\delta < 0$ is the distance between the sphere and the original film surface, and r is the capillary radius - with r much smaller than the radius of the sphere, R,[23] Fig. 6.14.

Because the meniscus force is linear with displacement, a *meniscus* spring constant can be introduced as:

$$-k_M = \frac{dF}{d\delta} \cong -2\pi \gamma_L \frac{R}{r} \tag{6.109}$$

which is assumed to be independent of both frequency and separation. The meniscus spring constant k_M corresponds to the effective spring constant, k_{eff} introduced above, and hence, can be determined at low frequencies and at large separations where viscous forces are negligible.

The viscous forces, $F_v = bv = b * dh/dt$ are measured at small surface deformations, where v is the relative velocity of the surfaces, $b = 6\pi R^2 \eta / h = k_L \eta / \alpha$ is the viscous drag coefficient (equation (6.55), *Moore*[24]) and equation

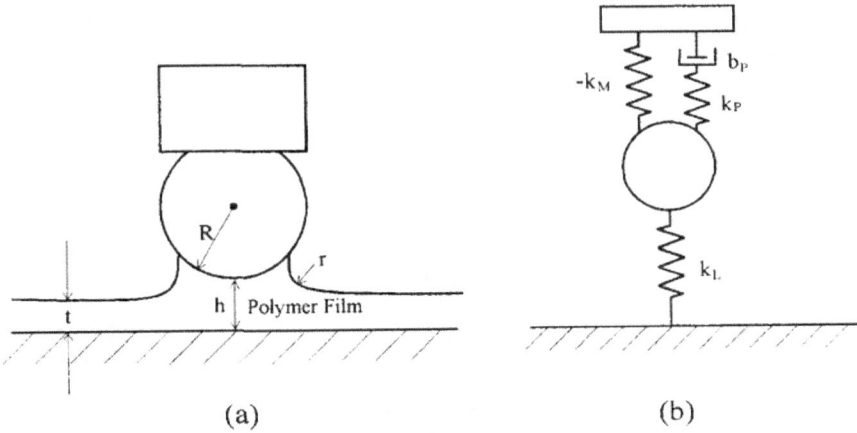

(a)   (b)

Figure 6.14: a) SFM approach for measurements of viscoelastic properties of liquid polymer films: (a) schematic, (b) mechanical model.

(6.103)), and h is the distance between the sphere and the wall. *Friedenberg* and *Mate's* measurements provide, as expected, a linear relationship between $a = k_L \eta / b$ and the separation distance, h, Fig. 6.15.

The damping coefficient, b, can be determined from

$$A = \frac{A_{out}}{A_{in}} = \frac{\omega b \sin \Phi - k_M \cos \Phi}{k_L - k_M - m\omega^2} \tag{6.110}$$

where A is the amplitude ratio, A_{out} is the response amplitude of the cantilever spring, A_{in} is the input modulation, Φ is the phase shift between input and response, k_M is the meniscus spring constant which is determined independently from the lever response at low frequency and large separation, k_L is the lever spring constant, and m is the mass of the cantilever probe[21]. The expression for the amplitude ratio reduces to

$$A = \frac{\omega b \sin \Phi}{k_L} \tag{6.111}$$

for very low modulation frequencies, i.e., $k_L - m\omega^2 \cong k_L$, and neglecting capillary forces.[n]

The sharp decrease of the reciprocal damping coefficient in Figure 6.15 is believed to arise from hard-wall contact, and hence, serves as the critical point

[n]Note that equations developed by *Montfort* and *Hadziioannou* for SFA, e.g., equation (6.97),

to define zero separation[21]. A viscosity of 430±40 cP could be determined from the linear extrapolation in Figure 6.15 and the radius of curvature of the sphere, and was found to be close to the bulk value of 350 cP[21].

One of the most fundamental questions in thin film applications is the interaction strength of films with their substrates. *Friedenberg* and *Mate* suggest using the capillary pressure, $P = \gamma_L/r$, which is assumed to be in equilibrium with the pressure of the liquid film on the substrate surface, as a quantitative measure for the interaction strength. Low pressure would indicate a weak interaction force. The capillary pressure is determined from the experimental determination of the stiffness of the meniscus, k_M, and the equation (6.109). In this particular example of PDMS, a disjoining pressure of 14 kPa for a 32 nm film, 5.8 kPa for a 92 nm film, and 2.5 kPa for a 128 nm film were found[21].

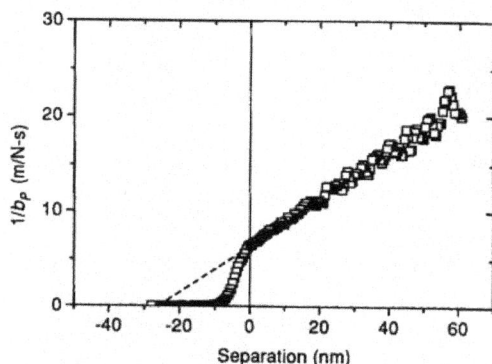

Figure 6.15: Reciprocal of the viscous damping coefficient, b, as a function of separation for a 92 nm PDMS film. The dotted line is a linear extrapolation. (From[21] with permission of C.M. Mate)

The frequency response of the amplitude ratio at large separation were found to be finite and non-zero at low frequency which was attributed to the capillary force[21]. As the frequency increases, the amplitude ratio was found to increase to unity because of enhanced viscous coupling between the cantilever and the polymer liquid[21]. *Friedenberg* and *Mate* claim that this is an important difference between SFM and SFA measurements[21].

---

and equations developed for SFM, e.g., equation (6.111), are related by

$$-A' \sin \Phi_1 = \frac{A}{\sin \Phi_2},$$

where Φ_1 and Φ_2 are the phase shift of the lower plate and the cantilever spring, respectively.

The significant difference between the SFM and SFA approach to confinement measurements of ultrathin films are the surface forces. In SFA measurements, capillary interactions are negligible compared to other forces due to the low curvature of the large contact area. This is however different for SFM measurements where the contact area is small and capillary forces at large separations are significant.

6.3.3 Viscous friction force measurements between lubricated surfaces

In the two previous paragraphs, viscous drag forces and interfering surface forces were discussed in experimental setups of two adjacent surfaces which are moving perpendicular to each other. In the event of lateral motions of two surfaces which confine a ultrathin film in between, typically referred to as lubricated frictional motion, it is expected that the overall friction is reduced and the surfaces slide smoothly relative to one another. Hence, the acting viscous forces for a Newtonian fluid can be expressed as $F = \eta v$ as already introduced above with the viscous damping coefficient for surface perpendicular motions.[o] It was, however, found that for very thin lubricant films the forces become much larger than those measured in the bulk liquid. Considering that the degree of freedom of motion in a confined "two-dimensional" fluid is significantly reduced in comparison to a three-dimensional bulk fluid, it is reasonable to assume that the characteristic relaxation times become orders of magnitude greater than those of the bulk[6]. Hence, increased viscous friction in an ultrathin film could be understood from the standpoint of entropy only, without considering the effect of surface interactions which will be discussed in more detail below.

Israelachvili and co-workers found that thin lubricant films can exhibit solid-like properties, including a critical yield stress and a dynamic shear melting transition, which can lead to stick-slip motions[4,25,26,27]. The generic shape of an overdamped stick-slip behavior is illustrated in Figure 6.16 which has been observed with SFA experiments as sketched in Figure 6.17[27]. The spring force, $F = -k(u - vt)$, results from the difference of the relative displacement of the block to the stationary lower surface, u, and the drive distance, vt, in conjunction with the spring constant k.

There are three possibilities which result in stick-slip motion of a measuring system which is based on springs:

---

[o]Considering the different lubrication regimes, neglecting surface deformations (this is possible for plane surfaces and mere lateral in-plane motions) and assuming a sliding distance h, the viscous friction force can be expressed by equation (6.55).

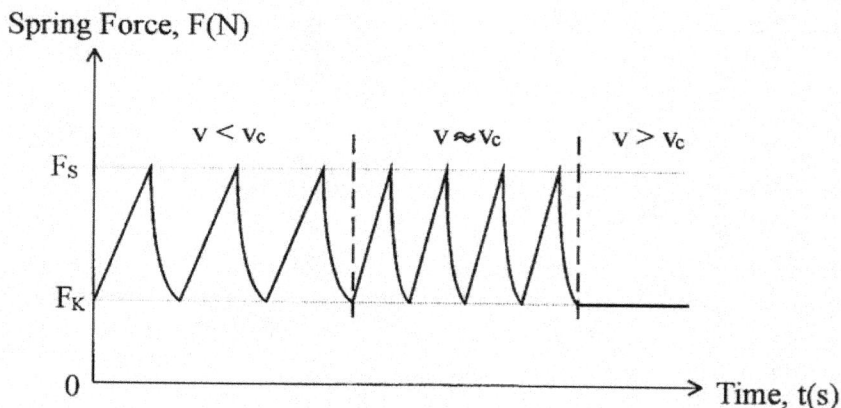

Figure 6.16: Illustration of a generic stick-slip motion of an overdamped spring system for increasing pulling velocities. Below the critical values for temperature, T_c, and shear rate, v_c, dominant stick-slip motions have been observed for hexadecane films. The stick-slip spikes disappear as the velocity is increased above v_c.

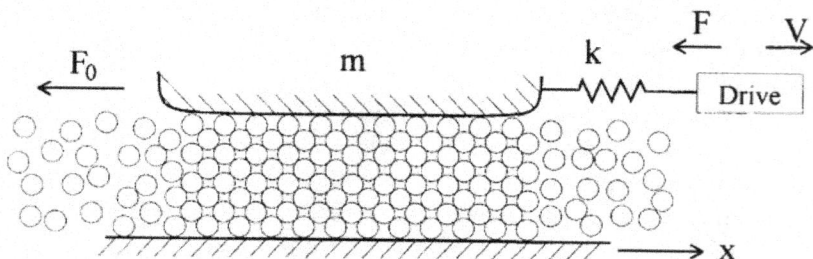

Figure 6.17: Mechanical analog of the SFA experimental setup used for measurements of friction forces Fo. A block of mass m which is confining a thin liquid film of hexadecane between atomically smooth mica is pulled laterally at velocity v. The lateral forces, F, are measured with an elastic spring of spring constant k [27]

(a) Spring instabilities occur due to the choice of soft measuring springs. It is not the liquid but the spring system which is measured.

(b) The confined liquid film is "frozen" and the measuring spring is stiff enough so that the yield of the confined film due to lateral displacement occurs before any spring instabilities. Further, it is assumed that the film material yields three-dimensionally. That process is called shear-melting. The force which is measured can be described by a viscous force.

(c) Due to the fact of confinement the exerted force is initially too small to deform the highly viscous film (stick regime). Yielding of the film occurs two-dimensionally along a slip plane either between the film and the solid mica surface or in between the film. This process - where again a stiff spring is assumed - is better described by solid friction.

Yoshizawa and *Israelachvili* claim that their observation of stick-slip behavior is associated with some sort of melting transition[27]. They support their statement by start-stop experiments with varying stopping times and sliding velocities above the critical stick-slip velocity, v_c. As mentioned above for sliding velocities, $v > v_c$, sliding occurs steadily. *Yoshizawa* and *Israelachvili* observed for hexadecane at $T = 17°C$, if sliding is discontinued for a time t_s smaller than a critical time t_c, the spring force F returns smoothly to the value it had prior to the stopping interval, while for a time $t \geq t_c$ a stiction spike is produced before the friction force returns to the value it had taken prior to the stopping interval. *Yoshizawa* and *Israelachvili* find the sharp onset for a stiction spike at $t_s = t_c$ sufficiently different from macroscopic dry friction experiments[28,29], which leads them to refer to t_c as the nucleation time of the frozen state. The generic behavior of friction at dry interfaces is that the stiction spikes increase logarithmically with time with no sharp transition time[30].

6.3.4 Theoretical shear simulations and mechanical models

In molecular dynamic (MD) simulations of *Gao, Luedtke* and *Landman*, it was found that the shear motion of a confined ultrathin hexadecane film can be visualized with the average motion of a single card in a *stack of cards*[31]. Previous drainage experiments of hexadecane films, performed experimentally and by MD simulation[32], confirm a layered density oscillation of surface-confined n-hexadecane[32]. The solvation force oscillations observed during normal compression are found to be very pronounced in n-hexadecane because of their strong repulsive and attractive regions which are due to the simple linear structural form of the molecules. *Landman* et al. showed that for a slightly more

complex structure of a branched alkane, such as squalane, the solvation forces
are mostly repulsive which exhibit a monotonic continuous decrease in the
number of confined segments contrary to the step-like variations observed with
n-hexadecane films. In friction experiments, the confined hexadecane film can
be pictured as being built up of discrete molecular layers which move statisti-
cally either in registry with the top or the bottom shear surface. The thicker
the film the higher are the statistical possibilities of shear motions so that after
five molecular layers no stick-slip motions can be found in the averaged lateral
shear stress. Note that *Landman's* MD simulation of shear does not suggest a
melting transition to be responsible for the stick-slip motion. The statistical
sliding between layers determines the overall friction force. Sliding occurs in
two-dimension, which is in contrast to viscous sliding where deformation occurs more or less isotropic in three dimensions depending on the complexity of
the fluid and interfacial interactions.

Various groups motivated by SFA or SFM experiments used simple mechanical models to investigate the stick-slip behavior[33,34]. It is important to
note that all of these attempts are phenomenological in nature by using differential equations where a spring and a viscous damping term are involved.
Depending on the *Ansatz* of the differential equation, the character of dissipation is already implied. For instance, an equation of motion of the simple
form

$$m\ddot{x} + kx + \eta\dot{x} = f(t) \qquad (6.112)$$

implies a Newtonian viscous dissipation.[P] *Carlson* and *Batista* proposed a phenomenological constitutive relation to describe the frictional forces as function
of the macroscopic variables, position, velocity and time[34]. Constitutive equations are referred to as rate and state laws, where the rate variable refers to
the sliding velocity, and the state variable is meant to capture all memory dependent effects. *Carlson* and *Batista* based their spring model calculation on
Yoshizawa and *Israelachvili's* friction measurement of hexadecane lubricated
surfaces as discussed above and sketched in Figure 6.17. Hence, it was assumed that the state variable represent the degree of "melting", as suggested
by *Yoshizawa* and *Israelachvili*. *Carlson* and *Batista* setup a single degree of
freedom equation of motion of the form

$$m\dot{U} = -k(U - vt) - F_o \qquad (6.113)$$

where U is the displacement of the block, and the other parameters are defined
as in Figure 6.17. The friction force F_o was introduced in its dimensional form,

---

[P]Dots denote time derivatives

containing the rate dU/dt and state Θ of the system, as follows[34]:

$$F_o = \Theta + \beta \dot{U} \qquad (6.114)$$

with

$$\dot{\Theta} = \frac{(\Theta - \Theta_{min})(\Theta_{max} - \Theta)}{t^*} - \alpha(\Theta - \Theta_{min})\dot{U} \qquad (6.115)$$

The state of the system has been arbitrarily constraint between Θ_{min} and Θ_{max}, corresponding to the "fully melted" and "frozen" states, respectively. The friction force F_o corresponds to the maximum static friction when $dU/dt = 0$, and to the dynamic friction when $dU/dt \geq 0$. Equation (6.115) describes the evolution of the state variable. A characteristic time t^* is introduced which makes the film freeze from any initial state $\Theta \neq \Theta_{min}$. The inverse of α plays the role of a characteristic melting length (i.e., slip distance over which the melting transition takes place). Like the experimental system, the model exhibits a transition from stick-slip to steady sliding at a critical velocity[34].

Carlson and *Batista* concluded that rate and state constitutive relations, while purely phenomenological, can provide some important guidelines for the design of mechanical systems, in terms of both material choices and operating conditions. More work remains to be done in the choice and the evaluation of particular constitutive relations. Considering the very recent MD simulations by *Landman* and co-workers, it seems reasonable also to consider rate and state laws for interfacial sliding in between liquid layers or at the liquid-solid interface.

In summary, it was found that even "simple liquids" such as short-chain alkanes behave quite complex under confinement. In recent studies it was also found that spherical molecules, such as octamethylcyclotetrasiloxane (OM-CTS) and low molecular polymer melts, cease to behave as bulk liquids if confined to a film of less than 2.5-5.0 nm in thickness[4,5,6,7].

6.4 Nanorheological and shear behavior of complex liquids

In the previous paragraphs, the shear and rheological properties of simple fluids have been discussed from the hydrodynamic regime to the boundary regime. Confinement has been introduced by bringing two molecularly smooth surfaces together separated by a thin compressed homogenous fluid. Complicating effects such as liquid film adsorption, interfacial interactions, and roughness have been neglected. It is the goal of the following paragraphs to discuss the behavior of complex liquids in the vicinity of interactive interfaces, the effect of roughness on the phase behavior of the "interfacial-liquid", and the rheological and shear properties of complex fluids which are confined between smooth and rough surfaces.

6.4.1 Rheological and shear properties of confined complex liquids composed of polymer brushes and solvent

Most of the liquids we encounter in nature and engineering are very complex, consisting of branched molecules, surface adhered brushes and colloidal particles. Fluids of that nature are called "complex liquids" to distinguish them from the "simple liquids" as discussed above. Extensive literature, both experimental and theoretical, can be found on these topics[27,35,36,37]. Surface adsorbed polymers which improve lubrication have been studied experimentally[17,38,39,40] and theoretically[19,20,41].

Since the 1950's, polymeric additives have been used extensively in engine oils to increase viscosity, decrease boundary friction, and reduce catastrophic failures such as shear thinning. Figure 6.18 shows the significant difference between (a) sliding of smooth surfaces immersed in toluene and (b) sliding of surfaces coated with polymer brushes in a good solvent. The effective friction coefficient meff could be reduced by three orders of magnitude from 0.6 to 0.005 by adding brushes to the system[42].

Acting forces between polymer coated surfaces have been extensively studied by *Klein* and co-workers[40]. Two regimes of lubrication forces are expected for two polymer-brush-coated surfaces, overlapping and non-overlapping polymer regimes.

For separation distances much larger than the thickness of the two polymer films (non-overlapping regime), the liquid flow in (or out) of the inter-surface gap is described by Reynolds hydrodynamic theory which leads for a plane-sphere geometry to equation (6.74). Reynolds theory can be extended by considering the hydrodynamic thickness, L_H, of the surface-adsorbed polymer films. The hydrodynamic thickness is defined as the distance from the solid interface where the velocity gradient of the liquid reduces to zero. The hydrodynamic force expressed in equation (6.93) can be extended as follows to fluid flow between polymer films or brushes:

$$F_H = -\frac{6\pi R^2}{(D - 2L_H)}\eta\frac{dD}{dt} \qquad (6.116)$$

It is important not to confuse that the hydrodynamic thickness L_H with the thickness of the polymer film, L. The relation between the two terms is given by $\delta H = (L - L_H)$, where δH represents the hydrodynamic penetration length of

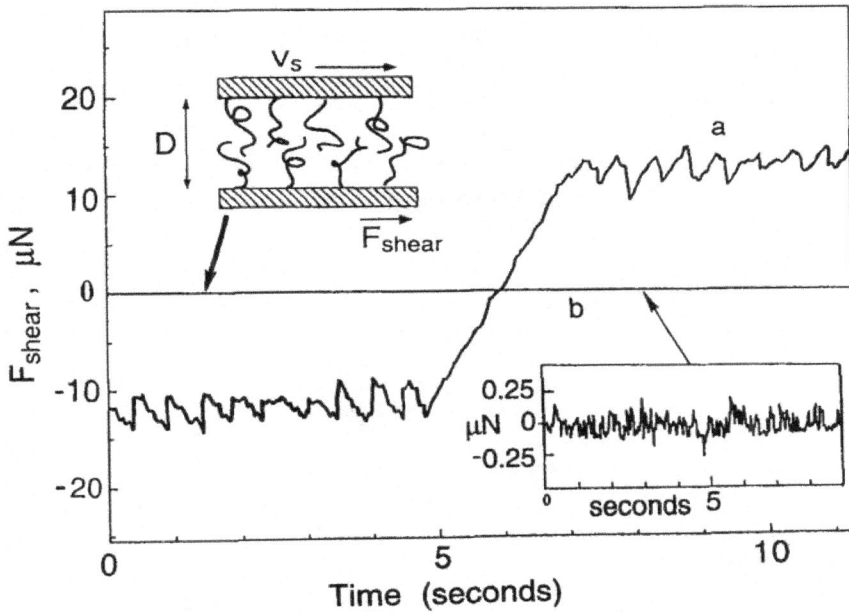

Figure 6.18: Friction force measurements (forward and reverse sliding direction, turning point at about 5s) (from Reference[42]). (a) Typical stick-slip motion for sliding between two smooth bare surfaces immersed in toluene (sliding distance, $D =1.4$ nm, sliding velocity $v = 45$ nm/s, normal load $N = 19$ μ N). (b) Sliding along polymer brushes reduces the friction forces significantly (c.f., magnification in inset) ($D =37$ nm, sliding velocity $v = 15$ nm/s, normal load N $= 19$ μ N).

the fluid flow field into the polymer layer. The penetration δH is an important parameter in the context of motion-dependent dissipative forces. The energy dissipation W per unit time and unit area caused by one polymer film is approximated by (6.52)

$$W \cong \frac{\eta_s(v(L)^2\delta_H)}{\langle \xi^2 \rangle} \qquad (6.117)$$

where $v(L)$ is the velocity at the film edge, η_s is the viscosity of the fluid (solvent), and $\langle \xi^2 \rangle$ is the mean square pore size or correlation length associated with the polymer network in the penetrated zone δ_H. The shear stress is then given by $\tau = W/v$.

For separation distances, D, much smaller than the thickness of the two polymer films (overlapping regime), the liquid has to flow through polymer segments. Energy dissipation occur mainly from friction experienced by the chains as they are dragged through the overlap region. The friction coefficient may be estimated as the ratio between the shear force required to slide the compressed brushes and the normal force required to compress them to a separation D[40]. The effective friction coefficient has been approximated for moderate compression and interpenetration of the polymer brushes to be[40]

$$\mu_{eff} = \frac{(6\pi\eta_{eff}vs^2)}{\beta^{1/2}k_BT} \qquad (6.118)$$

where η_{eff} is the effective viscosity, v is the sliding velocity, $1/s^2$ is the surface chain density, $\beta = 2L/D$ is the compression ratio of the brushes, and k_B is the Boltzmann constant. It is interesting to note that the resistance of shear is reduced for higher compression and/or denser brushes.

The efficiency of polymer brushes in reducing shear forces is shown in Figure 6.19 where already after compressions of 0.2 shear forces are below the resolution of the measurement[42].

6.4.2 Rheological and shear properties of compressed polymer layers melts

Luengo and *Israelachvili* focused their attention on polymer melts to study the full range of dynamic properties of a complex fluid film, and to develop a general scaling formalism of constitutive relations that effectively describes the transition from pure bulk to pure frictional behavior[12]. They chose polybutadiene (PBD) as sample with a molecular weight above the molecular entanglement weight to ensure a viscoelastic behavior. A droplet of the PBD melt was immersed between two mica sheets and slowly compressed to ensure mechanical

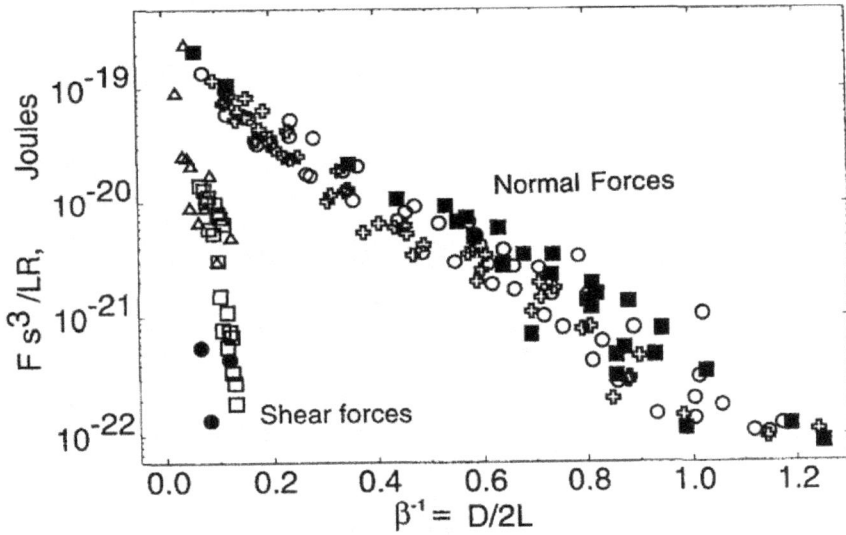

Figure 6.19: Normal and shear forces F between compressed polymer brushes, covering a wide range of molecular weights, brush heights, and surface densities. The surface separation axis and force axes are scaled to reduce the normal forces for the different brushes to a single curve. s, L, and R are the interanchor spacing, the brush, height, and the curvature of the polymer-bearing substrate, respectively (from Reference[42]).

equilibrium before shearing. It is important to note that true thermodynamic equilibrium of such high molecular weight molecules is not attained. This is in contrast to SFM measurements where thermodynamic equilibrium can be assumed because of the contact area which is on the order of six smaller than in SFA measurements. The surface forces are found to be steeply repulsive, as expected from the MD discussion above, which demonstrates with squalane and hexadecane the complexity of the molecular structure for being responsible for repulsive solvation forces[32]. It is also in correspondence with equilibrium type of interaction proposed by *de Gennes* for "pinned polymer melts"[44]. Pinning of polymer melts at interactive interfaces that alter the phase behavior of polymeric boundary layer to become more rubbery or glassy is known from SFA[6] and SFM[45] measurements, and will be discussed in more detail below. *Israelachvili* and co-workers found three dynamic regimes during oscillatory shear motion as a liquid film progressively thins[12]: the bulk rheological regime, the transition regime, and the tribological regime.

In the tribological regime, the surfaces are separated by $h < 15$ nm and stabilized by the steric "hard-wall" repulsion between them, the shearing surfaces are highly flattened and have a Couette flow geometry[12]. The kinetic friction force was found to be constant over sliding velocities of 0.01-1 μm/s which implies an effective viscosity $\eta_{eff} \propto (U/h)^{-1}$, where U is the shear velocity and h the separation distance, as discussed above by introducing the Stribeck curve. This indicates a pure two-dimensional frictional sliding which is in contrast to the three-dimensional viscous drag process of a bulk liquid.

In the transition regime, the fluid still behaves like a rheological fluid but the rheological properties are no longer described by the bulk values. The transition regime was found in PBD (MW = 7000 g/mol) between a separation distance of about 40 nm and 15 nm as shown in Figure 6.20. The effect of the transition regime (also called hydrodynamic layers) is extended over 2-4 R_g from each surface which suggests a surface-pinned polymer melt. The adsorbed or pinned polymers alter the bulk rheological properties[17].[q]

The bulk rheological regime has been discussed above in the previous section of that chapter.

---

[q] R_g stands for radius of gyration and is a size measure of an averaged polymer coil in the matrix or melt

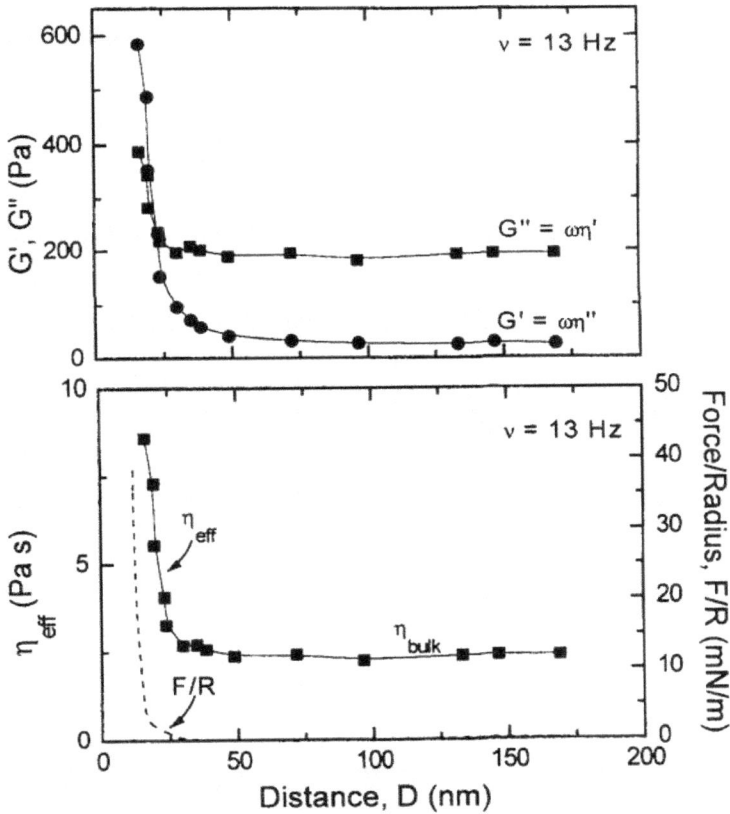

Figure 6.20: Shear moduli and viscosity measurement of PBD as function of the separation distance D covering the bulk ($D >$50nm), transition ($D =$15-50 nm) and friction regime ($D <$15nm) (from Reference[12]).

6.4.3 Film thickness variations of compressed polymer layers under shear

Georges, Loubet and co-workers studied rheological and frictional properties of adsorbed layers of poly-isoprene (PI) in a molecular tribometer, Fig. 6.21[39].

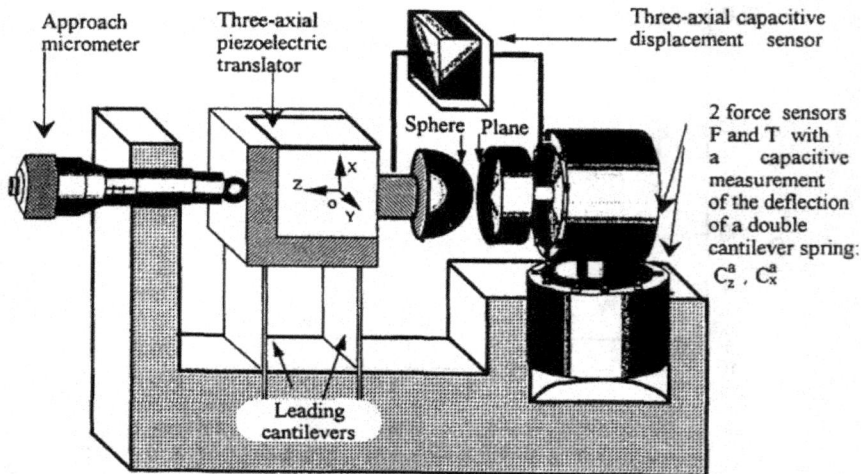

Figure 6.21: Schematic representation of a *"molecular tribometer"*. Differences to conventional surface forces apparatus (SFA) are the ball-plane geometry, the adaptable solid material which can be used, and the capacitance sensor detection scheme (from Reference[39]).

A droplet of PI solution (solvent 2,4, dicyclohexyl-2-mehylpentane (DCHPM)) was confined between the two metallic cobalt coated solid surfaces of the tribometer. A small amplitude oscillatory motion of about 0.1 nm RMS is superimposed to the steady approach motion of the sphere toward the plate in the tribometer. The thickness of the adsorbed polymer layer is determined from force profiles during a first compression while the hydrodynamic thickness is determine from damping profile measurements, Fig. 6.22. Based on Reynolds theory of lubrication (equations (6.92) and (6.116)) and assuming Newtonian behavior in the dilute regime (i.e., for surface separation distances larger than 25 nm), the damping coefficient A can be introduced as:

$$A = -\frac{F_H}{\dot{D}_{eff}} = \frac{6\pi R^2}{D_{eff}}\eta; \quad D_{eff} = D - 2L_H \tag{6.119}$$

where R is the radius of the sphere, η is the viscosity of the dilute polymer solution (i.e., the inverse of the slope of the damping profile) and L_H is the hydrodynamic layer thickness determined from the extrapolation of the damping profile in Figure 6.22(a).

Friction forces and PI film thickness have been investigated under various compression regimes. In very compressed films, where the sphere indents the two adsorbed PI films of thickness $2L$, a Hertzian circular contact area is found. The measured tangential compliance is the sum of the tangential compliance of the Hertzian contact (πa^2) and the tangential compliance of the confined film. A shear modulus G of about 250 MPa was obtained with

$$G = \frac{D}{\pi a^2 C_f},$$ (6.120)

where $D = 2L$, $a = 2.4$ μm, and the tangential compliance of the film $C_f = 4.4 \times 10^{-6}$. *Georges* et al. found that at this compression rate the PI layers behave as a solid in a rubbery state. The film thickness which was initially found to be equal to $D = 19.3$ nm, decreases as the lateral displacement X increases, Fig. 6.23. No immediate relaxation is observed when sliding is stopped. The variations of the film thickness is attributed to creep in the layer. Opposite behavior is observed with an increase in the film thickness when the sliding speed is increased. *Georges* et al. interpret that results as a change in the interpenetration regime which is assumed to increase for lower sliding speeds[39].

Two regimes of sliding could be differentiated for very compressed films - the pinning and non-pinning regime. Pinning could be observed up to a sliding velocity of about 20 nm/s (or a transit time larger than of 0.025 s), Fig. 6.24. *Georges* et al. claim that the non-pinning regime is due to the vibrations of the molecular groups which define a potential energy valley at the interface.

In conclusion, *Georges* et al. found that in a highly compressed polymer film at low lateral shear velocities the interpenetration of the molecules lead to an increase of the friction force and consequently to a shear alignment which manifest itself in a stabilized friction force. At high sliding velocities, a limiting friction coefficient was found corresponding to the minimum pressure necessary to achieve a "hard wall" contact between the two polymer layers, Fig. 6.24.

6.4.4   Nanorheological properties of interfacially confined films

Tribometer and SFA measurements provide valuable insight into the shear behavior of highly confined ultrathin liquids and polymer layers between two solid surfaces. The confinement due to normal compression accounts for deviations from the bulk rheological properties as long as the liquid can be assumed to be

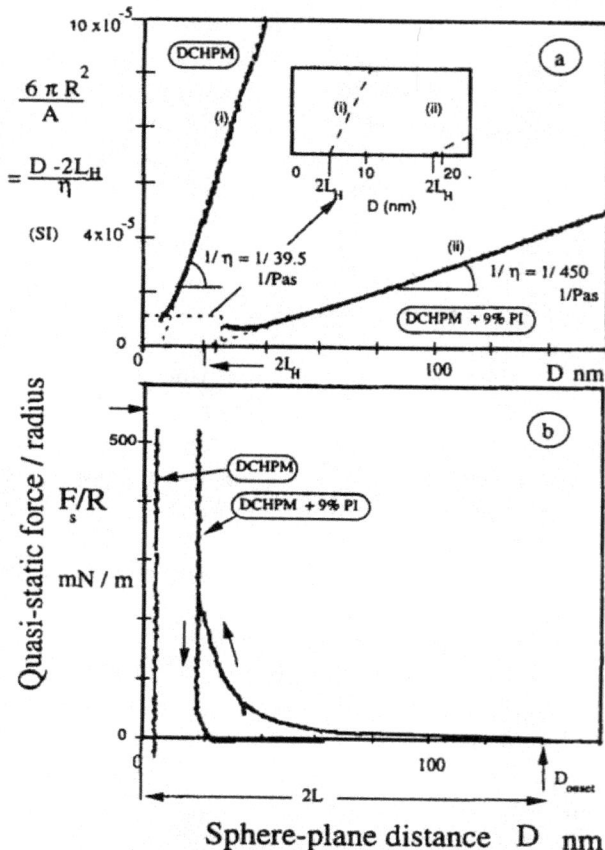

Figure 6.22: (a) The effect of the semi-dilute polymer surface (i.e., the hydrodynamic thickness) is determined from damping during oscillatory normal motions of the sphere against the plate in the tribometer. (b) Quasi static force (F_s) versus distance (D) profiles between the sphere of radius R and the plane in pure DCHMP and in PI solution (9% w/w). The force profiles are normalized to yield the interaction energy per unit area in the Derjaguin approximation[46]. The thickness of the layer is determined from the onset of repulsive forces at D = 135±20 nm which yields an adsorbed polymer layer thickness of $D/2$ = 68±10 nm (from Reference[39]).

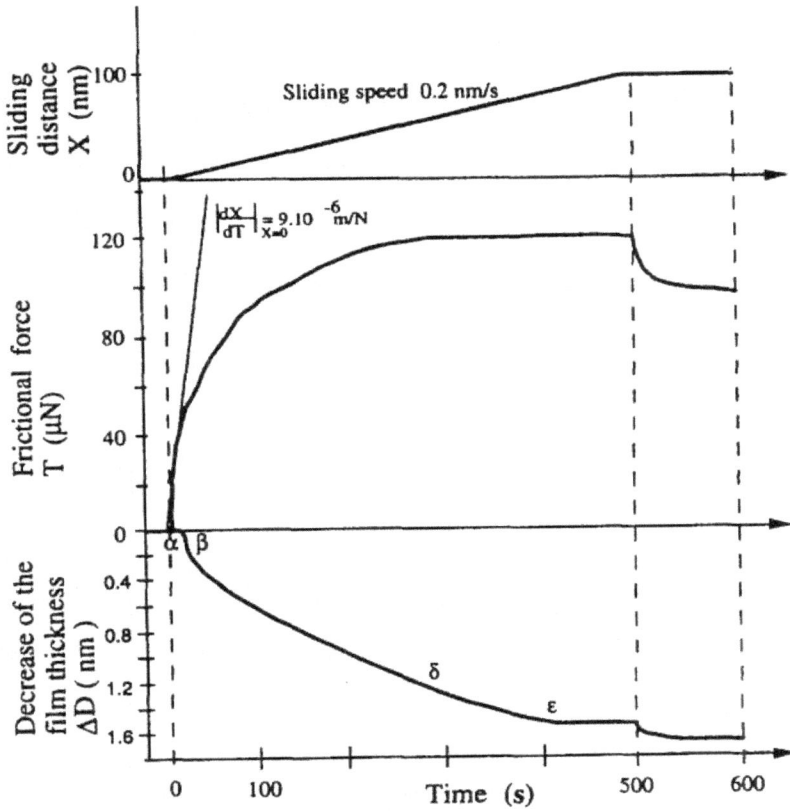

Figure 6.23: Friction and film thickness as function of the sliding distance for highly compressed PI polymer layers under a normal load of 508 μN which corresponds to a pressure of 28 MPa. A stabilized friction force is established after about 300 s (from Reference[39]).

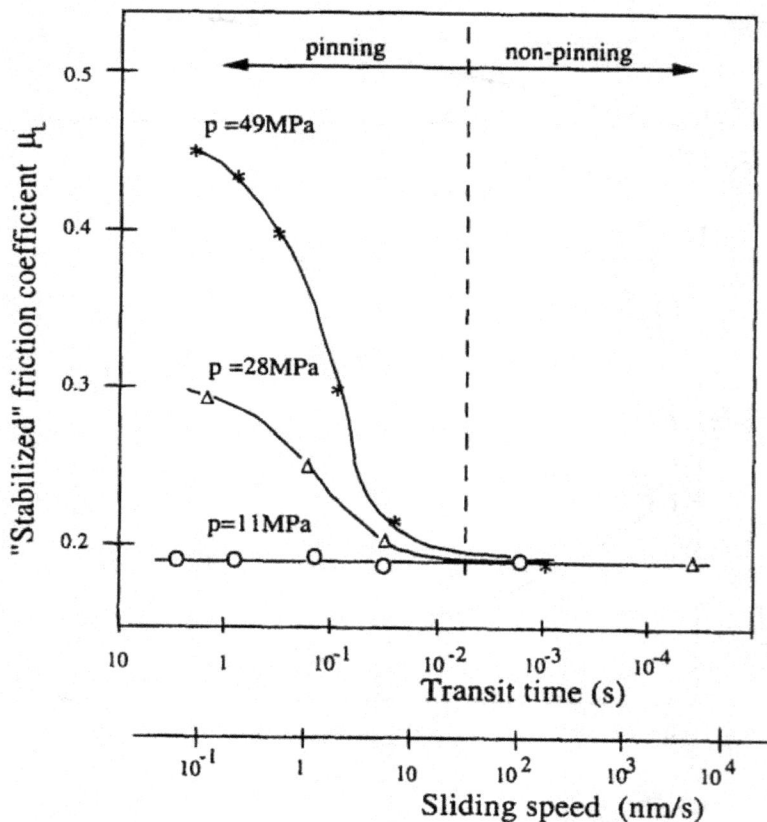

Figure 6.24: Stabilized friction coefficient versus transit time (From Reference[39]).

homogeneous in its properties. In complex systems, such as grafts and surface adsorbed polymer layers, the experimental outcome and the interpretation become more complex as it was illustrated above. Even in the case of well-adhered polymer layers, where sliding was assumed to occur at the interface of the two layers at high compression, the interpretation leaves open many questions and asks for further investigations. Questions which might, for instance, arise are concerned with the glass transition temperature at the interface during compression, the interpenetration regime between the two polymer layers, the free surface properties, and the effect of interfacial interactions on the mechanical properties.

That interfacial interactions can dramatically alter the shear behavior of liquid-like[b] polymer systems was found by *Rafailovich, Overney* and coworkers who have investigated the free surfaces of ultrathin polymer system adsorbed at interactive solid surfaces[22,47]. They made a very interesting observation on how the mechanical properties of a high-molecular-weight polymer, poly(ethylene-copropylene) (PEP, M_w 290k) could change continuously the phase from a more liquid-like polymer to a more glassy-like polymer. The chosen substrate surfaces on which the PEP films were spin-casted were hydrophobized silicon and polyvinyl pyridine (PVP). In the case of silicon, it was found that the surface mechanical properties measured by SFM is constant as long as the PEP film had a thickness larger than a critical value of about 100 nm. In closer vicinity to the silicon substrate, i.e., a film thickness below 100 nm, the mechanical properties of PEP were found to change linearly with the film thickness, Fig. 6.25. Lateral shear force measurements conducted with the SFM (discussed in more detail below) contain information about dissipation and reversible shear properties. In the case of PEP the dissipative part of the lateral forces was found to be independent of the film thickness. Hence, changes in lateral forces reflect changes in the shear mechanical properties of the sample. The easier a material is to shear (i.e., the "softer" it behaves in lateral direction) the higher is the lateral force under these conditions. Hence, the results of lateral force measurements presented in Fig. 6.25 reveal the existence of a PEP/Si boundary layer which continuously changes in the mechanical properties from a rubbery phase to a more glassy phase the closer the layers are to the silicon surface.

That indeed the silicon surface is responsible for the formation of an interfacial boundary layer has been tested with a low interaction interface, PVP. It is well known in polymer science that for two polymers, such as PEP and PVP,

---

[b]We will understand under "liquid-like" polymers, polymers investigated at temperatures far above the bulk glass temperature.

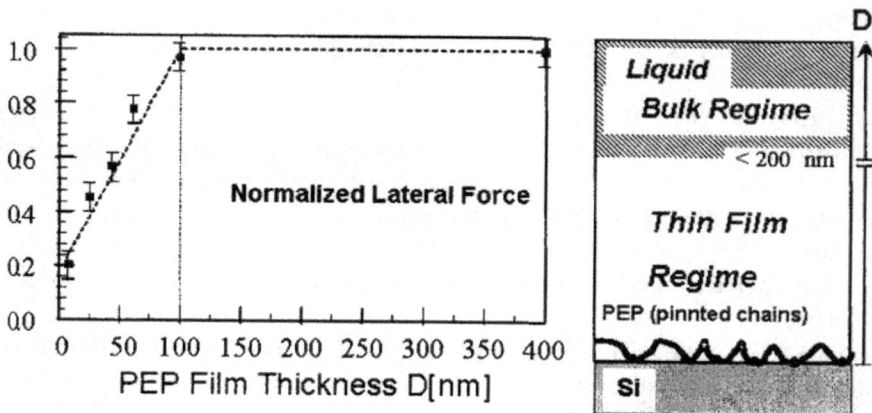

Figure 6.25: Interfacially confined PEP system. A continuous change in surface mechanical properties of ultrathin PEP films on silicon is observed. The lateral forces are a measure of the surface stiffness and are decreasing for ultrathin film PEP films on silicon. A mechanical boundary layer is formed in closest vicinity to the silicon interface (from Reference[45]).

with negative spreading coefficient S the interface can be considered sharp with barely any diffusion from one layer to the other. Shear measurements by SFM and dewetting revealed, as expected, no measurable dependence of the surface mechanical properties on the PEP film thickness in the case of a PEP/PVP system[45], Fig. 6.26.

Although there is no measurable effect in the surface mechanical properties, even for a film which is only 10 nm thick, that does not however imply that there is no interdiffusion zone. On the contrary, an interdiffusion zone of up to 10 nm thickness has been found between in non-wetting binary polymer system by secondary ion mass spectrometry (SIMS)[45]. Consequently, there must be a difference between a polymer-polymer interdiffusion zone of non-wetting polymer components, and the interfacial boundary layer of a wetting polymer and a solid substrate. As noticed experimentally by *Granick*[6] and *Georges*[39], and theoretically by *de Gennes*[44], surface pinned polymer melts change the mechanical properties of polymers over a distance of a few radii of gyration (R_g). In the case of PEP ($M_w = 290k$), R_g is about 4 nm which explains why the interdiffusion layer between PEP and PVP could not be detected with a PEP film thickness above 10 nm. Unexpected and unexplained by *de Gennes* theory[44] is the SFM result of a wetting polymer on a solid surface where the effective boundary layer exceeds by nearly an order of magnitude the radius of gyration.

Rafailovich, *Sokolov* and *Overney* extended their work to polymer-polymer interfaces with positive spreading coefficient and confirmed the PEP/Si result[45]. They conclude that high molecular weight polymer melts at wetting surfaces form mechanically altered boundary layers of significant thickness on the order 100 nm - far thicker than expected from the radius of gyration. This is a very important finding and may not be neglected in SFA measurements where it is, unfortunately, very difficult to deconvolute the effect on shear forces and rheological properties of *interfacial confinement* from the *confinement due to compression*.

It can be concluded that liquids between two solid interfaces are locally confined at each surface due to surface interactions and roughness, and globally by the normal compression (the effect of roughness will be discussed in more detail below). If the film thickness, D, is larger than two times the thickness of the mechanically altered boundary layer at each surface, L_B, SFA shear measurements results are directly related to the global confinement. However if, $D < 2L_B$, mechanical boundary layer properties should be included in the interpretation. It is important to note that there is not necessarily a sharp transition between the boundary layer and the bulk, and hence, assumptions of Hertzian contacts, as made by *Georges* et al. for polymer solutions, might not apply for polymer melts of higher molecular weight.

6.4.5 Lateral confinement of simple liquids

In the previous paragraphs, we have learnt that liquids under global confinement between two smooth surfaces exhibit shear and rheological properties very different from classical bulk measurements. Already simple liquids which are assumed to be homogenous and rate independent show rheological behaviors only known from solids or viscoelastic material. Layering was observed under extreme compression to a few remaining molecular layers in the gap. The observations of a decrease in entropy are due to the reduction in dimensions from bulk to ultrathin film. Three dimensional bulk liquids are confined to nanometer thicknesses with pressures in the MPa regime so that molecular surface corrugation are like "imprinted" into the liquid. *Gao* and *Landman* found in a MD simulation that if the commensurability of the molecular surface corrugation of the two solids around the confined liquid of spherical shaped molecules is altered (e.g., incommensurable), the amplitude of the oscillatory solvations forces are reduced[48]. The following questions arise if the confinement area is drastically reduced from micrometers to nanometers, or perfect molecularly ordered surfaces (like mica) are replaced with amorphous surfaces:

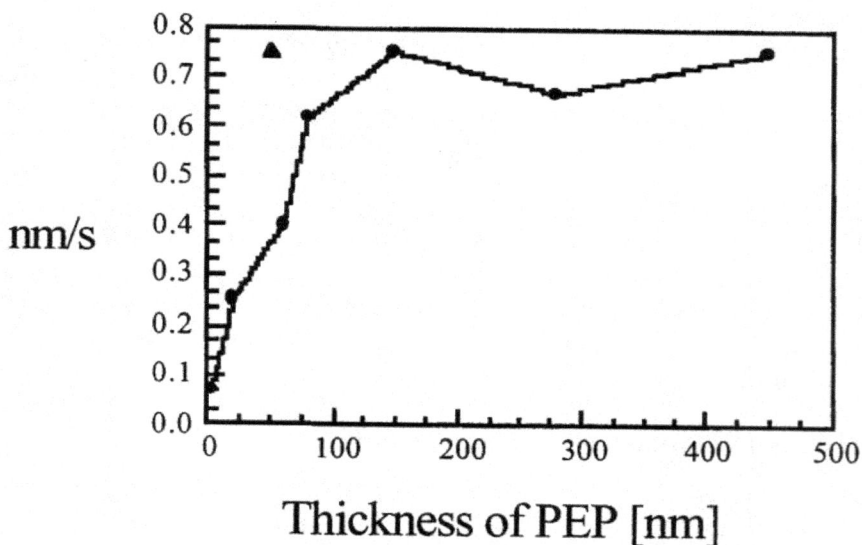

Figure 6.26: Indirect shear force measurements of PEP by measuring the dewetting velocity of a top layer of polystyrene (PS, M_w 220k, 13 nm). This setup is in analogy to SFA experiments, with the difference that shear can be applied without normal compression. Hence, the shear results can be directly related to the interfacial confinement of PEP on PS (•) or PVP (△). Dewetting results are in correspondence with SFM lateral force measurements on PEP/Si, Figure 6.25. No deviations from the surface-bulk properties have been found on PEP/PVP down to a PEP thickness of 15 nm (from Reference[45]).

(a) Is a solid-like behavior still existing under such confinement? And if not, how relevant are SFA studies with perfect mica surfaces for an improved understanding of lubrication where the solids are usually oxidized and rough with asperity contact radii smaller than micrometers?

In a very recent scanning force microscopy (SFM) study of *Meyer*, it was found that a liquid can show structure without a large scale compression area, Fig. 6.27[49]. The experiments show that the simple liquid, hexadecane, exhibits structured features on a gold surface without being normally confined. The structured areas appear only in local "depression" zones of the gold surface, Fig. 6.27. In other words, the liquid appears to be laterally confined. This is an interesting aspect where roughness is responsible for structuring a simple liquid and has analogs in the theory of granular material considering multiphase systems. There is something special regarding the gold surface which may not be neglected. The gold was (i) thermally evaporated as a thin film on a mica sheet, (ii) glued to a silicon wafer and (iii) mechanically stripped off. This method is known as *template-stripped gold method*[50]. It can be assumed that the mica-peeled gold surface was imprinted by the large molecular structure of mica, and then, the wetting hexadecane film picked up that structure from the imprinted gold surface. The exact origin of hexadecane's structure is still under investigation

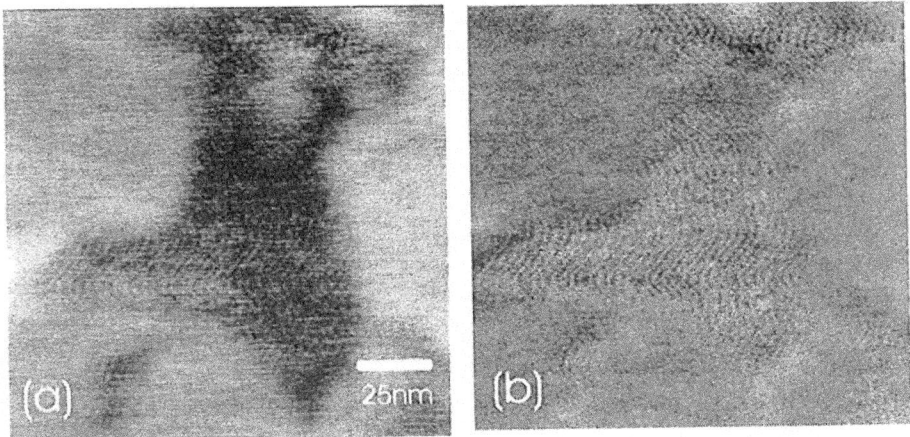

Figure 6.27: Contact scanning force microscopy (SFM) measurements of laterally confined liquid hexadecane on gold. Scan regime 100 nm. (a) Topography image. (b) Lateral force image. Bright contrast refers to a high value in both measurement modes. Hexadecane shows two phases - a structured and unstructured phase in topographical lower areas (depression zones) and higher areas , respectively (from Reference[49]).

We have seen that the fundamental study of liquid confinement started by SFA is very relevant considering the extraordinary experimental result of *Meyer*, and today's needs of lubrication for new engine designs with higher compression cycles and efficiency, and for new electronic devices of critical dimensions on the nanometer scale. Confinement of fluids is - as we saw - manifold and not just restricted to normal compression. It also occurs in the form of interfacial confinement due to surface interactions as discussed in the last paragraph, and lateral confinement.

6.4.6   Measurements of interfacial and lateral confinement of low viscosity liquids

In recent SFM studies, *Overney* and co-workers have attempted to measure the viscosity gradient of liquids due to interfacial interactions on the nanometer scale[22,51]. The investigation was split into three parts. First, it had to be established that the SFM is sensitive enough to probe changes in viscosity of low viscosity liquids. Second, a method had to be found to satisfactorily calibrate SFM results so that they could be compared with conventional rheological measurements. And third, the SFM procedure had to be applied to *model systems where the outcomes were predictable.*

Considering a contact area of about 10 nm^2, which corresponds to an average conventional SFM probe, and normal compression loading of about 10 nN, pressures on the orders of Gigapascal can be easily achieved. That such pressures pose no problem for crystalline systems had been shown previously[52], but not for non-structured material. Crystalline material can disperse energy very fast over a large area (or bulk) and hence avoid local plastic deformation. Amorphous material or liquids have response times to stresses which are in most cases significantly slower than the ones in crystals. The consequence is that the amorphous or liquid materials plastically yield to stress and this rate dependent. In a high viscous material, yielding causes visible plastic deformation. In a low viscous material, such as a simple liquid, the consequence of yielding would be that the SFM probe, which is basically a spring, would not feel a restoring force. As we have seen above, every material is, to a certain amount, elastic or viscous. It just depends on the rate the experiment is conducted. At sinusoidal modulation of frequencies of above 100 Hz and amplitudes on the order of nanometers, any liquid would cause, due to inertia, a restoring force to act against the SFM probe. The question now would be if a small change in the viscosity causes a change in the modulation response, significant enough to be detected?

Overney and *Rafailovich* modified the SFM in a way which allowed the application of AC and DC force-displacement curves as previously applied by *Salmeron et al.*[22,53], Fig. 6.28. The AC approach was performed by superimposing on the DC approach a small modulation amplitude. The displacement forces and modulation amplitude (plus phase relation with the input modulation) of the cantilever system were monitored simultaneously during the approach. A simple liquid, such as toluene, showed the behavior of a homogenous liquid until the last nanometers before the cantilever probe hit the silicon substrate, Fig. 6.28a. Different is the response, however, for an engineered complex liquid which was composed of a monodisperse diblock copolymer of poly-4-vinylpyridine (PVP) (polymerization index $N = 20$) and predeuterated poly(styrene) (PS) ($N=200$) attached on silicon and immersed in toluene, Fig. 6.28.b. The PVP block adheres strongly to the silicon surface forming a PS brush which extends into the toluene phase which is known to be a good solvent for PS[54].

The goal of this first study was to measure the height of the polymer brushes and to compare the SFM result with neutron reflectivity of the same system. Both measurements provided independently a brush film height of about 24 nm which suggests the SFM to be a useful tool to measure the rheological properties of simple or complex liquids with a resolution in three-dimensions on the nanometer scale. Because of the very small contact area, it can be assumed that material properties measured are in equilibrium which is an important addition to the non-equilibrium measurement with SFA.

The SFM data could be quantified with the theory by *Montfort* and *Hadziioannou* (see above section 6.3.1). Hence, the viscous loss is:

$$\eta' = \frac{h_o k_o}{6\pi R^2} A' \frac{\sin \Phi}{\omega}, \tag{6.121}$$

and the storage shear capability (moduli) is:

$$J = \omega \eta'' = \frac{h_o}{6\pi R^2} [\left(\frac{k_o}{A' \cos \Phi} - 1 \right) + k_{eff}], \tag{6.122}$$

where $k_o = (k_L - m_L \omega^2)$, k_L and m_L are the spring constant and the mass of the cantilever, respectively, ω is the modulation frequency, $A' = A_{in}/(A_{in} - A_{out})$, h_o is the uncompressed film thickness, k_{eff} is the effective spring constant determined from static force displacement curves (see below), and Φ is the phase shift between input and output modulation.

Montfort and *Hadziioannou's* method of calibration is, however, restricted in SFM measurements to films or brushes where a zero compression height,

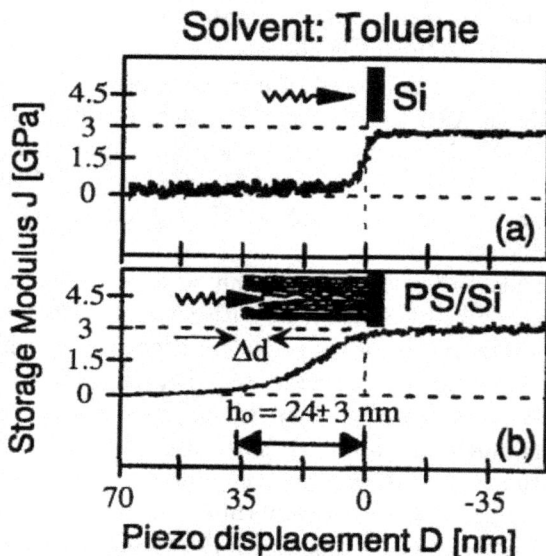

Figure 6.28: Static force and dynamic moduli measurements by SFM of (a) simple liquid (toluene) and (b) an engineered complex liquid (polystyrene brush). (a) A sharp transition in the elastic response between the simple liquid toluene and the solid silicon surface is observed. (b) A continuous change in the elastic response is found in the toluene dissolved PS chains which are end attached to the silicon substrate. A height length of about 24 nm was found which corresponds to neutron reflectivity data and suggests the SFM as a nanorheometer. (c) Illustrative sketch of the SFM setup (from Reference[22]).

h_o, is defined. *Overney* and *Drake* introduced a "blind" calibration method for liquids which is tested upon a continuous change of viscosity in the vicinity of interactive interfaces[55]. With this method, the observable of the dynamic SFM modulation measurements, the phase shift Φ, is compared to kinematic (zero frequency) viscosity values of standard test oils for rheometers. A first SFM calibration curve, which was established from standard calibration oils (measured in the bulk) of different viscosities, was tested with bulk squalane measurements performed at various temperatures by SFM and laser diffusion. The test showed positive as illustrated in Figure 6.29. It is important to note that the blind calibration method demands that subsequent measurements be conducted at the same frequency and cantilever-laser position as used during the calibration. The calibration is always established in the bulk liquid, i.e. far away from any interfaces.

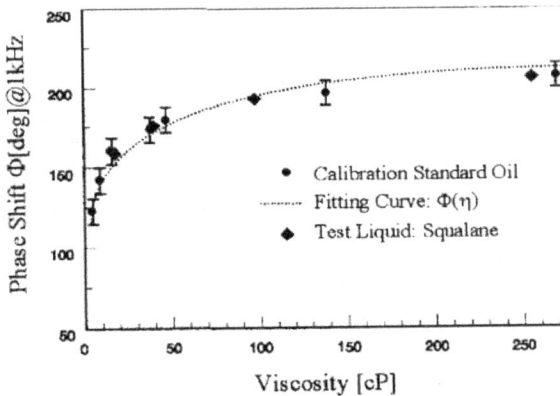

Figure 6.29: Calibration of phase and kinematic viscosity. Independent squalane measurements confirm the calibration procedure to be successful (from Reference[55]).

Figure 6.30 shows the calibration curve established from a set of bulk liquids. Water and ethanol has been chosen as simple liquids. Squalane, a branched alkane, was chosen for its strong viscosity dependence on temperature and its chemical neutral behavior at silicon surfaces. The target of the previous SFM test was the complex mixture of a basestock (150N) that is used as a solvent for various additives which form together with the basestock a fully formulated engine oil.

In dynamic force approach curves, the phase behavior was studied as a

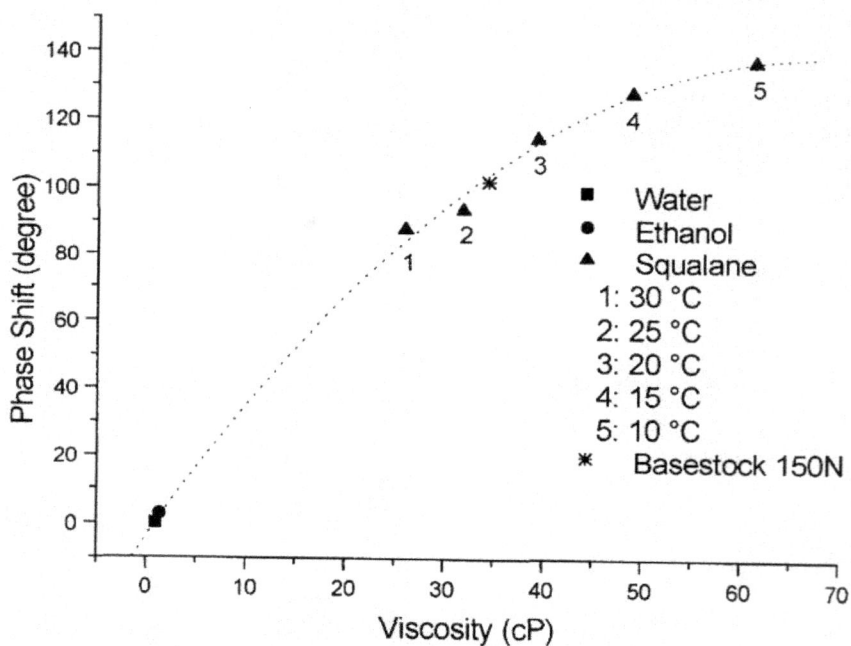

Figure 6.30: Bulk liquid calibration of various liquids - from simple liquids (water, ethanol) to more complex systems (squalane (branched alkane), basestock for motor oil) (from Reference[55]).

function of the distance to a silicon surface, Fig. 6.31. Water and ethanol showed, as expected, a sharp transition in the phase shift from the liquid to the solid interface. The difference in the phase is a direct measure of the viscosity of the two liquids using the calibration curve of Figure 6.30. Squalane showed a continuous change in the phase from the liquid to the solid over a distance of about 20 nm. Changes in viscosity from 20 cP to 60 cP of squalane due to temperature changes did not affect or extend the qualitative transition behavior of the phase. The basestock, with the highest molecular complexity of the tested liquids and a viscosity of 32 cP in the range of squalane, showed a significant larger transition region from its bulk value to solid interface. Further interpretation of these measurements are still in progress.

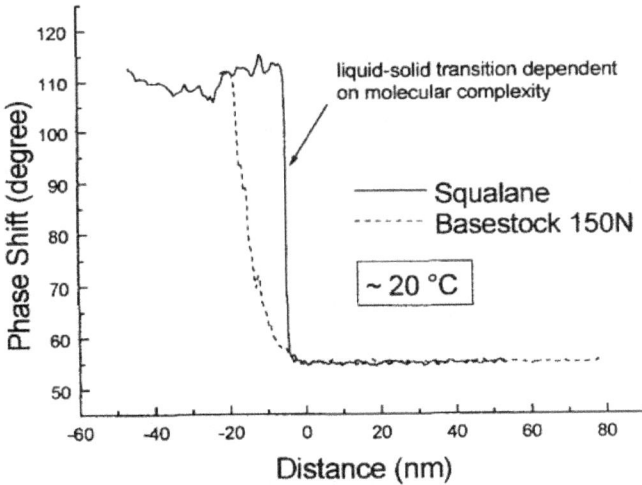

Figure 6.31: Phase relation between input and response modulation during dynamic force approaches (from[55]).

Measurements on the nanometer scale of viscosity gradients and the investigation of interfacially confined boundary layers are just beginning. Much new insight is expected and this will not only be important for the field of lubrication. Many processes, e.g., spreading, dewetting, etc., depend on the mechanical properties of material at interfaces.

6.4.7 Dewetting-shear-apparatus

An unconventional way to study the shear properties of an interfacially confined polymeric systems was introduced by *Rafailovich* and *Overney*[45,47] who used as a shear apparatus a hydrophobized silicon surface and a sample-non-wetting high glass temperature polystyrene layer of 13 nm thickness. The sample is placed in between the "dewetting shear apparatus". Shear forces are applied by increasing the temperature in a vacuum oven above the glass temperature of PS. Because the two polymer systems are chosen to have a negative spreading coefficient, dewetting occurs. The shear results due to dewetting are measured in-situ by optical microscopy, Fig. 6.32, and ex-situ by SFM. The sample used in the study of *Rafailovich* and *Overney* was a low glass temperature polymer, polyethylene-copropylene (PEP) film which has been discussed above. The mechanical properties of a thin film of PEP are known to show a strong dependence on its thickness and the choice of the substrate[45]. For a silicon substrate, it can be expected (see above) that the shear behavior of PEP is different for a PEP film thickness exceeding the boundary layer thickness of about 100 nm from a ultrathin film of less than 100 nm. In Figure 6.32, the shear results for a 4 nm and 400 nm film are documented and graphically illustrated.

A sharp transition interface is expected from two non-wetting systems. That dewetting can occur inside the liquid phase, as illustrated in Figure 6.32, of a polymer system is quite an astonishing finding considering that liquid/liquid dewetting with a negative spreading coefficient should occur at the interface of the two liquids only[56]. However, the spreading coefficient, S, does not consider interfacial entanglements or roughness. It is defined for none wetting liquids as

$$S = \gamma_2 - (\gamma_{12} + \gamma_1) \qquad (6.123)$$

where γ_1=40±0.5 dyn/cm and γ_2=30.9±0.1 dyn/cm are the surface tensions of the PS and PEP film, respectively, and γ_{12}=1-2 dyn/cm is the interfacial tension. SIMS measurements revealed an interdiffusion zone of up to 10 nm between PS and PEP films before dewetting at 110 °C[47]. It was assumed to be the cause for the PS/PEP interaction that is necessary to rupture the liquid-like film of PEP. The abrupt strain release of the dewetting PS film and the inertia of the highly viscous PEP film were claimed to be the cause of the liquid fracture of the PEP film[47].

The rate of which the momentum is transferred from the PS layer to the PEP film, and the mechanical properties of PEP close to the interface, i.e., its critical response time to shear stress, is responsible for the location of the shear plane. If PEP is thick, i.e., liquid-like with a large critical response time, the PEP film is ruptured and the shear plane location is situated deep-inside the PEP phase, Fig. 6.32b. If PEP is thin, i.e., thinner than 100 nm where its mechanical behavior is more glassy, the response time is much faster and the dewetting shear plane is established closer to the PS/PEP interface, Fig. 6.32b. The dewetting velocity was observed to be smaller for thin PEP films and larger for thick PEP films, Figure 6.32, because of the varying mechanical properties which strongly depend on the shear plane location (i.e., the distance to the silicon substrate).

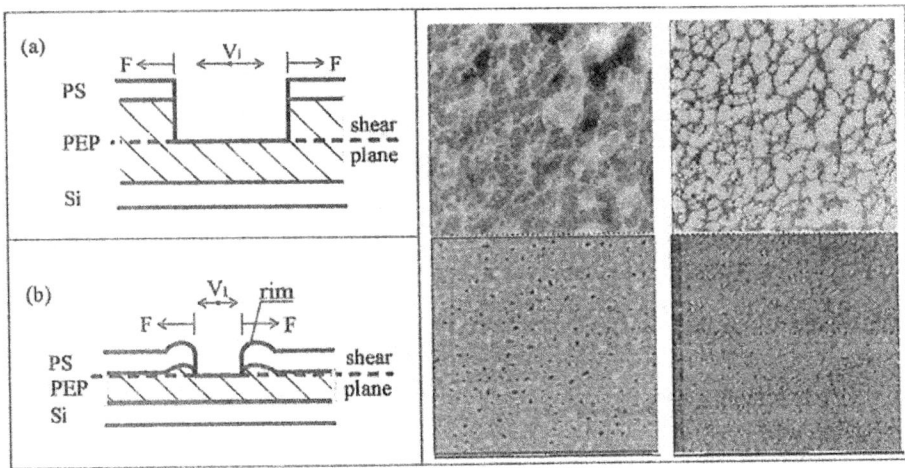

Figure 6.32: Dewetting-shear-experiment. The apparatus consists of a PS film of 13 nm thickness and a hydrophobized silicon surface. The sample is polyethylene-copropylene (PEP) with (a) a thickness of 400 nm and (b) of 4 nm. Shear is induced by a dewetting PS layer at 110°C. SFM graphs of $50 \times 50 \mu m^2$ (topography (middle) and friction (right)) show a snapshot of the dewetting process after 30 minutes. On the (left) the differences in the shear process are graphically illustrated (From[47])

6.4.8   A list and summary of distinct confinements

Influencing the interfacial confinement of ultrathin liquid films provide the possibility of controlling the shear behavior. The distinct confinements discussed in this paragraph are sketched in Figure 6.33. It was found that liquids confined by normal compression can exhibit very strong deviations from bulk properties. Solid-like shear behavior of simple liquids could be observed which led to new terms such as "shear melting", and consideration of slip conditions between liquid layers. The formation of surface adsorbed films has shown to strongly affect the shear behavior of complex systems. For instance, very low shear forces were measured in brush systems. Further, interfacial confinement can influence the viscous properties of low viscosity liquids, and the mechanical properties of polymer melts and rubber. And finally roughness has shown to be an addition to interfacial confinement (also called lateral confinement) and is partially responsible for the structuring of interfacially confined simple liquids. Lateral confinement was also responsible for momentum transport between non-wetting liquid interfaces.

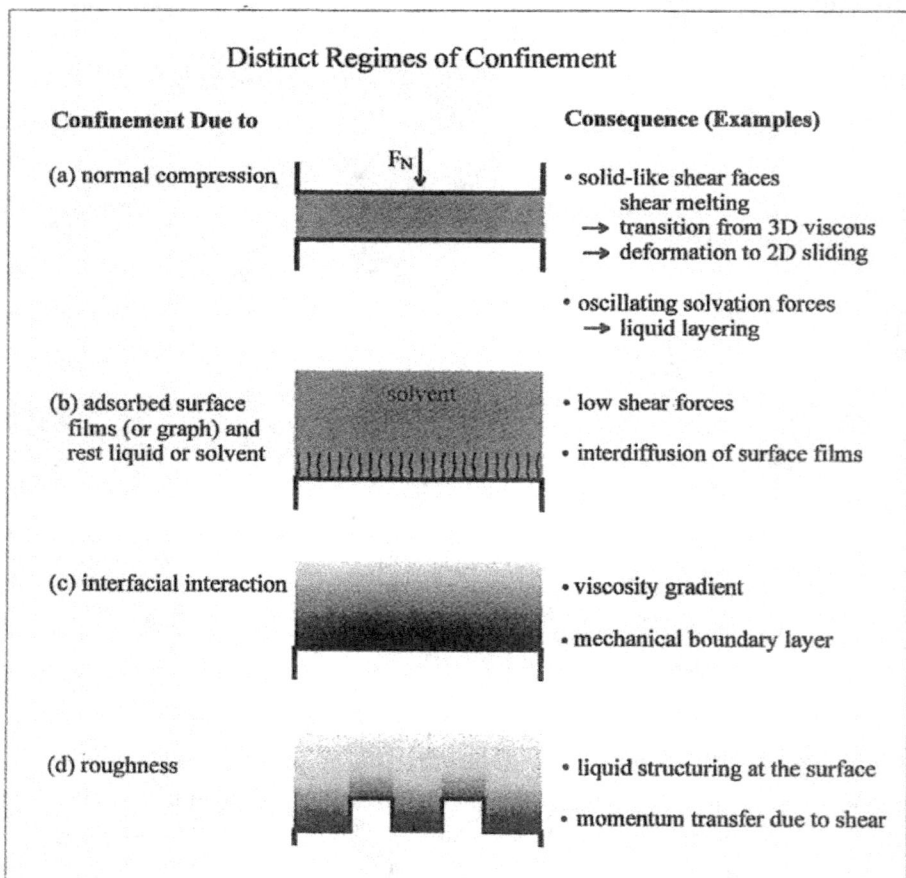

Figure 6.33: Distinct regimes of confinement

In the field of lubrication a combination of these distinct confinement regimes should be considered. The better we understand why a lubricant behaves in a distinct way, i.e., different from another lubricant or in another confinement, the more we can we influence and design a desired shear behavior. It is not the lubricant alone and its chemistry which makes good lubrication possible. The confinement in which the lubricant shears and moves is as important.

6.5 Nanorheological and mechanical properties of polymeric surfaces and thin films measured by SFM

In section five of Chapter 6, we are concerned about the measurements of nanomechanical properties in three-dimension of solid or liquid-like polymeric interfaces. In the past few years, the technique that showed most promising results and innovative new measuring modes is the scanning force microscopy (SFM), which is the focal point of the following discussion.

6.5.1 Introductory remarks

Since the inception of the scanning tunneling microscopy (STM) by *Binnig* and *Rohrer*[57], various nanoscale scanning devices evolved, summarized under the term of scanning probe microscopy (SPM)[58]. SPM techniques have been in strong demand due to their real space imaging capability on the micron and submicron scale. One of the early desires has been to analyze quantitatively surface properties on the nanoscale. In that respect, the *force modulated* SFM is unique and very promising[47,5359−76].

The SFM force modulation mode was introduced by *Maivald et al.* in 1991[59]. This new nanoscale contrast mechanism relies on variation in the surface mechanical properties. In the subsequent years, researchers have applied this technique successfully to *soft* materials such as polymers[22,47,53,59−62,64,65,67−78].

This paragraph intends to highlight some of the work accomplished by SFM where nanomechanical properties of polymer films and polymeric surfaces were investigated. Although the SFM force modulation mode opened the door for direct surface rheological studies on the nanometer scale, it is not its only measuring mode capable of delivering elastic, viscoelastic and/or hardness information. There are other SFM modes, which will be introduced too, that have been applied over the last few years to investigate nanomechanical properties[70,74,77,79−83].

The first few paragraphs of this section deal with static force measurements, models and calibration issues of sinusoidal elastic and viscoelastic perturbations on solid and liquid surfaces with a common SFM cantilever. They are followed by experimental applications of the force modulation mode to polymeric systems. Other SFM modes of operation which are applied to determine contact mechanical properties on the nanoscale are introduced. Finally, this section will end with some critical remarks about the SFM technique applied to *nano-contact-mechanics*.

6.5.2   Static deformations and sinusoidal perturbations

In the perturbation theory of small amplitudes of change, first order approximations are undertaken as discussed in detail at the beginning of this chapter. Hooke's law is a classical example where the resulting force on a material is assumed to change linearly with the applied distortion. In the lower dimensional case of an idealized spring, where it is assumed that any deformation occurs in one dimension only, the linear connection of the retaining force and the distortion is described by the spring constant.

In a very simplified model, *Maivald et al.* pictured the sample as a one-dimensional spring (sample spring constant k_s) during a small elastic indentation with a cantilever (spring constant k_l)[59]. The effective spring constant ksys (also called stiffness) of the entire system is then given by

$$\frac{1}{k_{sys}} = \frac{1}{k_s} + \frac{1}{k_l}.^c$$

(6.124)

Based on Hooke's law, the resulting force ΔF on the surface is

$$\Delta F = k_l \Delta z_l = k_s \Delta z_s = k_{sys}(\Delta z_l + \Delta z_s)$$

(6.125)

where Δz_l and Δz_s are the cantilever deflection and the sample deformation, respectively. In order to determine the cantilever deflection in units of a distance, it is necessary to calibrate the sensitivity of the cantilever detection scheme. In the setup of a laser beam deflection detector, it is the sensitivity of the photodiode which has to be determined. Based on equation (6.124), the detector is calibrated with a calibration sample with a stiffness exceeding significantly the stiffness of the cantilever[22]. The systems spring constant, k_{sys}^{cal}, determined from the calibration sample is measured with a typical force-displacement curve as illustrated in Figure 6.34. For stiff calibration samples

---

[c] Note that k_{sys} corresponds closely to the effective spring constant, k_{eff}, introduced above.

compared to the cantilever spring constant, the systems force deflection response, S_{sys}^{cal}, in volt/distance (i.e., the slope of the force-displacement curve in Figure 6.124) and, k_{sys}^{cal}, in N/m are related as follows:

$$S_{sys}^{cal} = \frac{1}{\lambda} k_{sys}^{cal} \cong \frac{1}{\lambda} k_l \Rightarrow \Delta F = \lambda S_{sys}^{cal} \Delta z_l \qquad (6.126)$$

with $\lambda [=]$ N/V as a unknown instrument depending factor. Hence, the ratio between the sample indentation and the deflection of the cantilever determined from equations (6.125) and (6.126) is:

$$\frac{\Delta z_s}{\Delta z_l} = \left(\frac{S_{sys}}{S_{sys}^{cal}} - 1 \right) \qquad (6.127)$$

where $k_{sys} = \lambda S_{sys}$, the system stiffness. Thus, the stiffness of the system, k_{sys}, and the stiffness of the sample, k_s, are

$$k_{sys} = k_{sys}^{cal} \frac{S_{sys}}{S_{sys}^{cal}} \approx k_l \frac{S_{sys}}{S_{sys}^{cal}}, \text{ and}$$

$$k_s \approx k_l \left(\frac{S_{sys}^{cal}}{S_{sys}} - 1 \right)^{-1}, \qquad (6.128)$$

respectively. Equations (6.128) are exact in the limit of an infinitely stiff calibration sample. This calibration procedure has been utilized in determining the stiffness of polymeric samples in dry nitrogen atmosphere[22]. The calibration samples were chosen to be silicon and ionic crystals in dry atmosphere, and in the case of silicon as the calibration sample, various liquid environments have been tested[22].

One problem with the static approach is the unfavorable signal-to-noise ratio due to naturally occurring fluctuations and variations in the system properties. Examples are temperature fluctuations, local contact position and size variations, and local sample imperfections. *Hamada* and *Kaneko* showed that a static approach for the hardness measurements on polycarbonate (PC), poly(methylmethacrylate) (PMMA) and epoxy (EP) surfaces differs substantially from macroscopic hardness measurements with a Vickers hardness tester[77]. They concluded that the operating regime is responsible for the discrepancy. While in the Vickers hardness test a large volume of material was tested, the SFM probed only the surface - and this very locally. Hence, imperfections in the material were concluded to be responsible for the discrepancy.

To improve the statistics of the local sample's property variations but also to decrease naturally occurring fluctuations, the entire surface is probed by

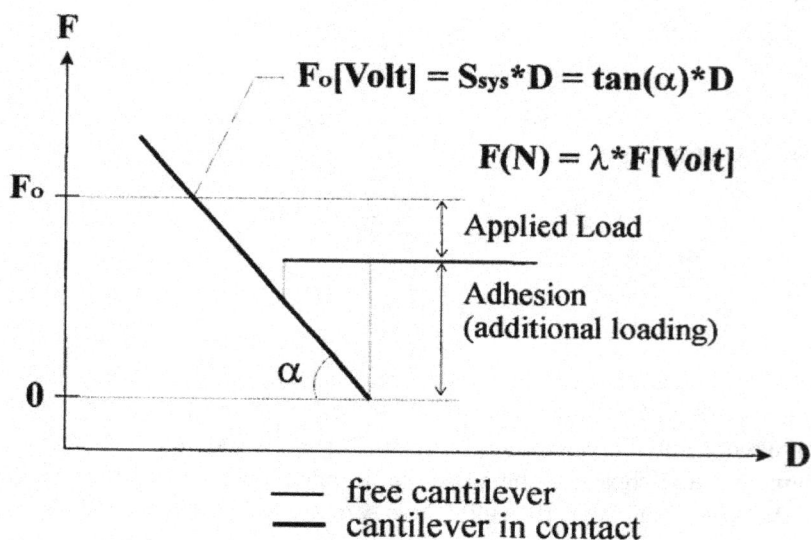

Figure 6.34: Calibration of loading forces and stiffness. The equilibrium load F_o is the sum of the applied load and the load given by adhesive forces. The slope of the force displacement (F-D) curve (bold) is mainly reflecting the convoluted stiffness of the sample and the cantilever. Signal-force conversion can be achieved with an ultrastiff calibration sample.

scanning a force modulated nanoprobe, Fig. 6.35. The force modulation serves as an artificially induced fluctuation which is known to dramatically improve the signal-to-noise level. By measuring the steady state response to such a non-equilibrium method, one creates momentum, energy and mass transport - the same procedure as used in classical rheological experiments. Usually the perturbations used are sinusoidal oscillating forces. Thus, the samples responses are proportional to the real and imaginary parts of the Fourier-Laplace transformed correlation functions at the applied frequency. Transport coefficients can be obtained by extrapolating to zero frequency.

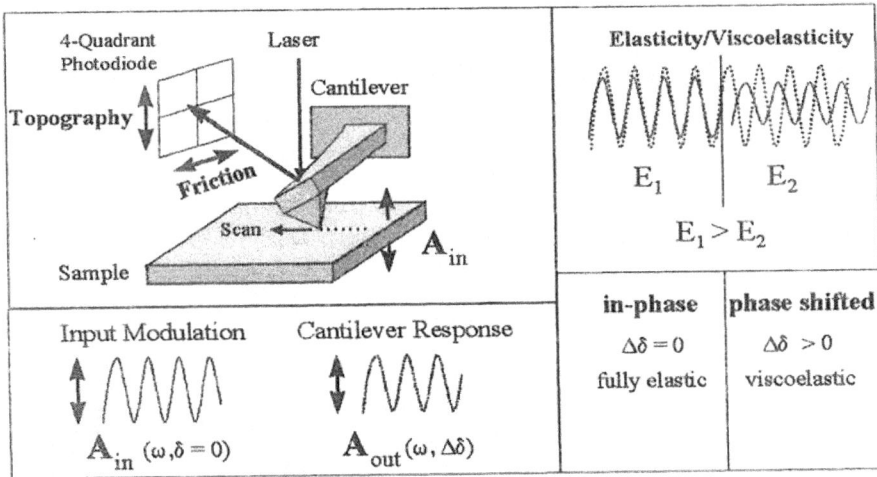

Figure 6.35: Principle of force modulation SFM (3-fold technique). A sinusoidal input modulation (A_{in}) can be introduced by a piezo on which the sample is mounted (as sketched), or by a piezo on which the cantilever is mounted. A cantilever that is in sample contact will locally indent the sample. Its measured deflection response (A_{out}) will be reduced by the indentation (δ). The elastic and viscoelastic components can be determined as described in the text.

Principle of force modulation SFM (3-fold technique). A sinusoidal input modulation (A_{in}) can be introduced by a piezo on which the sample is mounted (as sketched), or by a piezo on which the cantilever is mounted. A cantilever that is in sample contact will locally indent the sample. Its measured deflection response (A_{out}) will be reduced by the indentation (d). The elastic and viscoelastic components can be determined as described in the text.

The first time a SFM experiment used the perturbation theory in analyzing viscoelastic properties was by *Radmacher*, *Tillmann* and *Gaub* who studied the effect of UV polymerized ultrathin films[61]. The dynamic moduli E', G', G'' and E'' (see equation (6.84)) which represent the in-phase and out-off-phase strains and are named E or G depending on the modulation direction (normal (E), lateral (G)), are measured with a dual phase lock-in amplifier[22,61].

6.5.3 Elastic indentation models of surfaces

The classical Hertzian contact theory has been applied in most sinusoidal force modulated SFM studies[53,61,64]. According to this elastic theory and assuming a sphere-flat configuration, the Young's modulus of the system ($1/E = 1/E_1 + 1/E_2$, E_i (i=1,2) elastic constants of both bodies in contact) is given by

$$E = \left(\frac{k_{sys}^3}{6RF_o} \right)^{1/2} \tag{6.129}$$

with the equilibrium force F_o, the contact radius R ($1/R = 1/R_1 + 1/R_2$, R_i (i=1,2) radius of curvatures of both bodies in contact) and the system spring constant k_{sys}[64]. The system spring constant is determined (a) with the slope of force displacement curves for static elastic measurements (see above), or (b) by

$$k_s = k_l \left(\frac{A_{out}}{A_{in} - A_{out}} \right), \tag{6.130}$$

for sinusoidal force modulation measurements, where k_l is the cantilever spring constant, A_{in} the input modulation amplitude (perturbation) and A_{out} the response amplitude[64].

Adhesion or long range forces are not considered in the Hertzian theory. Therefore, the contact area between non-conforming elastic solids falls to zero when the load is removed[84]. Experiments however, like SFM studies, showed very impressively that measurable competing forces of attraction and repulsion between atoms or molecules in both bodies are acting within a certain separation distance[85-90]. *Johnson*, *Kendal* and *Roberts* considered in a continuums theory (JKR model) the aspect of adhesion in the elastic contact regime[91]. Hence, the equilibrium force F_o consist of two loading terms F_{appl} and F_{adh} with

$$F_o = F_{appl} + F_{adh},$$

$$F_{adh} = 3\pi RW + \sqrt{6\pi RWL + (3\pi RW)^2}, \tag{6.131}$$

where F_{appl} is the applied load, F_{adh} is the load resulting from adhesion[91], and W is the interaction energy per unit area between the sample and probing tip.

By adding a long-range potential to the continuum JKR theory to avoid edge infinities, a very complex and computer intensive theory, the DMT model (Derjaguin-Muller-Toporov) was developed[92]. DMT and JKR theories have been successfully tested in macroscopic experiments[84] and also on the microscopic scale with the SFM[93,94,95].

In general, it has to be noted that all acting forces (resp. their potentials) should be considered in a theoretical discussion of equilibrium forces and elastic constants. In force modulated SFM experiments, the equilibrium load can be determined with the force displacement curve as illustrated in Figure 6.34[22].

6.5.4 Static force measurements on polymeric systems

With *static* SFM, we define scanning a cantilever in contact with the sample without additional modulation. The principle on which it works is very similar to profilometry, where a hard tip is scanned across the surface and its vertical movements monitored. As a result of the miniature size of the SFM tip, which is mounted at the bottom end of a cantilever-like spring, it is possible to image the corrugation of the surface potential of the sample[52].

With static SFM, polymeric systems have been studied on the submicrometer and submolecular scale in predominantly two modes of operation: (a) topography[69] and (b) lateral force[70]. The following Tables 6.2 6.3 6.4 are intended to provide a brief overview of research domains where the static SFM has been successfully applied.

Static SFM has been successfully applied to study effects on polymer surface treatments such as UV irradiation and corona discharge[69,78,161]. *Goh* and co-workers investigated surface mobility at the molecular level by cross-linking experiments on poly(methylstyrene). The surface was exposed to UV light, Figure 6.36. Periodic deformation pattern were generated by repeated SFM scans. It was found that prolonged UV radiation hinders the patterning process[69].

SFM studies on excessive corona-discharge treatments showed deteriorative effects on uniaxial and biaxial stressed polypropylene (PP) films[78]. Together with peel-force experiments, GPC and attenuated total reflection measurements, it was found that the failure in the adhesive properties of corona treated PP surfaces is due to local surface melting, Fig. 6.37.

Hild et al. studied polymer films under stress conditions[81,142]. With a small stretch apparatus built on a SFM stage, they studied in-situ, the structural deformation of hard elastic PP films on the nanometer scale. With this approach, the quasi-elastic strain model of *Noether* and *Whitney* could experimentally be confirmed beyond an elongation of 40%, Fig. 6.38.

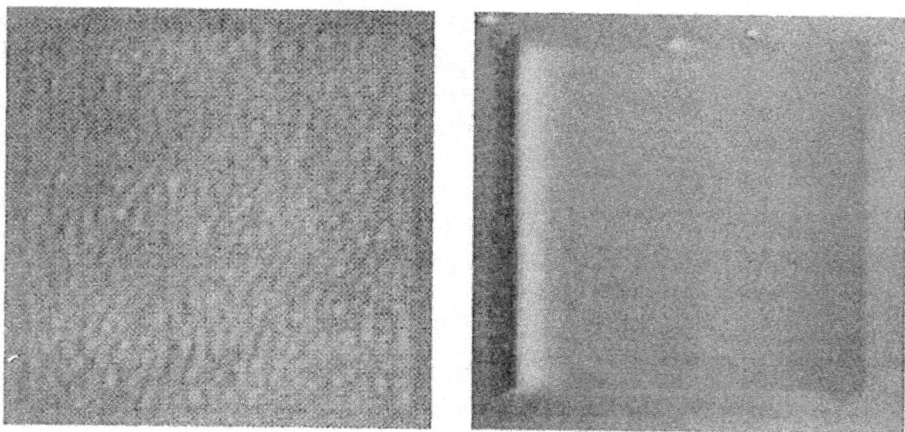

Figure 6.36: The effect of UV irradiation on the surface toughening tested by SFM scratch experiments. (Left) UV-untreated poly(methtylstyrene) surface is easily scratched by the SFM probe. (Right) UV irradiated surface hinders the patterning process (From[69]).

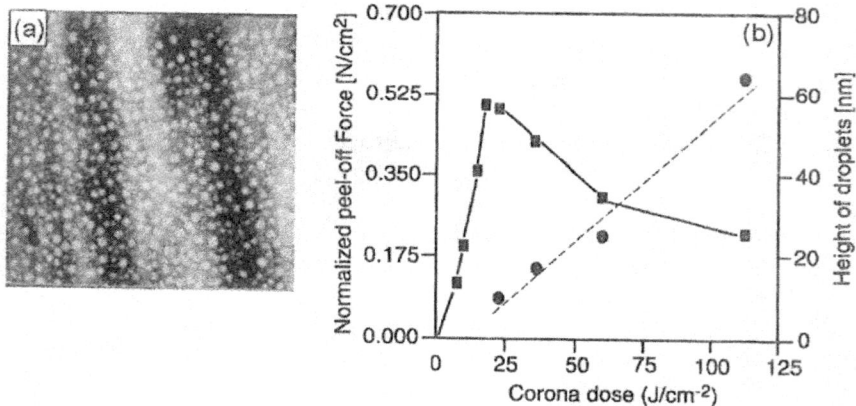

Figure 6.37: Excessive corona discharge treated PP surface (From[78,161]). (a) $20 \times 20\mu m^2$ topography scan on a striated uniaxial PP surface. The surface was corona-discharge treated with an energy dose of 112.5 J cm^{-2}. Local surface melting caused the formation of droplet-like protrusions. (b) Failure in the adhesive properties due to excessive corona discharge treatment occurred after a corona dose of about 18 J cm^{-2}. Droplets are formed after a corona dose of 18 J cm^{-2} with heights which are linearly increasing with increasing corona dose. From[70].

Surface Morphology	
homopolymers	Kevlar[96]
	polyaniline[97-99]
	polycarbonate[100]
	poly(epichlorohydrin)[101]
	polyethylene[96,102-105 106-110]
	poly(hydroxyaniline)[11]
	polyimide[112 113]
	polyoxymethylene[114]
	polypropylene[115-120]
	polystyrene[121-124]
	Teflon[100]
blends (phase separation)	polypropylene/polyurethane[125]
	polystyrene/poly(ethyleneoxide)[126]
	polystyrene/poly(methyl methacrylate)[126]
diblock copolymers	polyethylene oxide-polystyrene[69]
	polyisoprene-polybutadiene[127,128,129,130]
	polystyrene-polybutadiene[128]
heterostructures of hompolymer and block copolymers	polystyrene/polyethylene-propylene[47]
latex particle film	polystyrene[12]
	polybutyl methacrylate[131,132]

Table 6.2: Surface Morphology of Polymers

One of the research fields that is of fast growing interest since the inception of the scanning probe microscopy is Nanotribology[70]. It is strongly interconnected with force modulation measurements. The common approach in *Nanotribology* is based on lateral force measurements[70]. Static and dynamic friction can be studied on a molecular scale. Its origin has been found in stick-slip motions of the cantilever spring while scanning under constant contact-load over the sample surface[52]. The stick-slip motion is caused by instabilities of the cantilever's spring which cause energy dissipation, Fig. 6.39. The shape and the amplitude of a single stick-slip occurrence are measures for elastic properties of the sample, and the adhesion between sample and tip, respectively. The frequency of the stick slip motion has been found to be periodic for nearly static sliding, and erratic for faster sliding[52].

The first SFM experiment that measured wearless lateral forces and elasticity simultaneously on a submicrometer scale found the strong relation between the two sample properties confirmed[64]. It is, however, important to note that friction, the dissipative part of the lateral force measurements, is not an intrinsic property of the sample. It contains extrinsic information such as adhesion forces between sample and tip. If elastic properties dominate the lateral force response, or the lateral force signal is adhesion corrected, lateral force measurements can provide important information about the surface mechanical shear

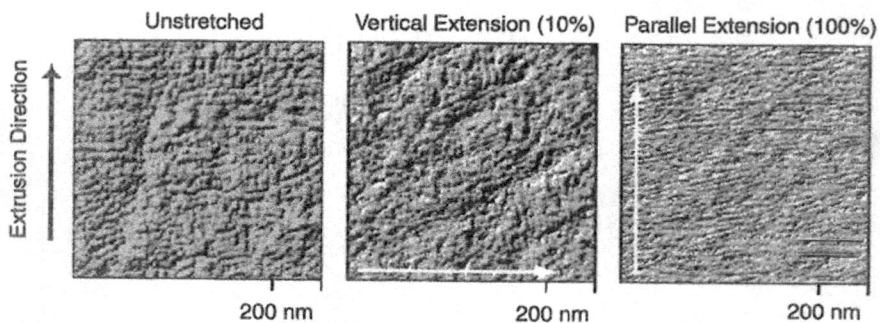

Figure 6.38: Structural deformation on hard elastic PP films. Interlamellar distances in the range of 30-50 nm were studied as a function of the vertical and parallel extension. (Modified from[142])

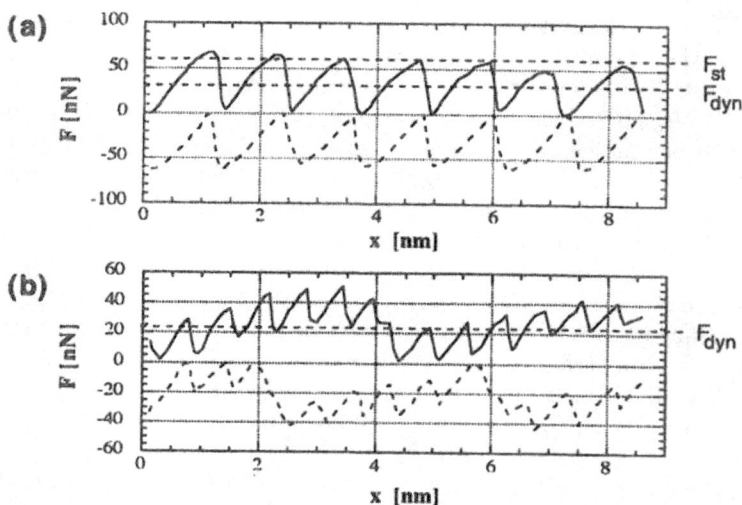

Figure 6.39: Molecular stick-slip behavior measured at a lipid film surface. (From[52]).

Mechanical Properties	
indentation	polycarbonate[77]
	poly(methyl methacrylate[77,133]
	polypropylene[133]
	poly(sebacic anhydride)[134]
friction (lateral forces)	polyoxymethylene[135]
	polypropylene[78]
	polypyrrole[136]
	polystyrene/polyethylene propylene[47]
	polystyrene/polyethylene oxide[137]
scratching, patterning	polycarbonate[138]
	poly(methyl methacrylate)[139]
	polystyrene[140]
shear deformation	poly(oxyphenylene)[141]
	polystyrene[141]
stretching	polypropylene[142]
wetting/dewetting	polyethylene -copropylene[143]
polystyrene/polyethylene-copropylene[47,130]	
Surface Treatments	
corona discharge	polypropylene[78]
UV irradiation	polyacenaphthalene[69]
	polymehtyl methacrylate[69]
	polystyrene[69]
	polymethyl styrene[69]

Table 6.3: Mechanical Properties and Surface Treatments of Polymers

properties of the sample[2]. An extensive study, where lateral forces have been compared with force modulation measurements will be summarized below in a paragraph about *three-fold measurement*.

6.5.5   Resolution limits of force modulation measurements

Many groups reported atomic or molecular resolution in lateral force and topography,[162] but only a few proved true atomic/molecular[52]. It is believed that most of the atomic or molecular resolved SFM measurements are not recorded with a single atomic tip. Multiple tip contacts that are commensurable with the sample lattice structure can provide unfortunately also periodic patterns (Moiré pattern). True atomic/molecular resolution is achieved if defect sites such as missing atoms or molecules, molecular boundaries or steps are found on the sample surface. An example of a molecularly resolved boundary, measured in air on a ultrathin organic film of 5-(4'-N,N-dihexadecylamino)benzylidene barbituric acid (lipid), is shown in Figure 6.40a.

How about atomic (molecular) resolution with force modulated SFM? Can elasticity be mapped on a molecular scale? First of all, it seems to be contradictory to compare a phenomenological defined property of a solid with the

Langmuir-Blodgett (LB) Films	
polymerized LB-films	144
polymer stabilized LB-films	64,65,145,146,147
Biopolymers	
morphology	actin[149]
filaments	neuro filaments[150]
	paired helical filaments[151]
fibers	collagen[152−154]
biodegradation	poly(DL-lactic acid)[155]
	DNA[156]
crystal growth	lysozyme[157]
force	DNA strands[158]
adhesion	biotin-streptavidin ligand[159]
phase behaviour	gelatin[160]

Table 6.4: Langmuir-Blodgett films and Biopolymers

discrete structure of matter. The elasticity of a single atom (molecule) is not defined, but the elastic flexibility of an atom in the lattice. *Overney* and *Leta* showed in 1995 that with a cantilever tip as sharp as the size of a single sample molecule, elasticity can be recorded on the molecular scale[163], Fig. 6.40b.

6.5.6 Procedure of scanning force modulation measurements

In the scanning force modulation approach, the modulation frequency is usually below the resonant frequency of the cantilever and the relaxation frequency of the sample. The sample or the cantilever is modulated in the z-direction (normal direction to the sample surface) at frequencies apart from resonant frequencies of the system, which includes the instrument, the sample, the tip and the physical properties of the contact zone. In the scanning mode, the modulation frequencies are set above the gain of the feedback-loop. Reasonable modulation frequencies are between 1-20 kHz while scanning, and 10-40 Hz in the static mode.[d] The response time of the electronics of the photo diode and system resonances sets limits in the choice of modulation frequencies.

Amplitude and phase of the response are measured simultaneously with a two-phase lock-in amplifier. The amplitude is proportional to the elastic properties of the sample and the phase is proportional to the phase shift caused by the sample, which is a measure for the viscoelastic flow[61].

Absolute values for viscoelasticity is obtained by normalizing the measurements to a modulation frequency of 10-40 Hz without scanning to eliminate electronic setup effects which occur at modulation frequencies above 1 kHz. In the elastic regime the Hertzian theory can be applied to relate the z-compliance to the Young's modulus, an intrinsic property of the sample (see equations 6.12

---

[d]Static mode of modulation is referred to as modulating without scanning).

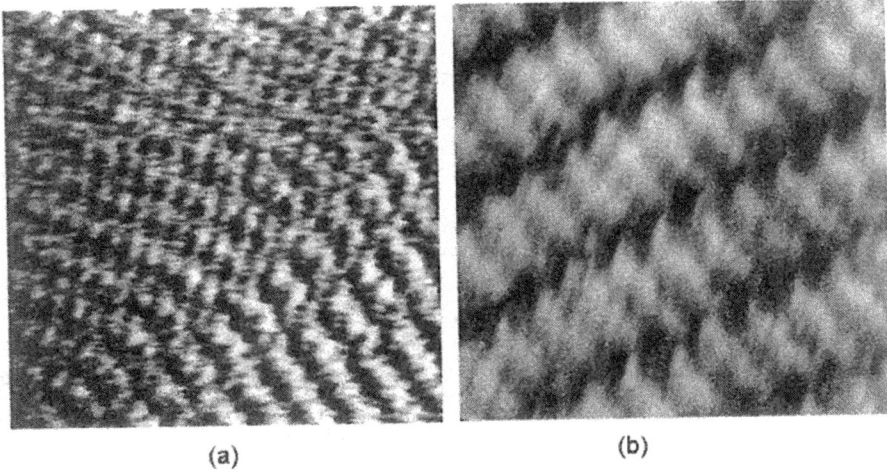

(a)   (b)

Figure 6.40: High resolution image of a lipid film surface. (a) 12×12 nm^2 scan (in lateral force mode) of a boundary between two domains. The upper limit of the contact area is determined to be on the molecular scale[52,163]. (b) 3.5×3.5 nm^2 elasticity scan. The resolution is due to the small contact area on the molecular scale (From[163], by permission of the publishers, Baltzer AG, Science Publishers).

and 6.13). This simplified elastic model does not account for adhesive forces and should therefore not be applied if the force-separation curve (FD curve) indicates local variations in adhesion.

An adsorbate-soiled tip has been observed to have the effect of rendering the viscoelastic measurements very frequency dependent, even inverting the contrast information[64]. Therefore, measurements must be performed very carefully, avoiding frequency-dependent modulation. Environmental conditions, such as humidity, have been found to affect the measurements. Therefore, caution is advised in reporting absolute values for measurements conducted in air. Instead, relative numbers or comparative absolute numbers, collected over a short time, are considered more valid.

If the spring constant of the cantilever is very soft compared to the sample, the cantilever will bend rather than elastically deform the sample. Only the spring of the cantilever will be measured and not the elastic properties of the sample surface. This is, however, not a serious problem for quantitative measurements on soft polymeric films. Viscoelastic contrast information can be achieved with a cantilever spring constant of 0.01-1.0 N/m on soft films. Reliable absolute measurements demand, however, stiffer cantilevers (0.1-10.0 N/m). Yet, wear problems with stiffer cantilevers on deformable films should be considered. An alternative method is described below with measurement in the fast modulation regime (GHz regime) where compliance measurements can be conducted with elastically soft probes on hard samples.

6.5.7   First promising measurement in force modulation

Maivald et al. applied in 1991 for the first time the force modulation SFM technique to measure mechanical properties of polymer surfaces[59]. The chosen sample, a carbon fiber and epoxy composite, showed a distinct difference in the surface elasticity. Based on the Hertzian mechanics (equation 6.129), surface moduli of 2.1×10^{11} Pa and 7.0×10^{10} Pa for the carbon fiber and epoxy, respectively, were calculated. The value of the fiber is in good agreement with bulk data. The value for the epoxy is an order of magnitude higher than the bulk value. The authors assume a tight packing effect in the composite material to be the reason for the discrepancy. They also point out the problem of determining the real contact area. A problem which has been encountered by everybody who tried to quantify force modulated SFM measurements.

6.5.8   Three-fold measurements: Topography, lateral force and force modulation

Langmuir-Blodgett (LB) films are often chosen as models of well structured films. The interplay between competing forces of polymer dynamics and carboxylate's intermolecular hydrocarbon interaction were studied on such models by a combined SFM technique of topography, lateral force and elasticity[64,65,147]. The studied films were either monolayer or bilayer structures of phase-separated hydrocarbons and fluorocarbons complexed with counterion polymers of polyallylamine (PAA) and poly(dimethyldiallylammonium chloride) (PDAA).

With the three-fold SFM approach, it was found that phase separated hydrocarbon and fluorocarbon areas showed local lateral force differences corresponding to the local mechanical response of each sample[64]. Lower lateral forces and higher Young's modulus were observed on hydrocarbon domains than on fluorocarbon domains. Young's moduli were determined by the Hertzian theory as described above in equation 6.129. The two-dimensional dispersion (i.e., the hydrocarbon agglomeration) of the two phases was determined by the counterion polymers, the hydrocarbons' chain length, and the subphase's pH value. In a systematic three-fold SFM study on film formation parameters, it was found that for chains longer than $C_{17}H_{35}COOH$, the hydrocarbons agglomerate two-dimensionally into circularly shaped islands. At shorter chain lengths, the islands deviate from the circular shape[165]. It was proposed that these differences in island formation result from a balance between attractive dispersion forces among hydrocarbon moieties and polymer (counterion) dynamics[64].

The phase dispersion was reported to depend strongly on subphase pHs[64]. At low pH, large hydrocarbon islands of about 300 and 1000 nm in diameter were found on top of a fluorocarbon sea. Intermediate pH showed much smaller circular islands of about 80 nm in diameter, Fig. 6.41. At high pH, circular interconnected islands were found. On films prepared at low pH, the Young' modulus has been reported to be twice as high (0.4 GPa) as on the samples prepared at higher pH[64]. It was concluded that the reason for the higher measured elastic response is due to higher molecular density. This was supported by the fact[166] that at lower pHs the density of molecules in the film is greater than at higher pHs because of a larger number of occluded neutral carboxylic acids.

Counterion polymers of different cross-sections have been found to affect the dispersion of the hydrocarbon and fluorocarbon phases, Fig. 6.41[64]. PAA and PDAA were chosen as counterions with different cross-sectional areas of 0.4 and 0.7 nm[2], respectively. It was found that an increase of the cross-sectional area caused a deviation from the circular shape of the hydrocarbon islands[65]. Furthermore, the sliding stability of the hydrocarbon domains was significantly reduced with PDAA. Randomly occurring (adhesive) wear was observed and analyzed by fractal analysis[65]. No significant changes in the elastic response were measured with increased cross-sectional area of the counterion polymer. Friction, however, increased as a result of occurring wear.

Topography, lateral force and elasticity measurements made it possible to differentiate between the sample's phases, and to achieve a better understanding of competing forces in complex polymeric systems. Wearless[e]lateral force measurements were for the first time introduced as a method to investigate nanomechanical properties of polymeric systems[64]. In the event of wear or local plastic deformation, however, lateral force measurements were found to be also a measure for adhesive forces[65]. This has been studied in more detail elsewhere[146].

6.5.9 Determination of mechanical properties of polymer blends

One of the questions which had been addressed by force modulation SFM is the phase separation of polymer blends at surfaces[167].

In *Takahara's* research group dynamic viscoelastic properties of octadecyltrichlorosilane (OTS) / [2 - (perfluorooctyl)ethyl]trichlorosilane (FOETS) and also polystyrene (PS) / poly(vinyl methyl ether) (PVME) blend films were

---

[e]The occurrence of microscopic wear can in most macroscopic friction experiments not be excluded. For a fundamental understanding of friction (or lateral forces), it is crucial to differentiate between lateral force observation with wear or without wear (i.e., "wearless").

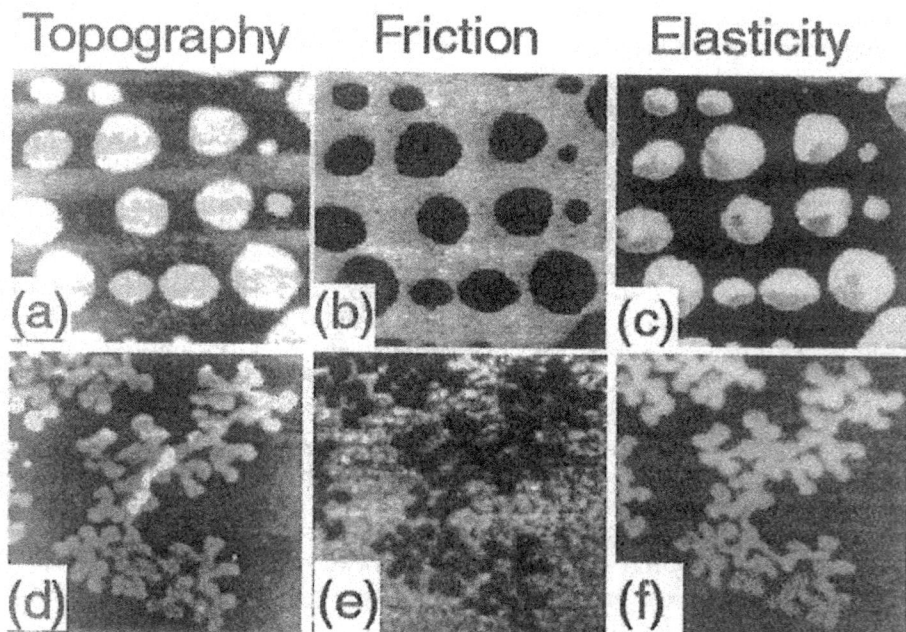

Figure 6.41: Threefold measurement of topography, lateral force and elasticity on phase separated polymeric stabilized hydrocarbon and fluorocarbon films. (a-c) 3×3 μm^2 scan on a phase separated mixture of behenic acid (BA) and PFECA complexed with PAA. Hydrocarbon "islands" are found on a fluorocarbon "sea". The hydrocarbon domains are stiffer and show less frictional resistance to sliding than the hydrocarbon substrate[145]. Mechanical surface properties dominate the lateral forces[64]. (From[70], by permission of the publishers, Elsevier Science Ltd.)
(d-e) 6×6 μm^2 scan on a phase separated mixture of arachidic acid (AA) and PFECA complexed with PDAA. Due to the larger cross-sections of the counterion polymer PDAA than PAA the hydrocarbon island boundaries are strongly deviating from the circular shape which was observed for counterions with smaller cross-sections. (From[147], Copyright ©1994 American Institute of Physics)

studied[167]. In the phase separated OTS/FOETS films with OTS islands, higher elastic modulus has been found in the OTS regions. The authors suggest that the difference in the elastic response between OTS and FOETS is due the crystalline state of OTS at 293°K and the amorphous state of FOETS. The PS/PVME ultrathin blend film was found to phase separate with circular PVME domains, 200 nm in diameter. Elasticity measurements revealed that the PS matrix is stiffer. Measurements of the complex modulus provided the authors with the possibility to calculate the tangents of the phase shift (c.f. equation 6.78). The magnitude of $\tan\delta$ for PVME and PS was reported to be approximately 0.25 and 0.1, respectively, measured at a frequency of 5 kHz[167]. This finding corresponds with the rubbery state of PVME at room temperature. However, it is important to note, that single frequency measurements (especially if conducted at higher frequencies above several hundred Hertz) are insufficient in determining the phase state of polymer surfaces. Nor are they useful in the discussion about differences between surface and bulk properties. It is necessary for force modulation SFM, i.e., nanorheological surface experiments, to extrapolate the results to zero frequency, as it is for state-of-the-art rheological bulk measurements.

As an example, force modulation SFM measurements of the elastic response, E', as function of the frequency for polystyrene (PS) and polyethylene-copropylene (PEP) are presented in Figure 6.42. These measurements were conducted at room temperature. The glass temperature, Tg, of PS and PEP are 100 °C and -62 °C, respectively. It is known that for temperatures above T_g, polymers behave very viscous with a low capacity to store energy. Below T_g the polymers act more like solids with increased shear elastic constants. If however, the perturbation frequency is high in comparison to the inertia of motion of very viscous polymers (PEP), the response can be as high as for high Tg polymers (PS) probed at the same frequency and temperature (see Figure 6.41 at 1kHz.).

6.5.10 Molecular mobility, interfaces and surface glass temperature

It is expected, and has been partially observed, that because of reduced dimensions from bulk to interfaces and surfaces, mechanical properties of polymers are different from the their bulk values[47]. In the dewetting study discussed in the last section in the context of interfacial confinement, it was found that interactions at interfaces can slow down the molecular mobility, and hence, change a liquid-like behavior of a polymer to become more solid-like. In the case of polyethylene-copropylene (PEP), for instance, the stress relief from a dewetting polystyrene (PS) layer which was mechanically interconnected with the PEP film, caused an abrupt momentum transfer. The PEP film fractured or yield depending on its available response time. The response time was found

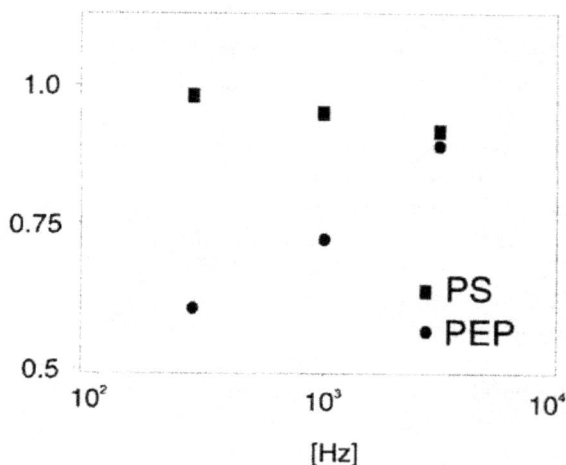

Figure 6.42: Normalized elastic response of PS and PEP measured at room temperature by SFM.

to rely significantly on the substrate distance within the mechanical boundary regime which was found to be an order of magnitude larger than the polymer's radius of gyration.

In a static SFM study, conducted by *Meyers et al.*, it was found that the surface of PS films (M_w 24K) was less glassy than reported for bulk PS[168]. Their finding was based on the recovery characteristics of initially scratched areas.

Thus, the determination of the surface glass temperature T_g as measure for increased or decreased molecular mobility at polymer surfaces is of great interest. *Kajiyama et al.* measured the T_g value at the film surface of a symmetric poly(styrene-block-methyl methacrylate) diblock copolymer (PS-PMMA) with temperature-dependent X-ray photoelectron spectroscopy and angular-dependent XPS[169]. Their measurements indicate that T_g at the PS-PMMA film surface is about 15-20% lower than expected for bulk PS or bulk PMMA.

No satisfying frequency and temperature dependent force modulation SFM "master curves" were published yet. It seems however, from Figure 6.42, that the SFM would be the perfect choice of an instrument in determining surface T_g values.

In a scanning rate dependent friction study on monodisperse PS films which was conducted simultaneously with force modulation measurements, *Takahara* and co-workers claim to have found for low molecular weight PS (< 30k) a glass-rubber transition state at the film surface at temperatures as low as $293°K^{170}$. The authors believe, that because a bare silicon substrate measured under the same conditions did not show a scanning rate dependence, adhesion can be neglected. In other words, the authors assume the same wear or plastic deformation behavior on low molecular PS films as on a bare silicon wafer. This is quite a questionable assumption, and if proven wrong, their finding above would be reduced to a mere scan rate dependent wear or plastic deformation behavior of friction as studied elsewhere[146].

6.5.11 Measurement of shear moduli

Young's modulus measurements are achieved by modulating the scanner (or the cantilever) in the surface's normal direction. But there are at least two intrinsic problems of the experimental set-up: (i) piezo shearing, and (ii) asymmetry between probing tip and sensor with the sample surface (i.e., the indentation of the tip is not perpendicular to the sample surface.

Overney and co-workers suggested to measure simultaneously the lateral displacement of the cantilever during surface asymmetric indentation[65]. It is accomplished by changing in a controlled manner the tilt angle between tip and sample. Such a set-up adds to the set-up illustrated in Figure 6.35 the possibility of recording the torsional modulation signal, too. Equipped with a second lock-in amplifier, the normal and shear moduli can be measured, and this, simultaneously with lateral force and topography while scanning.

In addition to the shear moduli information, the technique also provides the possibility to study the stick-slip regime which determines static and dynamic friction. Modulated stick-slip measurements experienced quite recently great attention in the field of surface forces apparatus (SFA) studies on polymers by *Granick* and co-workers[171].

6.5.12 Surface mechanical properties measured by lateral forces

In combined lateral force and viscoelastic studies, it has been found that lateral force measurements can provide important insight into surface mechanical properties. It is, however, important to insure no plastic deformations and no local changes in adhesion, i.e., no variations in the extrinsic part of the lateral force signal.

An example of lateral force measurements applied to surface mechanical properties is found with the copolymer resins study of *Nysten et al.*[67]. Toughening effects on i-polypropylene/(ethylene propylene)copolymer resins were investigated as function of the viscosity ratio between their xylen soluble and insoluble fractions and their impact resistance strength. In the first resin, resin A, (EP wt % 20, melt flow index 6.2, viscosity ratio 2.2, notched izod 4.7) the EP nodules are reported having irregular shapes and to contain multiple crystalline inclusions. In the second resin, resin B, (EP wt % 20, melt flow index 6.3, viscosity ratio 1.0, notched izod 3.3) circular EP resins with one crystalline inclusion were found. Larger friction forces were observed on the amorphous EP phase. Additional force modulation measurements confirmed the mechanical observation achieved by lateral force microscopy. Based on the lower friction values and stiffer response of the inclusions, the authors presume crystalline PE inclusions in the soft amorphous EP nodules. No qualitative differences could be found between the iPP matrix and the crystalline inclusions. Resin A, with its good impact resistance, showed stiffer EP nodules than resin B which is less resistant to impact. Thus, the authors claim that macroscopic observation of impact resistance can be understood by the stiffness of the EP nodules.

Haugstad, Gladfelter and *Jones* showed experimentally on the submicrometer scale a frictional "heating" induced rubbery behavior of a gelatin top surface layer[172]. Residual elevated friction and reduced elastic modulation response were imaged in regions which were initially scanned under pertubative scanning conditions (i.e., scanning above a critical contact force of 70 nN with low scan velocity on the order of 100 nm/s). The relaxation time was studied and found to be too slow to be caused by heat transfer only. Therefore, the authors presume that conformational metastable changes occurred at the surface. They were not visible in topography which was recorded simultaneously.

In another study by *Haugstad* and co-workers of surface stresses on a gelatin film, it was found that small indentations, with a SFM tip, induced local stiffening of the gelatin surface[173]. Reduced friction was recorded at the circular indentation spots, Figure 6.43. Simultaneously measured topography images revealed elevated heights at these spots which are assumed to be an indication of local delaminations of the gelatin film from the substrate[173]. In a detailed quantitative analysis, it was found that the time dependence of relaxation towards the friction of the unperturbed region is consistent with a "stretched exponential" function $(exp(-t/\tau)^{\beta}$, with fractional exponent $\beta)$[173]. This observation is often exhibited by disordered systems due to a distribution of relaxation times.

6.5.13 Surface stresses as indicators of surface instabilities

Surface or near-surface stresses have been observed by lateral force and force modulated microscopy during the onset of dewetting in the binary system PS/PEP[47], Fig. 6.44. Circles of square-micrometers in dimensions of low lateral forces were found around dewetting holes which were more than one order of magnitude smaller than the stress fields in the lateral force image. Topography revealed in the stress regime a small height deflation of less than 2 nm compared to a PS/PEP film thickness of 13nm/400nm. Lower lateral force and lower stiffness was measured in the stress regime.

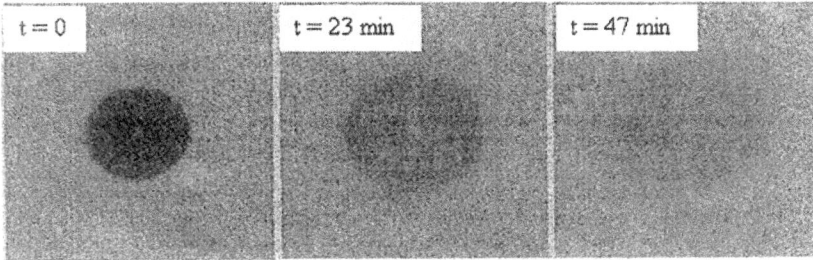

Figure 6.43: Three time-shots of 50×50 μm^2 lateral force images of a stress relaxing gelatin film. The effect of surface stiffening (dark circular spots) induced by a SFM tip at time zero (left image) is relaxing over time. (From[173], With permission of G. Haugstad).

Rafailovich and *Overney* proposed a dewetting cavity model, with a cavity beneath the surface[47]. With such a model in mind, the force modulation measurements are superimposed by the macroscopic flexibility of the film on top of the cavity. Hence, the overall measured stiffness is reduced. On the other hand, the SFM lateral force mode is particularly probing the stresses at the surface. Thus, the measured mechanical response determined by lateral force measurements is higher than the one measured by the force modulation method. Other explanations than the cavity model are still under investigations and are considering the complexity of the confinement during the instability is formed.

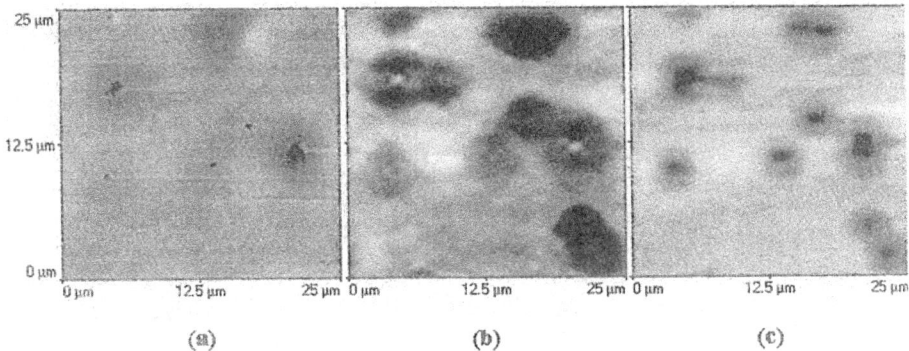

Figure 6.44: (25×25 μm^2 SFM scan images capturing mechanical stress fields during the onset of dewetting of PS/PEP. A dewetting plane about 25 nm below the surface is instantaneously established. The topography image reveals small holes about 200 nm in diameter. The material around the holes shows a very small height reduction (< 2 nm) but a very distinct change in the mechanical properties (From[47], Copyright ©1996 American Institute of Physics)

6.5.14   Static and dynamic force-displacement measurements

Force-displacement (FD) measurements represent a complementary approach for investigating nanomechanical properties. There are two modes of operation: (a) a static mode as illustrated in Figure 6.34, and (b) a dynamic mode as illustrated in Figure 6.28c.

In the static FD method, the system stiffness (or the Young's modulus) can be determined. The Young's modulus (equation 6.129) depends strongly on the cantilever 's normal spring constant and radius of curvature (area of contact). *Burnham* and *Colton* demonstrated in 1989 the applicability of the FD method to determine surface mechanical properties[79]. However, because of uncertainties of the surface contact area, it has been reported to be very difficult to quantify the results[174].

In a very recent work by *Corcoran et al.*, the sharp and not well defined SFM tips were replaced with spherical glass beads with diameters from 4 to 1000 μm[175]. Very stiff springs of 4500 N/m were used. Glass bead indentation measurements were performed on polycarbonate (PC) and polystyrene films. A strong decrease in the measured modulus of PC was observed over an increase in the sphere's diameter, Fig. 6.45[175]. This is in correspondence with the comparative study of *Hamada* and *Kaneko* between SFM and a Vicker's hardness tester[77]. *Corcoran et al.* claim the effect of adhesive forces and surface roughness to be responsible for the decrease in the modulus with increased contact area. Static FD measurements on thin polystyrene films with indentations on the order of more than a tenth of the film thickness show strong contributions from the substrate[175].

The purpose of the dynamic FD mode is to measure, besides surface mechanical properties, also mechanical or viscous properties of liquids in the vicinity of solid interfaces and polymer melts[22]. This technique combines the perturbation theory with the static FD measurement by adding a small sinusoidal modulation to the steady approach or retraction curves (FD-curves)[22]. This combined approached of dynamic FD measurements has been named ac-FD-approach[22]. Its modulation amplitudes are on the nanometer scale, and its steady approach velocity a few nanometer per seconds. From the measured force signal, the system stiffness, the elastic and viscous component of the modulus can be acquired simultaneously. The *ac-FD-approach* provides nanometer scale viscoelastic information of liquid samples as illustrated on polymer brushes in contact with solvents, Fig. 6.28[22]. This method showed very valuable in the studies of Overney and Rafailovic[22]. They found, for instance, that in poor solvents, e.g., water, the storage moduli of collapsed polystyrene chains, extrapolated to zero shear rate, reduces to 0.1 ± 0.03 GPa. This value is similar to that obtained for a bulk PS melt of comparable molecular weight at 135 OC. Additionally small erratic modulus fluctuations had been observed

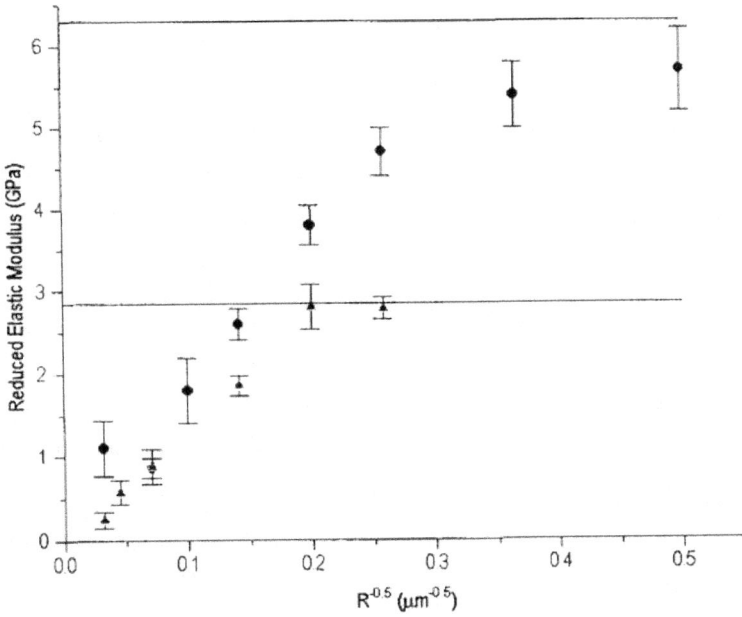

Figure 6.45: Reduced elastic modulus on PC as a function of inverse indentor radius. Sphere indentations with variable radius R are compared to results measured by a sphero-conical diamond microindentor with nominal radii of 20 μm (dashed line). (From[175], With permission of R. Colton).

on the collapsed PS brush system and interpreted as escape transitions (chain dynamics)[22]. Moduli measurements on PS brushes in a good solvent, toluene, which caused the PS chains to extend about 25 nm into the solvent, were reported to show an increase in the elastic response of 1-2 orders of magnitude compared to water as solvent[22]. The static FD measurements were compared to surface forces apparatus measurements and found to be qualitatively consistent within a 10 nm compression regime. A quantitative comparison was only possible by fitting a contact radius which is, as discussed above, not directly accessible with conventional cantilever tips.

6.5.15 Ultrasonic force method

An extension of the state-of-the-art force modulation mode is the SFM ultrasonic force mode (UFM) which was introduced by *Kolosov* and *Yamanaka* in 1993[63,66]. This SFM operation mode opens up the prospect of studying surface dynamic processes on the nanometer scale. In addition, viscoelastic properties with elastic constants, orders of magnitude higher than the one of the probing cantilever, and short range interfacial interactions such as van der Waals interaction can be studied. A very related SFM mode, the scanning local acceleration microscopy (SLAM), was introduced in 1996 by em Burnham, *Kulik*, and *Oulevey* with similar applications to surfaces as the UFM[176,177].

The principle of the UFM is very similar to the force modulation mode. A cantilever with a typical resonance frequency below 100 kHz is brought into contact with a sample which is mounted on a piezo transducer. The piezo transducer is driven, in contrast to the force modulation mode far above the cantilever's resonance, with vibration frequencies ranging from 1 MHz to more than 100 MHz. A nonlinear force-distance dependence between cantilever tip and sample, which is usually observed for soft cantilevers and attractive potentials in closest sample vicinity (<1 nm), will cause a shift in the average position of the lever and generate higher harmonics of the ultrasonic vibration. The force acting on the cantilever is then described by[63,66]:

$$F_m(z) = k(z - z_s), \text{ and}$$

$$F_m(z.a) = \frac{1}{2\pi} \int_0^{2\pi} F(z - A\cos x)dx, \tag{6.132}$$

where k is the lever spring constant, z_s is the displacement of the object surface out of the area of contact, and A is the vibration amplitude. The local system spring constant can be determined from the deflection-versus-ultrasonic-amplitude dependence. Thus, the Young's modulus can be found with equation 6.129.

The UFM SFM was applied to study the nanomechanical structure of floppy disks composed from metal-coated ferrite particles and soft polymer binder[66].

6.5.16 Pulsed force method

Stiffness and adhesion can be determined with the pulsed force SFM mode of *Marti, Schild* and *RosaZeiser*[178,179]. While the above discussed ultrasonic force mode is well suited for elasticity measurements of hard samples, the pulsed force mode is tailored for very soft samples where strong adhesion forces are present.

The pulsed force mode is based on very fast force-displacement curve measurements while scanning. Only discrete force values are recorded which admit higher modulation frequencies. Scanning even on very adhesive and soft material is possible because the tip is touching the sample only for a short moment. Scanning is further improved by drastically reduced shear forces.

The principle is as follows: The cantilever is sinusoidal modulated. Sample and lever interact over harmonic potentials and an adhesive potential The sample and the cantilever are described by their harmonic spring potential (a spring constant ks is assigned to the sample as introduced in equation 6.124). The effective potential is the sum of the three potentials. The modulated cantilever is approaching the surface and its amplitude will be damped by the adhesive potential until the attraction exceeds the spring constant and the cantilever jumps into contact. After the tip is in contact, a first maximum in the force curve occurs as a consequence of the harmonic and adhesive potentials. This maximum signal is used as feedback for the scan process. The shape of the force-displacement curve between initial contact and force maximum is a convoluted measure for the system's effective spring constant and the adhesive forces. An adhesive image can be obtained by recording the forces at the jump off point, and an elastic image at a fixed point between maximum force and the jump in or jump out point.

Schild and co-workers applied the pulsed force SFM mode on a polystyrene-isoprene blend. The polystyrene phases were as expected found to be stiffer than the isoprene phases. With a rather costly model of six fit parameters, the relative shear bulk modulus value between isoprene and polystyrene were confirmed within the same order of magnitude in the pulse force experiment, Figure 6.46.

6.5.17 Scanning static elastic method

A very different and intriguing approach has been undertaken by *Yang*[80]. He used the SFM to measure microscopic deformation zones of free-standing glassy polymers by varying the applied load from scan to scan. Free-standing films,

Topography   Local Stiffness   Adhesion

dark = hard   dark = low Adhesion

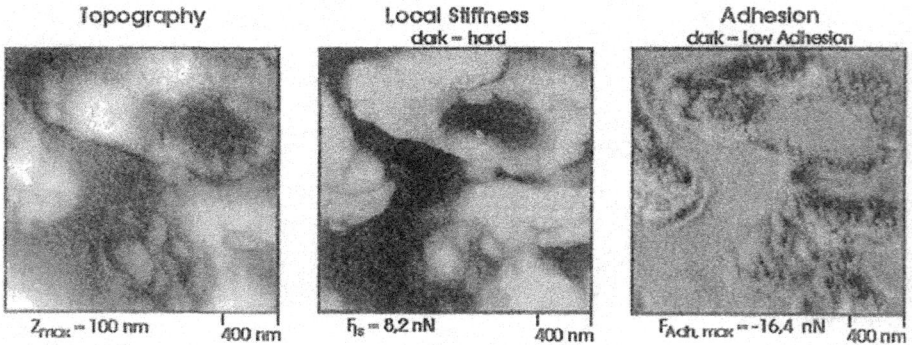

z_{max} = 100 nm   400 nm F_{ls} = 8,2 nN   400 nm $F_{Adh.\,max}$ = -16,4 nN   400 nm

Figure 6.46: Pulsed force mode provides simultaneous information about topography (slightly distorted), elasticity and adhesion. Local stiffness measurements identify the high spots (bright areas) in topography as polystyrene. Higher adhesion is measured on the isoprene areas (From (178) T. Schild, S. Hild, O. Marti, presented at the American Chemical Society Meeting in Orlando (1996)).

polystyrene (PS) and poly(2,6-dimethyl-1,4-phenylene oxide) (PPO), were spin cast from toluene solutions on glass slides and then floated from water onto TEM copper grids. Polymer films of various thickness (d_f) from 0.1 to 0.5 μm were tested. Each segment of the square mesh divided polymer film was modeled as edge-clamped free-standing film. Stresses in film segments were introduced by merely scanning a cantilever in contact over the surface. The force-constant *topography* contour lines were measured depending on the applied equilibrium load (c.f. Figure 6.34), and surface elastic constants extracted.

Based on work from von *Karman* and *Seewald*[180], global and local film deflections were considered. The elementary bending process of a uniform edge-clamped film is described by the global film deflection. Local effects which vanish very quickly from the SFM contact points are considered to be onsets of plastic deformations which lead to rupture (or fracture) of the film. The global deflection were considered constant for all loading points as long as scanning was performed in the center of a grid segment with small scan sizes, at least two orders of magnitude smaller than the size of the square segment[80].

The Young's modulus ratios E_z/E_f (subscripts z and f refer to the values of the material within and outside the deformation zone, respectively) were determined from loading dependent height variations of crazes and micro-shear zones appeared as large depressions on the polymer surface[181]. Hence, the measured total depth D of crazes is a convolution of the film thicknesses dz and df, and the local deformation δ_z and δ_f. Yang applied the following first order approximation to depth-loading measurements to calculate local elastic moduli[80]:

$$D = \frac{1}{2}(d_f - d_z) + (\delta_f - \delta_z) \cong \frac{1}{2}(d_f - d_z) + \lambda \left(\frac{d_f}{d_z} \frac{E_f}{E_z} - 1 \right) \frac{F_o}{(d_f E_f)} \quad (6.133)$$

where λ is determined by the choice of the geometrical model of the film segment. The ratio d_f/d_z is determined by extrapolation of zero loading forces. Bulk values were chosen for E_f.

Yang chose the elementary beam bending (EEB) analysis[182] to determine λ and found significant softening of (a) the glassy 0.1 μm thick PPO films in micro-shear zones of more than 4 μm width, and (b) the glassy 0.1 μm thick PS films in all grazes and independent of thermal aging at 75 °C. Because of the choice of the EEB analysis which accounts only for free film ends instead of clamped ends and models the film as a beam rather than a plate, *Yang's* calculation provides about 30 % lower values[183]. Nevertheless, it was found with *Yang's* scanning static elastic SFM method that shear yielding and crazing in glassy polymers cause a significant decrease in the Young's modulus. This finding supports the theory of strain softening during plastic deformation.

6.5.18 Summarizing critical remarks

In the past, various critical reviews have addressed the applicability of the SFM for the characterization of polymer surfaces[184,185]. Aspects that have been of particular interest are the SFM's limit of resolution and defect-free imaging. Both are strongly connected.

While in the case of well ordered monolayers (or multilayers) of self-assembled soft organic films plastic deformations caused by the cantilever tip can be imaged on a consecutive large scale scan[186], tip induced conformational changes on polymer surfaces are not as easy to confirm. The reason is the scale of deformation. In a van der Waals crystalline film, for instance, single deformations can be expected to occur on a scale of 10 or less square-nanometer leaving a topographically visible trace of deformation. Deformations of polymer surfaces, however, reach over microns although induced on the nanometer scale. A very nice example has been shown above in the paragraph about induced surface stiffening by *Haugstad et al.*[173]. Hence, lateral force and surface viscoelasticity measurements have been found to be useful in determining if the cantilever plastically interacted with the polymer sample surface. Topographical imaging alone was shown to be unconvincing[45,47,172,173].

Plastic deformations (i.e., conformational changes) at the surface limit the resolution of SFM measurements. There are four possible origins to be named for plastic deformations: loading pressure, polymer-tip wetting, scan speed, and shear forces. The loading pressure can be reduced by avoiding the formation of capillaries due to humidity and/or by using larger contact radii of curvature. Ambient capillaries can be diminished in a dry nitrogen gas environment or eliminated in a liquid environment. Cantilever tips of larger radii of curvature can be chosen with the disadvantage of lower imaging resolution. Tip wetting by the polymer sample can be reduced by the choice of (a) the right probing temperature, (b) the time in contact with a particular area, and/or (c) the suitable material for cantilever tip coating. The probing temperature and the time in contact are strongly influencing the diffusion of the polymers toward the tip. The right tip coating reduces the strength of the interaction. The probing time can be set by the scan speed, and/or by modulating normal to the sample surface. A disadvantage of increasing the scan speed is that it reduced feedback control resulting in high shear forces which can cause plastic deformation. To achieve a reduced contact time by mere modulation, a huge modulation amplitude might be necessary that results in high normal forces, and hence, can also cause plastic deformation.

The combination and weights of the mentioned procedures are crucial for reducing plastic deformation and gaining high resolution. The literature still lacks a comprehensive study of the optimization parameters of defect-free imaging of polymer surfaces, especially for polymers with low glass temperatures.

Multiphase and multicomponent systems are of particular interest in polymer science. Heretofore, it is desirable to identify the single components which are present at the sample surface.

It is obvious that a SFM operated in the topography mode does, only in very rare cases, provide insight into the chemical nature of multicomponent system. Other means of operation are in demand. In 1992, the lateral force mode (also called friction mode) was introduced as a nano-surface-tool suited to provide chemical distinctions of two component systems[145]. The lateral force values found on single component systems were compared with the lateral force values measured on the binary systems and depending on their correspondence the material was identified. In these early reported experiments[145] lateral forces were found to be dominated by the shear mechanical properties of the samples[65]. *Friesbie* and *Lieber* extended this viewpoint to chemical distinction by functionalizing the cantilever tip and thus selectively enhanced the acting interaction forces[187]. Because of their particular choice of the tip coating material, adhesion forces were considered to dominate the lateral forces.

In recent independent studies in the groups of *Colton*[82] and *Overney*[45], it was found that chemical distinctions based on mere lateral force measurements do not necessarily provide only information about the chemical nature of the sample surface. Both groups showed that the lateral force signal can significantly vary on chemically homogeneous surfaces just because of local differences in mechanical properties[45,82]. The ability of the material to shear can, therefore, not be neglected. Particularly for a functionalized tip, where a strongly curved monolayer film is very locally confined, scanning induced shear rates must be immense. Further adhesion measurements on chemically homogeneous samples conducted by the two groups of *Colton*[82] and *Overney*[45] showed strong variations due to local differences of the contact area. For instance, differences in grafting density in polyethylene acrylate polystyrene (PEA-xPS) block copolymer film show a significant effect on lateral forces and adhesion despite the fact that the surface is chemically uniform, Fig. 6.47. It was found that the chemically identical surfaces were wetting the tip differently depending on the backbone flexibility of graft copolymers.

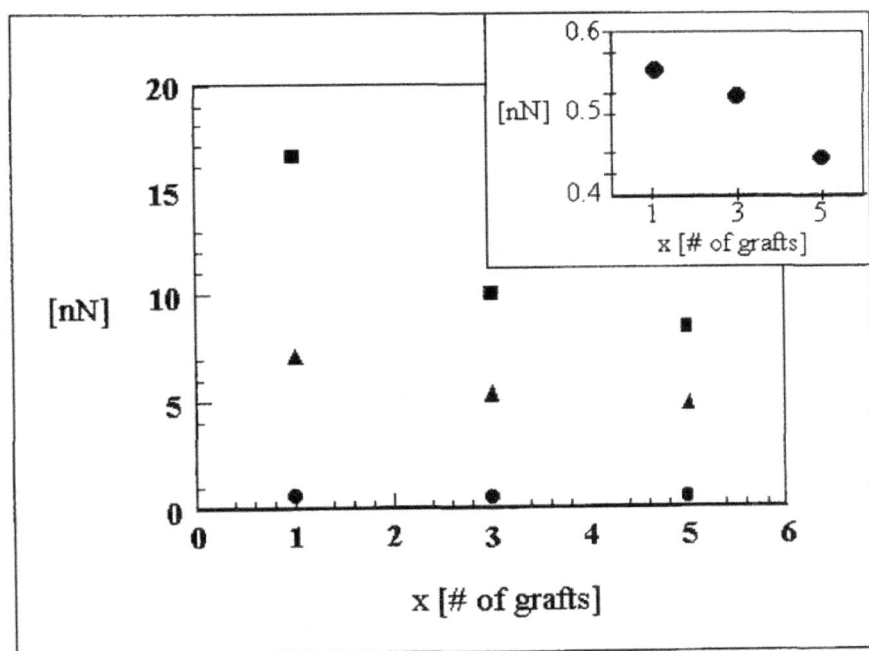

Figure 6.47: Backbone bending flexibility of PEA-xPS. □ Adhesion forces provide information about stiffening of graft copolymer surface with increased grafting density. △ Absolute lateral forces are adhesion influenced. ○ Lateral forces, adhesion corrected. Inset: Magnification of adhesion corrected lateral forces. (From[45] Copyright ©1997 Material Research Society).

A chemical distinction of multicomponent surfaces can be very difficult. The biggest challenge for functionalized tips is the determination of the contact area. As long as this value is not established, quantitative results, and even relative comparisons of interaction energies, are very difficult to accomplish.

The sample's response to force modulations can depend on various parameters. The most important ones are the equilibrium load, the amplitude of modulation, and the modulation frequency. As with all parameters there is an operational regime which is acceptable. Outside that regime, either the uncertainties do not tolerate a quantification of the results or, even worse, it comes to misinterpretation of the results.

Equilibrium loads or amplitudes of operation which exceed the elastic limit of the sample material set the experiment into a regime where quantifying theories are rare and the elastic theories, such as the linear viscoelastic theory, fail. Huge amplitudes of modulation can also cause unstable equilibrium loads during the scanning process due to feedback problems.

It is well known in polymer rheology that polymeric systems show in their viscoelastic response a very strong frequency dependence, Fig. 6.41. As long as the measurements are conducted outside any resonances of the system the results are valuable (if reported together with the frequency, amplitude and equilibrium load). However, and this is a very important point, one should avoid reporting measurements conducted at resonances. First of all, at resonances there are no amplitudes defined. Experimentalists seem, from time to time, tempted to neglect theories especially at instability points. Of course, there is no infinite amplitude for a realistic spring system. But the system is also no longer described by the same mathematical equations. In other words, the spring is no longer linear. With the loss of knowledge about the amplitude, the force modulation method at any resonance of the system seem to be reduced to a mere qualitative stiffness measuring tool. A very recent and interesting publication from *Liu's* group showed that at contact resonances an enhanced contrast force modulation response map can be recorded between the two components of octadecyltriethoxysilane (OTE) and mica, Figure 6.48[188]. A very nice contrast - but, and this is the disturbing aspect of resonance force modulation measurements, the contrast is reversed. Resonance measurements provide, therefore, only the information about the existence of heterogeneities in stiffness. They do not identify the stiffer (or softer) areas nor provide absolute or relative values. It is only a *"nice picture"* that remains and merely provides the operator with a hunch about the rheological properties of the sample.

In summary, the SFM has been shown to be a very versatile instrument in studying the nanomechanical and rheological properties of surfaces, interfaces and ultrathin films. Future work will certainly be concerned with the improved quantification of SFM measurements.

1. Sheth, K. C., Chen, M. J. & Farris, R. J. (1995) *Mat. Res. Soc. Symp. Proc.* **356**, 520-534.
2. Bird, R. B., Stewart, W. E. & Lightfoot, E. N. (1960) *Transport Phenomena* (John Wiley & Sons, New York).

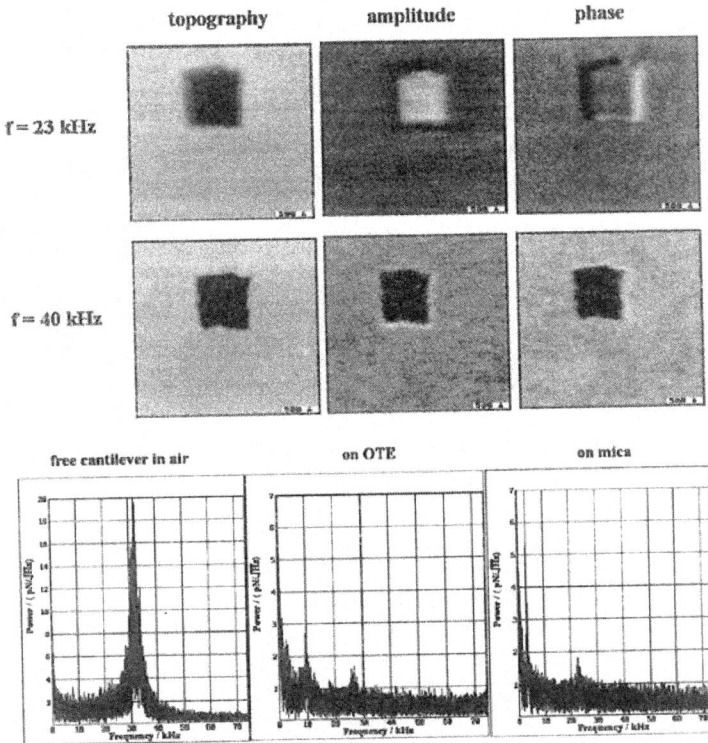

Figure 6.48: $25 \times 25 \ nm^2$ SFM scans of an OTS coated mica substrate. OTS was previously removed from the center (square hole (dark in topography)). It was reported that only at 23 kHz the hole area showed a higher elastic response (brighter contrast) than the OTS layer. Spectrum analysis (left) showed that the contact resonance frequency in the hole area (presumably mica) of 23 kHz is responsible for the inverted elastic contrast. (From[188], With permission of Gang-yu Liu).

3. Pinkus, O. & Sternlicht, B. (1961) *Theory of Hydrodynamic Lubrication* (McGraw-Hill Book Company, New York).

4. Gee, M. L., McGuiggan, P. M., Israelachvili, J. N. & Homola, A. M. (1990) *J. Chem. Phys.* **93**, 1895.

5. Hu, H.-W., Carson, G. A. & Granick, S. (1991) *Phys. Rev. Lett.* **66**, 2758.

6. Granick, S. (1991) *Science* **253**, 1374.

7. Hu, H.-W. & Granick, S. (1992) *Science* **258**, 1339.

8. Thompson, P. A., Robbins, M. O. & Grest, G. S. (1995) *Israel J. Chem.* **35**, 93.

9. Urbakh, M., Daikhin, L. & Klafter, J. (1995) *Europhys. Lett.* **32**, 125.

10. Rabin, Y. & Hersht, I. (1993) *Physica* **A 200**, 708.

11. Urbakh, M., Daikhin, L. & Klafter, J. (1995) *J. Chem. Phys.* **103**, 10707-10713.

12. Luengo, G., Schmitt, F.-J. & Israelachvili, J. (1997) *Macromolecules* **30**, 2482-2494.

13. Montfort, J. P. & Hadziioannou, G. (1988) *J. Chem. Phys.* **88**, 7187-7196.

14. Chan, C. Y. C. & Horn, R. G. (1985) *J. Chem. Phys.* **83**, 5311-5324.

15. Israelachvili, J. N. (1986) *J. Colloid Interface Sci.* **110**, 263-271.

16. Israelachvili, J. N. (1988) *Pure & App. Chem.* **60**, 1473.

17. Klein, J., Kamiyama, Y., Yoshizawa, H., Israelachvili, J. N., Fredrickson, G. H., Pincus, P. & Fetters, L. J. (1993) *Macromolecules* **26**, 5552.

18. Georges, J. M., Millot, S., Loubet, J. L. & Tonck, A. (1993) *J. Chem. Phys.* **1993**, 7345.

19. Milner, S. T. (1991) *Macromolecules* **24**, 3704.

20. Fredrickson, G. H. & Pincus, P. (1991) *Langmuir* **7**, 786.

21. Friedenberg, M. C. & Mate, C. M. (1996) *Langmuir* **12**, 6138-6142.

22. Overney, R. M., Leta, D. P., Pictroski, C. F., Rafailovich, M. H., Liu, Y., Quinn, J., Sokolov, J., Eisenberg, A. & Overney, G. (1996) *Phys. Rev. Lett.* **76**, 1272-1275.

23. Mate, C. M. & Novotny, V. J. (1991) *J. Chem. Phys.* **94**, 8420.

24. Moore, D. F. (1972) *The Friction and Lubrication of Elastomers* (Pergamon Press, New York).

25. Israelachvili, J. N. (1988) *Science* **240**, 189.

26. Yoshizawa, H., McGuiggan, P. & Israelachvili, J. N. (1993) *Science* **259**, 1305.

27. Yoshizawa, H. & Israelachvili, J. N. (1993) *J. Phys. Chem.* **97**, 11300.

28. Dieterich, J. H. (1979) *Pure Appl. Geophys.* **116**, 790.

29. Baumberger, T., Heslot, F. & Perrin, B. (1994) *Nature* **367**, 544.

30. Beeler, N. M., Tullis, T. E. & Weeks, J. D. (1994) *Geophys. Res. Lett.* **21**, 1987.
31. Landman, U. (1997) private communication .
32. Gao, J. P., W.D., L. & U., L. (1997) *J. Chem. Phys.* **106**, 4309-4318.
33. Zhong, W. & Tomanek, D. (1991) *Europhys. Lett.* **15**, 887.
34. Carlson, J. M. & Batista, A. A. (1996) *Phys. Rev.* **E 53**, 4153-4164.
35. Israelachvili, J. N., Kott, S. J. & Fetters, L. J. (1989) *J. Polym. Sci., Phys. Ed.* **27**, 489.
36. Montfort, J. P., Tonck, A., Loubet, J. L. & Georges, J. M. (1991) *J. Polym. Sci., Polym. Phys. Ed.* **29**, 677.
37. Hatzikiriakos, S. G. & Dealy, J. M. (1992) *J. Rheol.* **36**, 703.
38. Luckham, P. & Klein, J. (1985) *Macromolecules* **18**, 721-728.
39. Georges, J.-M., Tonck, A., Loubet, J.-L., Mazuyer, D., Georges, E. & Sidoroff, F. (1996) *J. Phys. II France* **6**, 57-76.
40. Klein, J. (1996) *Annu. Rev. Mater. Sci.* **26**, 581-612.
41. Sens, P., Marques, C. M. & Joanny, J. F. (1994) *Macromolecules* **27**, 3812.
42. Klein, J., Kumacheva, E., Mahalu, D., Perahia, D. & Fetters, L. J. (1994) *Nature* **370**, 634.
43. deGennes, P.-G. (1979) *Scaling Concepts in Polymer Physics* (Cornell Univ. Press, Ithaca, NY).
44. deGennes, P. C. R. (1987) *Acad. Sci. Paris* **305 II**, 1181.
45. Overney, R. M., Guo, L., Totsuka, H., Rafailovich, M., Sokolov, J. & Schwarz, S. A. in *Dynamics in Small Confining Systems IV*, eds. Drake, J.M., Klafter, J., Kopelman, R., *Mat. Res. Soc. Symp. Proc.* **464**, 133-144 (1997)
46. Israelachvili, J. N. (1992) *Intermolecular and Surface Forces*, 2nd Ed. (Academic Press.
47. Overney, R. M., Leta, D. P., Fetters, L. J., Liu, Y., Rafailovich, M. H. & Sokolov, J. (1996) *J. Vac. Sci. Technol.* **B 14**, 1276-1279.
48. Gao, J. P. & Landman, U. (1997) *Phys. Rev. Lett.* **79**, 705-708.
49. Meyer, E. (1997) private communication .
50. Hegner, M., Wagner, P. & Semenza, G. (1993) *Surface Science* **291**, 39-46.
51. Overney, R. work in progress.
52. Overney, R. M., Takano, H., Fujihira, M., Paulus, W. & Ringsdorf, H. (1994) *Phys. Rev. Lett.* **72**, 3546-49.
53. Salmeron, M., Neubauer, G., Folch, A., Tomitori, M., Ogletree, D. F. & Sautet, P. (1993) *Langmuir* **9**, 3600-11.
54. Taunton, H. J., Toprakcioglu, C., Fetters, L. J. & Klein, J. (1988) *Nature* **332**, 712-14.
55. Overney, R. M. & Drake, J. M. work in progress.

56. Brochard-Wyart, F., Martin, P. & Redon, C. (1993) *Langmuir* **9**, 3682.
57. Binnig, G. & Rohrer, H. (1982) *Helv. Phys. Acta* **55**, 726.
58. Bottomley, L. A., Coury, J. E. & First, P. N. (1996) *Anal. Chem.* **68**, 185R-230R.
59. Maivald, P., Butt, H. J., Gould, S. A. C., Prater, C. B., Drake, B., Gurley, J. A., Elings, V. B. & Hansma, P. K. (1991) *Nanotechnology* **2**, 103-106.
60. Radmacher, M., Tillmann, R. W., Fritz, M. & Gaub, H. E. (1992) *Science* **257**, 1900.
61. Radmacher, M., Tillmann, R. W. & Gaub, H. E. (1993) *Biophys. J.* **64**, 735-42.
62. Salmeron, M. B. (1993) *Mat. Res. Soc. Bull* **18**, 20-5.
63. Kolosov, O. & Yamanaka, K. (1993) *Jpn. J. Appl. Phys.* **32**, L1095 - L1098.
64. Overney, R. M., Meyer, E., Frommer, J., Guentherodt, H.-J., Fujihira, M., Takano, H. & Gotoh, Y. (1994) *Langmuir* **10**, 1281-1286.
65. Overney, R. M., Takano, H. & Fujihira, M. (1994) *Europhys Lett* **26**, 443-447.
66. Yamanaka, K., Ogiso, H. & Kolosov, O. (1994) *Appl. Phys. Lett.* **64**, 116.
67. Nysten, B., Legras, R. & Costa, J. L. (1995) *Journal Of Applied Physics* **78**, 5953-5958.
68. Hoper, R., Gesang, T., Possart, W., Hennemann, O. D. & Boseck, S. (1995) *Ultramicroscopy* **60**, 17-24.
69. Goh, M. C. (1995) *Advances In Chemical Physics* **91**, 1-83.
70. Overney, R. M. (1995) *TRIP* **3**, 359-364.
71. Heuberger, M., Dietler, G. & Schlapbach, L. (1995) *Nanotechnology* **6**, 12-23.
72. Heuberger, M., Dietler, G. & Schlapbach, L. (1996) *Journal Of Vacuum Science & Technology* **B 14**, 1250-1254.
73. Tsukruk, V. V., Bliznyuk, V. N., Hazel, J., Visser, D. & Everson, M. P. (1996) *Langmuir* **12**, 4840-4849.
74. Yamada, R., Ye, S. & Uosaki, K. (1996) *Japanese Journal of Applied Physics Part 2 - Letters* **35**, L846-L848.
75. Chen, J. T. & Thomas, E. L. (1996) *Journal of Material Science* **31**, 2531-2538.
76. Nie, H. Y., Motomatsu, M., Mizutani, W. & Tokumoto, H. (1996) *Thin Solid Films* **273**, 143-148.
77. Hamada, E. & Kaneko, R. (1992) *Ultramicroscopy* **A 42**, 184-90.
78. Overney, R. M., Guentherodt, H. J. & Hild, S. (1994) *J. Appl. Phys.* **75**, 1401-1404.

79. Burnham, N. A. & Colton, R. J. (1989) *J. Vac. Sci. Technol.* **A 7**, 2906-13.
80. Yang, A. C.-M. (1995) *Materials Chemistry and Physics* **41**, 295-298.
81. Hild, S., Gutmannsbauer, W., Luethi, R., Fuhrmann, J. & Guentherodt, H.-J. (1995) *Journal of Polymer Science Part B* **34**, 1953-1959.
82. Koloske, D. D., Barger, W. R., Lee, G. U. & Colton, R. J. (1997) *Mat. Res. Soc. Symp. Proc.* **464**, 377.
83. Corcoran, S. G., Colton, R. J., Lilleodden, E. T. & Gerberich, W. W. (1997) Phys. Rev. B **55**, 16057-16060.
84. Johnson, K. L. (1985) *Contact Mechanics* (Cambridge University Press, Cambridge.
85. Blackman, G. S., Mate, C. M. & Philpott, M. R. (1990) *Phys. Rev. Lett.* **65**, 2270-3.
86. Meyer, E., Heinzelmann, H., Gruetter, P., Jung, T., Hidber, H. R., Rudin, H. & Guentherodt, H. J. (1989) *Thin Solid Films 181*, 527-544.
87. Hartmann, U. (1991) *Adv. Mater.* (Weinheim, Fed. Repub. Ger.) **2**, 594-7.
88. Weisenhorn, A. L., Maivald, P., Butt, H.-J. & Hansma, P. K. (1992) *Phys. Rev.* **B 45**, 11226-11236.
89. Gauthier-Manuel, B. (1992) *Europhys. Lett.* **17**, 195-200.
90. Burnham, N. A. & Colton, R. J. (1993) in *Scanning Tunneling Microscopy and Spectroscopy*, ed. Bonnell, D. A. (VCH Publishers, New York), pp. 191-249.
91. Johnson, K. L., Kendall, K. & Roberts, A. D. (1971) *Proc. Royal Soc. A* **324**, 301.
92. Derjaguin, B. V., Muller, V. M. & Toporov, Y. P. (1975) *J. Coll. Interace Sci. 53*, 314.
93. Burnham, N. A., Dominguez, D. D., Mowery, R. L. & Colton, R. J. (1991) *Phys. Rev. Lett. 64*, 1931-4.
94. Thomas, R. C., Houston, J. E., Crooks, R. M., Kim, T. & Michalske, T. A. (1995) *J. Am. Chem. Soc.* **117**, 3830-3834.
95. Meyer, E., Luethi, R., Howald, L., Bammerlin, M., Guggisberg, M. & Guentherodt, H.-J. (1996) *J. Vac. Sci. Technol.* **B 14**, 1285-1288.
96. Snetivy, D., Vancso, G. J. & Rutledge, G. C. (1992) *Macromolecules* **25**, 7037-7042.
97. Nyffenegger, R., Gerber, C. & Siegenthaler, H. (1993) *Synth. Met.* **55**, 402-7.
98. Avlyanov, J. K., Josefowicz, J. Y. & MacDiarmid, A. G. (1995) *Synth. Met.* **73**, 205-8.

99. Kugler, T., Rasmusson, J. R., Osterholm, J. E., Monkman, A. P. & Salaneck, W. R. (1996) *Synthetic Metals* **76**, 181-185.
100. Magonov, S. N., Kempf, S., Kimmig, M. & Cantow, H. J. (1991) *Polym. Bull.* (Berlin) **26**, 715-22.
101. Singfield, K. L., Klass, J. M. & Brown, G. R. (1995) *Macromolecules* **28**, 8006-15.
102. Patil, R., Kim, S. J., Smith, E., Reneker, D. H. & Weisenhorn, A. L. (1990) *Polym. Commun.* **31**, 455-7.
103. Magonov, S. N., Qvarnstrom, K., Huong, D. M., Elings, V. & Cantow, H. J. (1991) *Polym. Mater. Sci. Eng.*.
104. Magonov, S. N., Qvarnstrom, K., Elings, V. & Cantow, H. J. (1991) *Polym. Bull.* (Berlin) **25**, 689-94.
105. Snetivy, D., Yang, H. & Vancso, G. J. (1992) *J. Mater. Chem.* **2**, 891-2.
106. Annis, B. K., Reffner, J. R. & Wunderlich, B. (1993) *J. Polym. Sci., Part B: Polym. Phys.* **31**, 93-97.
107. Magonov, S. N., Sheiko, S. S., Deblieck, R. A. C. & Moller, M. (1993) *Macromolecules* **26**, 1380-6.
108. Patil, R. & Reneker, D. H. (1994) *Polymer* **35**, 1909-14.
109. Jandt, K. D., Buhk, M., Miles, M. J. & Petermann, J. (1994) *Polymer* **35**, 2458-62.
110. Eng, L. M., Jandt, K. D., Fuchs, H. & Petermann, J. (1994) *J. Appl. Phys.* **A 59**, *Solids Surf.* 145-50.
111. Bowman, G. E., Cornelison, D. M., Caple, G. & Porter, T. L. (1993) *J. Vac. Sci. Technol.* **A 11**, 2266-8.
112. Nysten, B., Roux, J. C., Flandrois, S., Daulan, C. & Saadaoui, H. (1993) *Phys Rev B-Condensed Matter* **48**, 12527-12538.
113. Kim, Y. B., Kim, H. S., Choi, J. S., Matuszczyk, M., Olin, H., Buivydas, M. & Rudquist, P. (1995) *Mol. Cryst. Liq. Cryst. Sci. Technol.* **262**, Sect. A 1377-86.
114. Snetivy, D. & Vancso, G. J. (1992) *Polymer* **22**, 422.
115. Lotz, B., Wittmann, J. C., Stocker, W., Magonov, S. N. & Cantow, H. J. (1991) *Polym. Bull.* (Berlin) **26**, 209-14.
116. Dorset, D. L. (1992) *Chemtracts: Macromol. Chem.* **3**, 200-1.
117. Schoenherr, H., Snetivy, D. & Vancso, G. J. (1993) *Polym. Bull.* (Berlin) **30**, 567-74.
118. Snetivy, D. & Vancso, G. J. (1994) *Polymer* **35**, 461.
119. Snetivy, D., Guillet, J. E. & Vancso, G. J. (1993) *Polymer* **34**, 429-431.
120. Stocker, W., Schumacher, M., Graff, S., Lang, J., Wittmann, J. C., Lovinger, A. J. & Lotz, B. (1994) *Macromolecules* **27**, 6948-55.
121. Li, Y. & Lindsay, S. M. (1991) *Rev. Sci. Instrum.* **62**, 2630-3.

122. Jandt, K. D., Eng, L. M., Petermann, J. & Fuchs, H. (1992) *Polymer* **33**, 5331-3.
123. Eppell, S. J., Zypman, F. R. & Marchant, R. E. (1993) *Langmuir* **9**, 2281-2288.
124. Kumaki, J., Nishikawa, Y. & Hashimoto, T. (1996) *J. Am. Chem. Soc.* **118**, 3321-3322.
125. Reifer, D., Windeit, R., Kumpf, R. J., Karbach, A. & Fuchs, H. (1995) *Thin Solid Films* **264**, 148-52.
126. Motomatsu, M., Nie, H.-Y., Mizutani, W. & Tokumoto, H. (1994) *Jpn. J. Appl. Phys.*, Part 1 **33**, 3775-8.
127. Schwark, D. W., Vezie, D. L., Reffner, J. R., Thomas, E. L. & Annis, B. K. (1992) *J. Mater. Sci. Lett.* **11**, 352-5.
128. Annis, B. K., Schwark, D. W., Reffner, J. R., Thomas, E. L. & Wunderlich, B. (1992) *Makromol. Chem.* **193**, 2589-604.
129. Collin, B., Chatenay, D., Coulon, G., Ausserre, D. & Gallot, Y. (1992) *Macromolecules* **25**, 1621-2.
130. Liu, Y., Rafailovich, M. H., Sokolov, J., Schwarz, S. A., Zhong, X., Eisenberg, A., Kramer, E. J., Sauer, B. B. & Satija, S. (1994) *Phys. Rev. Lett.* **73**, 440-3.
131. Wang, Y., Juhue, D., Winnik, M. A., Leung, O. M. & Goh, M. C. (1992) *Langmuir* **8**, 760.
132. Goh, M. C., Juhue, D., Leung, O. M., Wang, Y. & Winnik, M. A. (1993) *Langmuir* **9**, 1319-22.
133. Boschung, E., Heuberger, M. & Dietler, G. (1994) *Appl. Phys. Lett.* **64**, 1794-6.
134. Shakesheff, K. M., Chen, X., Davies, M. C., Domb, A., Roberts, C. J., Tendler, S. J. B. & Williams, P. M. (1995) *Langmuir* **11**, 3921-7.
135. Nisman, R., Smith, P. & Vancso, G. J. (1994) *Langmuir* **10**, 1667.
136. Li, J. & Wang, E. (1994) *Synthet. Metal* **66**, 67-74.
137. O'Shea, S. J., Welland, M. E. & Rayment, T. (1993) *Langmuir* **9**, 1826-35.
138. Jung, T. A., Moser, A., Hug, H. J., Brodbeck, D., Hofer, R., Hidber, H. R. & Schwarz, U. D. (1992) *Ultramicroscopy* **42**, 1446-51.
139. Mamin, H. J. & Rugar, D. (1992) *Appl. Phys. Lett.* **61**, 1003-5.
140. Leung, O. M. & Goh, M. C. (1992) *Science* **267**, 64-6.
141. Yang, A. C. M., Kunz, M. S. & Wu, T. W. (1993) *Mater. Res. Soc. Symp. Proc.* **308**, 511-6.
142. Hild, S., Gutmannsbauer, W., Luthi, R., Haefke, H. & Guentherodt, H. J. (1994) *Helv. Phys. Acta* **67**, 759-760.
143. Zhao, W., Ravailovich, M. H., Sokolov, J., Fetters, L. J., Plano, R., Sanyal, M. K., Sinha, S. K. & Sauer, B. B. (1993) *Phys. Rev. Lett.* **70**, 1453-6.

144. Arisawa, S., Fujii, T., Okane, T. & Yamamoto, R. (1992) *Appl. Surf. Sci.* **60-61**, 321.

145. Overney, R. M., Meyer, E., Frommer, J., Brodbeck, D., Luethi, R., Howald, L., Guentherodt, H. J., Fujihira, M., Takano, H. & Gotoh, Y. (1992) *Nature* (London) **359**, 133-5.

146. Overney, R. M., Takano, H., Fujihira, M., Meyer, E. & Guntherodt, H. J. (1994) *Thin Solid Films* **240**, 105-109.

147. Overney, R. M., Bonner, T., Meyer, E., Ruetschi, M., Luethi, R., Howald, L., Frommer, J., Guentherodt, H.-J., Fujihira, M. & Takano, H. (1994) *J. Vac. Sci. Technol.* B **12**, 2227-30.

148. Haugstad, G. (1995) *TRIP* **3**, 353-59.

149. Manne, S., Hansma, P. K., Massie, J., Elings, V. B. & Gewirth, A. A. (1991) *Science* **251**, 183.

150. Karrasch, S., Heins, S., Aebi, U. & Engel, A. (1994) *J. Vac. Sci. Techn.* B **12**, 1474-7.

151. Pollanen, M. S. (1995) *Nanotechnology* **6**, 101-3.

152. Chernoff, A. G. & Chernoff, D. A. (1992) *J. Vac. Sci. Techn.* A **10**, 596-9.

153. Baselt, D. R., Revel, J. P. & Baldeschwieler, J. D. (1993) *Biophys. J.* **65**, 2644-55.

154. Gale, M., Markiewicz, P., Pollanen, M. S. & Goh, M. C. (1995) *Biophys. J.* **68**, 2124-8.

155. Shakesheff, K. M., Davies, M. C., Roberts, C. J., Tendler, S. J. B., Shard, A. G. & Domb, A. (1994) *Langmuir* **10**, 4417-19.

156. Radmacher, M., Fritz, M., Hansma, H. G. & Hansma, P. K. (1994) *Science* **265**, 1577-9.

157. Durbin, S. D. & Carlson, W. E. (1992) *J. Crystal Growth* **122**, 71-9.

158. Lee, G. U., Chrisey, L. A. & Colton, R. J. (1994) *Science* **266**, 771-3.

159. Lee, G. U., Kidwell, D. A. & Colton, R. J. (1994) *Langmuir* **10**, 354.

160. Haugstad, G., Gladfelter, W. L., Weberg, E. B., Weberg, R. T. & Weatherill, T. D. (1994) *Langmuir* **10**, 4295.

161. Overney, R. M., Luethi, R., Haefke, H., Frommer, J., Meyer, E., Guentherodt, H. J., Hild, S. & Fuhrmann, J. (1993) *Appl. Surf. Sci.* **64**, 197-203.

162. Proc. International Conference on Scanning Tunneling Microscopy (STM'95), S. V. C., USA, July 23-28, 1995) *J. Vac. Sci. Technol.* B **14**.

163. Overney, R. M. & Leta, D. P. (1996) *Tribology Letters* **1**, 247-52.

164. Overney, R. M., Takano, H., Fujihira, M., Overney, G., Paulus, W. & Ringsdorf, H. (1995) in *Forces in Scanning Probe Methods*, eds. Guentherodt, H.-J., Anselmetti, D. & Meyer, E. (Kluwer Academic Publishers, London), Vol. 286, pp. 307-12.

165. Meyer, E., Overney, R., Luethi, R., Brodbeck, D., Howald, L., Frommer, J., Guentherodt, H. J., Wolter, O., Fujihira, M. & et. al. (1992) *Thin Solid Films* **220**, 132-7.
166. Nishiyama, K., Kurishara, M. & Fujihira, M. (1989) *Thin Solid Films* **179**, 477.
167. Kajiyama, T., Tanaka, K., Ohki, I., Ge, S.-R., Yoon, J.-S. & Takahara, A. (1994) *Macromolecules* **27**, 7932-34.
168. Meyers, G. F., DeKoven, B. M. & Seitz, J. T. (1992) *Langmuir* **8**, 2330-5.
169. Kajiyama, T., Tanaka, K. & Takahara, A. (1995) *Macromolecules* **28**, 3482-3484.
170. Kajiyama, T., Tanaka, K. & Takahara, A. (1996) *Macromolecules*, in press.
171. Granick, S. (1996) *Mat. Res. Soc. Bulletin* **21**, 33-6.
172. Haugstad, G., Gladfelter, W. L. & Jones, R. R. (1996) *J. Vac. Sci. Technol.* **A 14**, 1864-8.
173. Haugstad, G. (1996) in Material Research Society (submitted to Langmuir 1997).
174. Hues, S. M., Draper, C. F. & Colton, R. J. (1994) *J. Vac. Sci. Technol.* **B 12**, 2211.
175. Corcoran, S. G., Hues, S. M., Colton, R. J., Schaefer, D. M., Draper, C. F., Meyers, G. F., DeKoven, B. M. & Webb, S. C. (1997) to be published .
176. Oulevey, F., Burnham, N. A., Kulik, A. J., Gallo, P. J., Gremaud, G. & Benoit, W. (1996) *J. de Physique IV* **6**, 731-734.
177. Burnham, N. A., Kulik, A. J., Gremaud, G., Gallo, P. J. & Oulevey, F. (1996) *J. Vac. Sci. Technol.* **14**, 794-799.
178. Schild, T. (1996) Ph.D. Thesis, University of Ulm.
179. RosaZeiser, A., Weilandt, E., Hild, S. & Marti, O. (1997) *Meas. Sci. Technol.* **8**, 1333-1338.
180. von Karman, T. & Seewald, F. (1927) *Abh. Aerodynam. Inst. Technol. Hochschule*, Aachen **7**.
181. Yang, A. C.-M. & Kunz, M. S. (1993) *Macromolecules* **26**, 1767.
182. Timoshenko, S. P. & Goodier, J. N. (1969) *Theory of Elasticity* (Mc-Graw Hill).
183. Overney, R. M. (1997) to be published.
184. Cantow, H. J. (1993) *Rev. Roum. Chim.* **38**, 1021-6.
185. Miles, M. J. (1995) *J. Proc. Charact. Solid Polym.* , 17-55.
186. Overney, R. M., Meyer, E., Frommer, J., Guentherodt, H. J., Decher, G., Reibel, J. & Sohling, U. (1993) *Langmuir* **9**, 341-6.

187. Frisbie, C. D., Rozsnyai, L. F., Noy, A., Wrighton, M. S. & Lieber, C. M. (1994) *Science* **265**, 2071-2074.
188. Kiridena, W., Jain, V., Kuo, P. K. & Liu, G. (1997) SIA , *Surf. Int. Anal.* **25**, 383-389.

Chapter 7

Generation of ultrasonic waves in sliding friction

Klaus Dransfeld
Fakultät für Physik, Universität Konstanz
D-78464 Konstanz
Germany

7.1 Abstract

It is well known, that phonons are excited in dry friction between insulating solids, and that these phonons are finally converted into heat. But less is known about the frequency spectrum of the phonons initially excited. In this paper several processes are discussed which do not directly lead to the emission of thermal phonons of energy kT but instead generate ultrasonic waves having frequencies between 10 MHz and 10 GHz. For relatively flat surfaces transverse acoustic waves of about 100 MHz are expected to be generated while for rougher surfaces the predicted frequencies are higher. Furthermore, the higher the load force the lower the expected ultrasonic frequency. Vice versa, if strong ultrasonic waves are applied to the sliding interface near frictionless sliding should occur. In the same sense, the strong build-up of ultrasonic phonons excited by friction, which is expected in small particles, may also be responsible for the low friction force observed for solid lubricants like graphite and MoS_2.

7.2 Introduction

Friction is one of the most common physical processes, of considerable practical importance, but at the same time one of the least understood problems of mechanics. In hydrodynamic friction, described very early by the well known

Stokes Law, a spherical particle which moves slowly in a viscous fluid experiences a drag force which is proportional to the velocity v.

If a flat solid block slides on another flat solid substrate the magnitude of the friction force F_f depends strongly on the lubrication in the presence of a fluid film between both solids. For a thick enough fluid film, thick by comparison with the roughness of the two moving surfaces, the surfaces are effectively kept apart and cannot interact directly, (provided the viscosity of the fluid film is high enough). In this case the friction force is small and of hydrodynamic origin[1]: It hardly depends of the surface roughness but increases with the relative velocity similar to Stokes Law of fluid mechanics.

We are, however, only concerned here with non-lubricated friction between dry or almost dry solids. If no lubricant is used at all, or if the lubricating fluid (of relatively low viscosity) is squeezed out almost completely between the two solids by a strong load force F_n, the two surfaces come into mechanical contact with one another and no sliding occurs for a very small applied friction force F_f. In this case of the so-called "dry friction" (or nearly dry friction) the lateral force has to exceed the force of static friction F_{stat} before sliding sets in. F_{stat} increases with the load force F_n, and the coefficient of static friction $\mu_{stat} = (F_{stat}/F_n)$ has for most dry solids in air values between 1.0 (Cu/Cu) and 0.1 (diamond/diamond)[2]. Under high vacuum conditions the coefficient of static friction can dramatically increase (to $\mu_{stat} > 100$ for Cu/Cu)[3]. Vice versa, in the presence of a lubricating film mentioned above μ_{stat} is usually smaller than 0.1.

Once sliding has started after overcoming F_{stat} a considerably smaller force F_f (the force of kinetic friction) is necessary to keep up the sliding motion. In contrast to the behavior of viscous forces in hydrodynamics the force of kinetic friction F_f between solids does in general not depend on the sliding velocity (at least not for small velocities). But like the force of static friction it increases with the load force F_n which presses both solids against each other.

In their pioneering first friction experiments on an atomic scale, using a tungsten tip sliding across a graphite surface, Mate et al.[4] demonstrated that the origin of the force of kinetic friction during the sliding motion was a stick-slip process on an atomic scale in analogy to the macroscopic stick-slip process between a moving bow and a violin string which is periodically plucked by the bow. This stick-slip mechanism originally suggested by Tomlinson[5] as the cause of dissipation in friction has lately been studied by many groups both experimentally[6] and theoretically[7].

Let us consider the energy conversion in friction: The mechanical work performed during the sliding motion (sliding velocity v under a friction force F_f) is $(F_f \cdot v)$ per second. In certain cases the process of friction leads to plastic deformation and/or to wear of the two participating surfaces. In many other cases, however, no plastic deformation or wear occurs at all during the process of friction. This has first been beautifully demonstrated by Israelachvili and Tabor[8] in their experiments with the bend-mica-technique as well as by Meyer and Frommer[9] with the atomic force microscope. We are concerned here only with this latter case of (dry) friction without plastic deformation and without wear, and here the principal question is, how the mechanical work of friction is converted into other excitations and - of course - finally into heat.

For insulating solids the mechanical work can only be converted into phonons, either thermal phonons of frequency $(kT/h) \approx 10^{13}$ Hz, or ultrasonic waves of lower frequencies (i. e. from MHz to GHz). In semiconductors and metals a conversion of the friction work into electronic excitations at the interface is also possible[10]. In this paper we will discuss only the process of dry friction between insulators, where phonons are the only excitations. The exact mechanism of their excitation and their frequency distribution is not well known. The discussion of this question is the main purpose of this paper.

As will be explained below, we believe that in the process of friction ultrasonic waves in the frequency range from 10^6 - 10^{11} Hz are generated. We will also discuss the existing experimental evidence and, finally, suggest new experiments in order to study the intensity and frequency distribution of the ultrasonic waves to be excited in the process of friction.

7.3 The stick-slip process between flat surfaces with adsorbed soft molecules

Fig. 7.1a shows schematically two single crystals pressed against each other by a load force F_n. The lower crystal is assumed to be absolutely rigid, while the surface atoms of the top slider are held only by relatively weak bonds to the remainder of the very rigid slider. Thus under the influence of the interaction forces between both solids it is only the surface atoms of the top slider which can elastically move up and down or sideways. If the slider is forced to move to the right with an avarage velocity v and if all surface atoms of the slider move coherently, the friction force F_f varies periodically in time dependent on the position of the slider as indicated in Fig. 7.1b. While the weakly bound surface atoms of the top slider are sticking to the lower solid the time dependent elastic deformation of the springs holding the top atoms leads to a friction force growing in time until the limit of static friction F_{stat} is reached and all surface atoms slip back by a slip distance (equal to the atomic distance a in Fig. 7.1). Thereby the previously stored potential energy is converted into vibrational energy of the weakly bound surface atoms (vibrating in the frequency range of 10^{13} Hz) which decays within about 10^{-11} s into heat. The weakly bound surface atoms mentioned above may be physisorbed or chemisorbed molecules. The existence of stick-slip processes on an atomic scale is in good agreement with the first observations in friction force microscopy. More details of this very simplified model[4] and information about further experiments[6] as well as theoretical studies[7] can be found elsewhere.

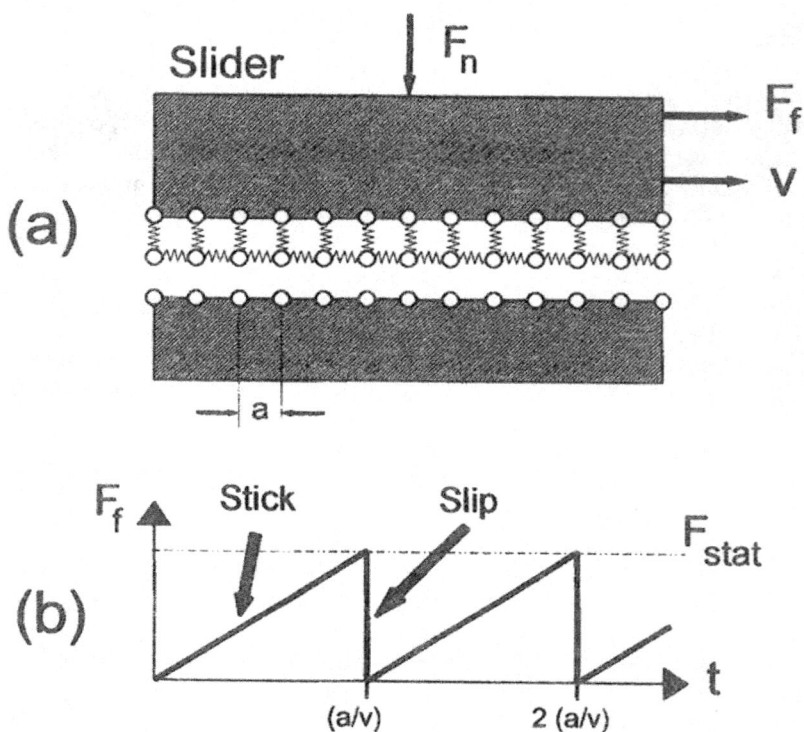

Figure 7.1: (a) Two crystals in sliding motion at velocity v under the action of a load force F_n and a friction force F_f. Both solids are very rigid with the exception of the weakly bound atoms on the surface of the slider. (b) The friction force as a function of time. During the sticking interval the elastic deformation takes place while in the slipping process (for $F_f = F_{stat}$) the previously stored energy is converted into molecular vibrational phonons. The slip length is about one atomic distance a.

The shortest slipping length is the atomic distance a, but in principle the slider can also slip by several units of a before it gets stuck again. If the slipping distance is only one lattice spacing and if the slider moves at a speed v there are $(v/a) = f_a$ stick-slip processes per second. For a speed of $v = 1$ cm/s the frequency f_a is of the order of 30 MHz. (This frequency has been referred to by Sokoloff[11] as the washboard frequency.) At this rate the burst of vibrational energy are released, and therefore at this rate local temperature fluctuations and thermal expansions are expected to occur.

In this very simplified description of the stick-slip process we have assumed the elastic deformation to occur only within one layer of molecules weakly bound to the surface of the top slider. We furthermore considered the bulk material of the two solids to be infinitely rigid, and only for this reason we could neglect the elastic deformation within the sliding block and substrate. Because of this simplifying assumptions only the vibrational energy of the surface molecules (frequency 10^{13} Hz) could be excited in the in the sliding process. In our view, this restriction to vibrations of the surface molecules only is rather artificial. It eliminates the emission of ultrasonic waves which will be discussed below.

7.4 Stick-slip processes between ideally flat surfaces without adsorbed soft molecules

In the absence of soft adsorbed molecules the stick-slip will lead to an elastic deformation of a large part of the solid as described by Persson[12,1] and by Dransfeld et al.[13]. In this case low frequency transverse acoustic waves are emitted both into the sliding block and into the substrate as explained below and shown in Fig. 7.2.

Let us consider a homogeneous large sliding block having a thickness d of several cm. Its lower surface is able to slide across the surface of the bottom solid. Both solids are single crystals of the same material and of the same orientation. The lower substrate is for simplicity assumed to be much more rigid than the top slider. Thus we can restrict ourselves to considering only the elastic deformations of the top sliding block during the process of friction. As before, the two sliding surfaces are pressed together with a load F_n which determines the force of static friction F_{stat}. If now the slider - held at its top part - is forced to move to the right at constant velocity v and if at the same time its lower surface is still sticking to the substrate B a shear deformation is set up in the sliding block near the interface: As shown in Fig. 7.2 (top left) the deformed region is first created at the interface and propagates with the velocity c of acoustic shear waves into the interior of the sliding block until - as the interface bonds break, the slip phase begins and in the absence of external forces the elastically deformed region adjacent to the lower surface of the sliding block is accelerated to the right until there is no relative motion between both surfaces any more and they stick together again. The distance a covered during the slip phase may be one or several interatomic distances. By repeated stick-slip processes at a rate f acoustic shear waves of the frequency f are emitted into the sliding block as shown in Fig. 7.2. For a slipping distance of $a = 3$ Å, a velocity v of 3 cm/sec the stick-slip frequency is $f = (v/a) =$ 100 MHz, which is the frequency of the transverse ultrasonic waves emitted into the slider. The corresponding acoustic wavelength λ (shown in the top of Fig. 7.2) is $(c/f) = 30\mu$m. Rough surfaces and a coparison with experimental observations will be discussed in section 5 and 6, respectively.

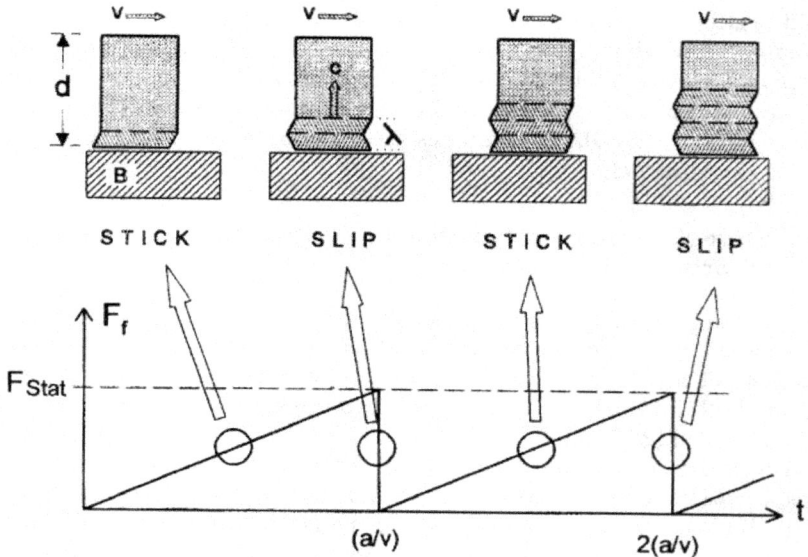

Figure 7.2: The excitation of transverse ultrasonic waves into the bulk of a sliding block S is schematically shown during a stick-slip process. This process is characterized by the sliding velocity v and a slipping distance a (of atomic dimension or a multiple thereof). The interface is assumed to be ideally flat (over several mm). For simplicity the lower substrate B is considered to be hard enough so that its elastic deformation can be neglected. The friction force shown in the lower part of Fig. 7.2 is acting at the interface between substrate and slider. This time dependent friction force leads to a shear deformation which propagates at the velocity of sound c into the interior of the sliding block. When a slip occurs the interface force drops to zero and under the influence of the previously build-up shear stress the sliders surface is accelerated into the reverse direction until it sticks to the substrate again, and so on. λ is the acoustic wavelength of the transverse waves emitted into the interior of the sliding block.

Next we will describe an example of practical importance, the excitation of standing transverse acoustic waves (i.e. of transverse vibrations) in a thin solid film serving as the top slider.

7.5 Excitations of transverse acoustic vibrations in thin films by stick-slip processes

If the top slider is replaced by a thin elastic film of thickness d plated on a rigid holder (see top part in Fig. 7.3), and if this compound slider under the influence of load F_n a friction force F_f is moving to the right with velocity v the stick-slip process of friction leads to the excitation of transverse acoustic vibrations of the film as shown in Fig. 7.3. Since the film is elastically softer than the neighboring solids (shaded in Fig. 7.3) it is able to perform shear vibrations at its fundamental resonance frequency of $f_d = (c/4d)$, with c being the velocity of acoustic shear waves in the film material.

For $d = 1$ mm the resonance frequency f_d is about 1 MHz, and correspondingly higher for thinner films. This mode of vibration excited by a stick-slip process of friction is similar to the excitation of a violin string by the moving bow. The vibrational amplitude A (equal to the slip distance) is $A = (v/f)$. For $d = 1$ mm and $v = 1$ cm/s we find $A = 100$ Å. (The slip distance can become considerably larger than the lattice constant.)

Sokoloff[14] pointed out a very interesting and probably related case when "small particles" (which may also be thin platelets) are located between the sliding surfaces. If the phonons excited within these small particles by stick-slip processes cannot escape fast enough, their population builds up, there can be no further dissipation and consequently the friction force must disappear. We have the surprising case of near frictionless sliding under such circumstances. It is well known that solid flakes of graphite, MoS_2 and similar lamellar structures are excellent solid lubricants[15]. Their presence between sliding surfaces reduces the friction force to very small values. Perhaps by stick-slip processes resonant ultrasonic shear vibrations are excited to high amplitudes in these thin sliding plates because of their weak acoustic coupling to the surrounding. For 1 μm thick MoS_2 plates the excited resonant ultrasonic vibrations would exhibit frequencies around 1 GHz, or multiples thereof. It is their strong excitation, which may be responsible for the lubricating effect of these layer compounds. Inelastic light scattering experiments may give valuable information about the validity af this assumption.

7.6 Excitation of ultrasonic waves by friction between rough surfaces
Theoretical considerations

In real friction experiments we must take into account that the solid surfaces, which are in relative motion, are not ideally flat single crystals but have a certain roughness. Therefore they are touching each other only at a few "spots". The small "contact area" of each of these spots is known[16] to increase linearly

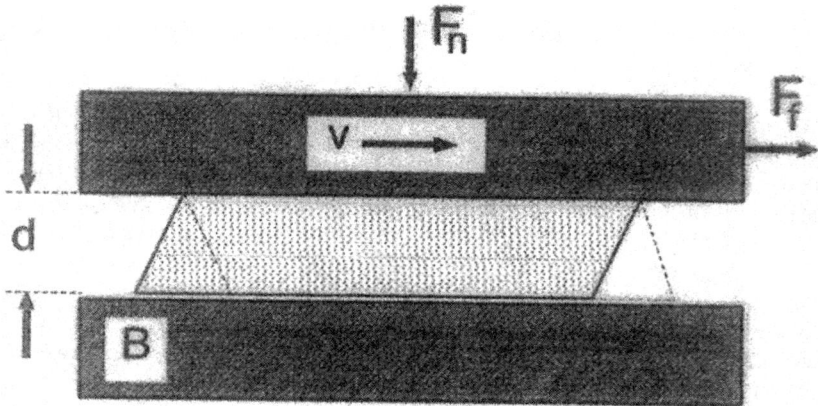

Figure 7.3: The excitation of a resonant shear vibration. The upper rigid block carries a thin film of an material which is less stiff. This film can oscillate in a shear vibration if the thickness is equal to a quarter of the acoustic wavelength. If the slider (pressed with a load force F_n against the bottom solid B is forced to a sliding motion by the friction force F_f shear vibrations will be excited in a stick-slip process of friction.

with the load F_n, but its total area of contact is only a small fraction of the geometrical bottom surface of the slider exposed to the substrate. In Fig. 7.4 only one elevated spot of the substrate is shown to be in direct contact with the slider and the diameter of the contact area is L. As the slider moves to the right with velocity v stick-slip processes are again assumed to take place. Whenever the friction force exceeds the force of static friction F_{stat} a new slip occurs. Consequently the force of friction is expected to be independent of the sliding speed (for small speeds). (The lower substrate may also be replaced by the tip of an atomic friction microscope with a correspondingly smaller contact area)

For a slip distance of atomic dimensions $a = 3\text{Å}$ and a sliding velocity of $v = 3$ cm/sec the stick-slip rate is of the order of 100 MHz. The corresponding acoustic wavelength $\lambda = 30$ μm is much larger than the diameter of the contact area L, which may only be 1000 Å or less.

During the sticking phase at least half of the elastic deformation energy is stored in a small volume of the slider (of diameter L and volume L^3) adjacent to the contact area. Since the diameter of this heavily strained volume is much smaller than the acoustic wavelength λ the deformation can be considered as a quasi static process. The local deformation itself is rather complex: The

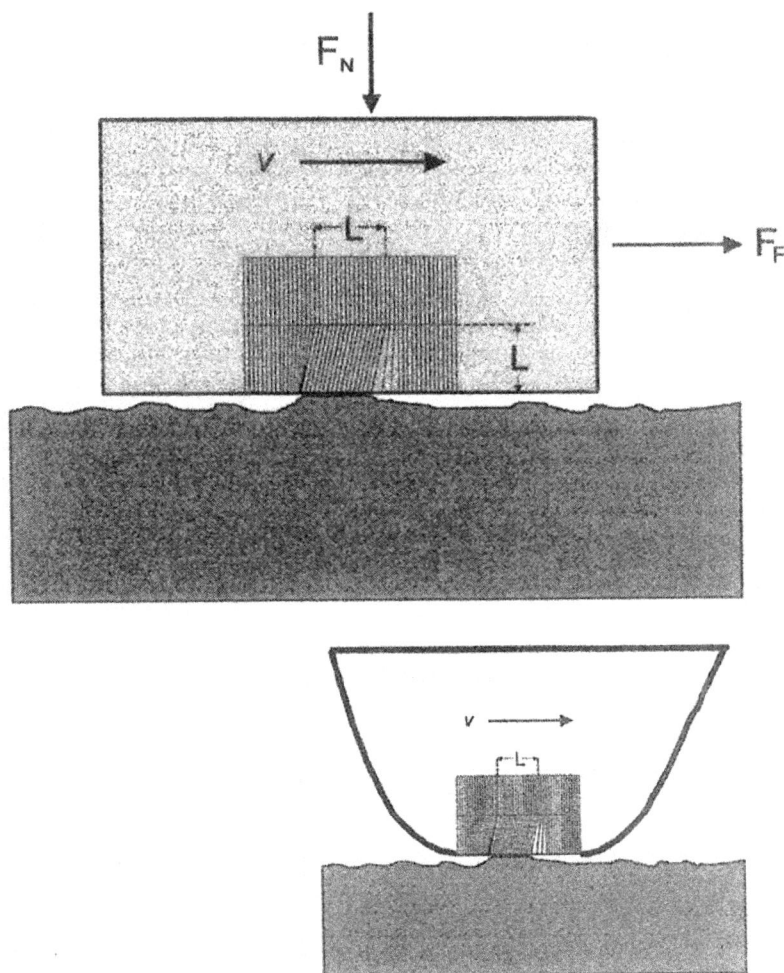

Figure 7.4: The process of dry friction between two rough surfaces is represented in a simplified model by a flat sliding block (top) in contact with a rough but rigid substrate. Mainly in the immediate vicinity of the area of contact the lattice of the slider is being elastically deformed during the sticking phase of the stick-slip process. L is the diameter of the contact area and at the same time the diameter of the elastically deformed volume of the sliding block. v ist the sliding speed, F_n and F_f are the load and friction force respectively. In reality a similiar localized deformation (not shown here for simplicity) is generated in the lower substrate near the contact area. Lower insert: Here the top slider is the tip of a friction force microscope or of a vibrating shear sensor for SNOM applications, instead of a sliding block. The physical arguments are, however, the same in all cases: The elastic deformation occurs mainly in a volume of size L^3 near the contact area. The force of friction arises from the generation of ultrasonic waves.

dominant shear deformation at the center of this volume (see fig 7.4) is accompanied by an elastic compression (left) and dilation (right) at the sides. But it is important to note that during the quasi static deformation nearly half of the elastic energy remains concentrated in the volume L^3 near the contact as already mentioned.

As soon as the slip sets in, the stored elastic strain of the small volume is relaxed and the area previously in contact with the substrate is accelerated into the opposite direction (to the right in Fig. 7.4) until sticking occurs again. During the slipping process the locally stored elastic energy propagates as ultrasonic wave with the speed of sound c into the interior of the slider. The distance of slip (slip length) is one atomic distance or a multiple thereof. The dominant wavelength of the emitted ultrasonic wave is about equal to the diameter L of strained volume, and the corresponding ultrasonic frequency is accordingly $f = (c/L)$, which for $L = 1000$ Å is of the order of 10 GHz.

The contact area (L^2) is known [16] to increase for higher load pressures F_n in proportion to the load. Therefore also the wavelength of the emitted ultrasonic waves ($\lambda \approx L$) increases with rising load ($\lambda \approx \sqrt{F_n}$).

At low sliding speeds when the stick-slip rate is smaller than (c/L), i.e. smaller than 10 GHz for L=1000 Å, the radiated acoustic power P is proportional to the stick-slip rate f and therefore to the sliding speed v, ($P = const \cdot v$). On the other side, for full conversion of the friction work into ultrasonic waves the emitted acoustic power is equal to $P = F_f \cdot v$. Therefore the force of friction is expected to be independent of the sliding velocity v. However, at high sliding speeds v, when the stick-slip frequency $f = (v/a)$ exceeds the geometrical resonance frequency (c/L), the ultrasonic power $(=F_f \cdot v)$ radiated at frequency f will become proportional to f^2 and hence to v^2. Consequently, the force of friction will increase linearly with the sliding speed v in this case.

7.7   Previous experimental studies of acoustic emission

There are many experimental investigations of the acoustic emission (AE) in manufacturing processes such as turning, milling, grinding and metal forming. Usually acoustic frequencies between 50 kHz and 2 MHz are studied with the pupose of controlling and optimizing these technical processes from an engineering point of view (for details see[17]). There is little known about the physical mechanism by which the acoustic waves are generated, whether, for example, by the process of friction alone or mainly by wear processes accompanying friction. In the technical literature it is often assumed that ultrasonic waves are emitted during friction by a rapid release of the strain energy, by plastic deformation or by fracture and the formation of micro-cracks. In most experimental studies on AE acoustic waves were only observed for frequencies up to about 2 MHz.

The piezoelectric transducers for the detection of AE were either mounted on a pin rubbing against a rotating disk or they were located on a block in frictional contact with a rotating ring. The observations, mostly at high sliding speeds of a few meters per second, are rather diverse: Some autors[18,19] report that the intensity of AE depends on the degree of wear, while others[20,21] found that the observed AE-signal was independent of the degree of wear but increased instead proportional to the frictional work. Even in the absence of detectable wear other destructive processes such as plastic deformation and the generation of microcracks were suggested as sources of AE[22]. For steel blocks having rough surfaces the AE signal was observed to grow with increasing load, but - unexpectedly - to decrease with load for smoother surfaces[23,24].

In summary, the previous AE studies at frequencies up to 2 MHz on steel and other technical materials were useful to optimize manufacturing processes but they give only little and at times even conflicting information on the physical nature of the acoustic emission process in the absence of plastic deformation or damage.

Unfortunately there are no experiments known to the author for the study of ultrasonic waves at frequencies above 2 MHz generated in the process of dry friction between well defined crystalline surfaces. In view of these uncertainties new experiments of the following type seem very desirable.

7.8 Proposed experiments for the detection of high frequency ultrasonic waves generated by friction

When using in an friction experiment a mobile slider and a substrate which are optically transparent, it should be possible to detect any ultrasonic waves excited in the process of friction by the Brillouin scattering of light. Fig. 7.5 shows a possible experimental setup. A mobile slider pressed down on a substrate both made of very flat single crystalline quartz is periodically moved in the horizontal direction by a mechanical drive (for example by a loudspeaker drive) imposing a sliding speed v of a few cm/s. The excited ultrasonic waves propagate from the interface into both solids. For flat samples with relatively large areas of contact (diameter of contact area $L >$ wavelength of emitted soundwave) the ultrasonic waves are expected to be plane shear waves of frequency $f = v/a$ (which is of the order of 100 MHz with a being the atomic distance). For rougher surfaces with smaller contact areas (of smaller diameter L, for example, of only 1000 Å) the ultrasonic waves have higher frequencies $f =$ (velocity of sound / L) (about 10 GHz for $L = 1000$ Å) and they propagate more diffusely into several directions.

The Brillouin scattering of light is a very powerful technique to detect phonons in this high MHz to GHz frequency range. In particular, the use of a high contrast Brillouin spectrometer consisting of multiple pass Fabry-

Perot filters (type "Sandercock") allows the detection of microwave phonons with extreme sensitivity, even if their number only slightly exceeds the thermal occupation at room temperature. Once the ultrasonic waves are detected it would be interesting to study also their frequency and intensity of as a function of load force, sliding speed and the state of the sliding surfaces.

The excitation of ultrasonic waves at microwave frequencies by the process of friction should also lead to an enhanced noise in the microwave range if a metallic tip is moved laterally across a conducting surface

7.9 On the possible reduction of friction by ultrasonic waves

If ultrasonic waves are excited by friction it seems also plausible that - vice versa - the force of friction may be influenced by irradiating the sliding interface by strong external ultrasonic waves. The stick-slip process of friction can only be effective in creating the kinetic force of friction if after each slipping phase complete sticking occurs again reliably and fast enough. This fast sticking is no problemm at sufficiently low temperature ($kT <$ binding energy) and in the absence of strong acoustic waves. At high temperatures, however, or under intense acoustic irradiation of the sliding interface the sticking (including also the solidification of thin fluid films between slider and substrate) does no longer occur, and therefore the force of kinetic friction is expected to decrease or nearly disappear. A convenient method to excite strong shear waves at microwave frequencies at the sliding interface is to plate a ferromagnetic film about 1000Å thick onto the substrate. If now ferromagnetic resonance is excited, by applying a dc magnetic field (perpendicular to the film) and an ac magnetic field at right angle to it, the surface of the ferromagnetic film is undergoing intense shear vibrations at the ferromagnetic resonance frequency which can be tuned by the dc magnetic field[25]. In the presence of strong shear vibrations near frictionless sliding is expected. In this way the friction force can be electrically modified for certain applications.

A similar near frictionless sliding may occur if the phonons excited in friction cannot escape rapidly enough from the interface. Thereby a build-up of a non-thermal high phonon population is created, which prevents further sticking processes and thus acts like a lubricant. Perhaps this type of an acoustic lubrication, first discussed by Sokoloff[14] in general terms, is responsible for the relatively low friction force of small particles of layered structures, such as the "solid lubricants" graphite or MoS_2.

7.10 Conclusions

It is well known, that phonons are excited in dry friction between insulating solids, and that these phonons are finally converted into heat. But less is

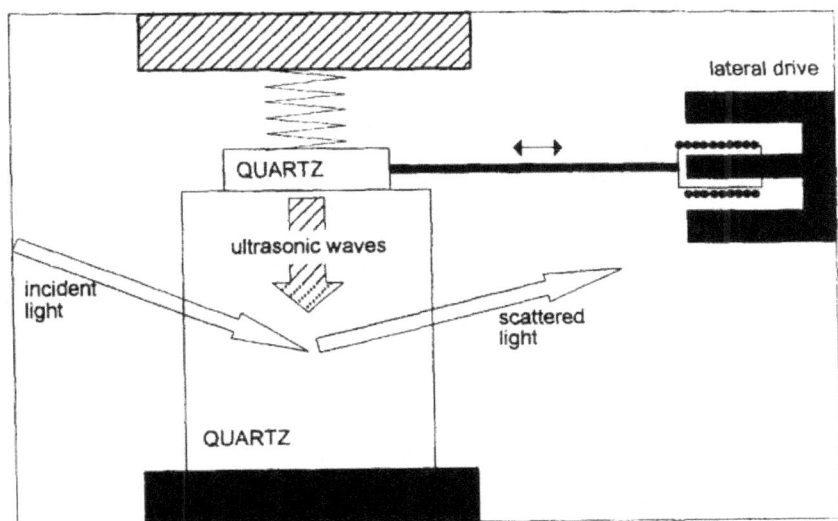

Figure 7.5: Setup for the optical detection of ultrasonic waves generated by friction between a mobile slider and a fixed substrate both made of flat single crystals of quartz.

known about the frequency spectrum of the phonons initially excited. In this paper several processes are discussed which do not directly lead to the emission of thermal phonons of energy kT but instead generate ultrasonic waves having frequencies between 10 MHz and 10 GHz. For relatively flat surfaces transverse acoustic waves of about 100 MHz are expected to be generated while for rougher surfaces the predicted frequencies are higher. Furthermore, the higher the load force the lower the expected dominant ultrasonic frequency. The actual build-up of the microwave phonon population depends on the ultrasonic absorption of the particular substrate and slider material. A strong build-up occurs also in small particles, from which phonons may not escape fast enough. Such a build-up may be responsible for the low friction force observed for solid lubricants like graphite and MoS_2. Vive versa, if strong external ultrasonic waves are applied to the sliding interface also a near frictionless sliding should occur.

7.11 Acknowledgements

I am grateful to W. Arnold at Saarbrücken for his advice on the previous acoustic emission experiments as well as to B. N. J. Persson (Jülich) and to

J. P. Sokoloff (Boston) for their valuable information concerning their own published work on friction.

7.12 References

1. B. N. J. Persson, Sliding friction of lubricated surfaces, *Comments Cond. Mat. Phys.* **17**, 281 - 305 (1995)
2. See for example: D. Buckley, *Surface effects in adhesion, friction, wear and lubrication*, Elsevier Scient. Publ. Comp., (1981)
3. F. P. Bowden and D. Tabor.*Friction and lubrication in solids*, Oxford University Press (1950)
4. C. M. Mate, G. M. McClelland, R. Erlandson and S. Chiang, *Phys. Rev. Lett.* **59**, 1942 (1987)
5. G. A. Tomlinson, A molecular theorie of friction, *Phil. Mag.* **7**, 905 (1929) See also: D. Dowson, *History of tribology*, Longman Publ., (1979)
6. See for example: R. Erlandson, G. Hadziioannou, C. M. Mate. G. M. McClelland and S. Chiang, *J. Chem. Phys.* **89**, 5190 (1988)
   S. R. Cohen, G. Neubauer and G. M. McClelland, *J. Vac. Sci. Technol.* **A8**, 3449 (1990)
   O. Marti, J. Colchero and J. Mlynek, *Nanotechnology* **1**, 141 (1990)
   G. Meyer and N. M. Amer, *Appl. Phys. Lett.* **57**, 2089 (1990)
   G. M. McClelland and S. R. Cohen; Tribology at an atomic scale, in Springer Series on Surface Sciences, Vol. 22, R. Vanselo et al., eds., Springer (1990)
   G. M. McClelland and J.N. Glosli, Friction at the atomic scale, in FUNDAMENTALS OF FRICTION: MACROSCOPIC AND MICROSCOPIC PROCESSES, I. L. Singer and H. M. Pollack, eds., p. 405 (1992)
   E. Meyer, R. Overney, L. Howald, D. Brodbeck, R. Lüthi and H. J. Gütherodt, in FUNDAMENTALS OF FRICTION, I. L. Singer et al. eds., Kluwer Academic Publisher, p. 427 (1992)
   E. Manias, G. Hadziioannou, I. Bitsanis and G. TenBrinke, *Europhys. Lett.* **24**, 99 (1993)
   H. Yoshizawa and J. Israelachvili, *J. Phys. Chem.* **97**, 11300 (1993)
7. See for example: W. Zhong and D. Tomànek, *Phys. Rev. Lett.* **64**, 3054 (1990)
   D. Tomanek, in SCANNING TUNNELING MICROSCOPY III, R. Wiesendanger and H. J. Güntherodt, Eds., Springer (1993), p. 269 - 292
   T. Gyalog, M. Bammerlin, R. Lüthi, E. Meyer and H. Thomas, Mechanism of atomic friction, *Europhysics Lett.* **31**, 269 (1995)

B. N. J. Persson, Theory of friction: Dynamical phase transitions in adsorbed layers, *J. Chem. Phys.* **103**, 3849 (1995)

B. N. J. Persson, Theory of friction: on the origin of stick-slip motion of lubricated surfaces, to be published (1996)

B. N. J. Persson, *Physics of sliding friction*, Springer Verlag (1997)

8. J. N. Israelachvili and D. Tabor, The shear properties of molecular film, *Wear*, **24**, 386 (1973)

9. E. Meyer and J. Frommer, Forcing surface issues, *Physics World* (4) **4**, 46 (1991)

10. B. N. J. Persson and A. I. Volokitin, Electronic friction of physisorbed molecules, *J. Chem. Phys.* **103**, 8679 (1995)

11. J. B. Sokoloff, *Surf. Science*, **144**, 267 (1984). See also: *Phys. Rev.*, **B 42**, 760 (1990) and: *Phys. Rev. Lett.*, **66**, 965 (1991)

12. B. N. J. Persson , Theory of friction: The role of elasticity in boundary lubrication, *Phys Rev.* **B 50**, 4771 (1994), see also ref. [5]

13. K. Dransfeld and Jie-Li, in Forces in Scanning Probe Methods, H. J. Güntherodt, D. Anselmetti and E. Meyer (eds), NATO ASI Series E: Applied Sciences, Vol. 286, Kluwer Academic Publisher (1995), p. 273 - 283

14. J. B. Sokoloff, Microscopic mechanism for kinetic friction: Possible near frictionless sliding for small particles, *Phys. Rev.* **B 52**, 7205 (1995)

15. For further details see for example: I. L.Singer: Solid lubrication processes, in Fundamentals of Friction: Macroscopic and Microscopic Processes, I. L. Singer and H. M. Pollock, eds., NATO ASI Series, Vol. 220 (1992), p. 239

16. J. F. Archard , Elastic deformation and the laws of friction, *Proc. Roy. Soc., London* **A 243**, 190 (1959)

17. C. L. Jiaa and D. A. Dornfeld, Experimental studies of sliding friction and wear via acoustic emission signal analysis, *Wear*, **139**,403 (1990)

18. T. Masaki, Use of atomic emission for the study of wear, Thesis, MIT (1986)

19. Yu. A. Fadin, A. M. Leksovskii, B. M. Ginzburg and V. P. Bulatov, Periodicity of acoustic emission with dry friction between steel and brass, *Tech. Phys. Lett.* **19**, 136 (1993)

20. S. Linggard and K. N. NG, An inestigation of acoustic emission in sliding friction and wear of metals, *Wear* **130**, 367 (1989)

21. S. H. Carpenter, C. R. Heiple, D. L. Armentrout, F. M. Kustas and J. S. Schwartzberg, Acoustic emission produced by sliding friction and its relationship to AE from machining, *J. Acoustic Emission*, **10**, 83 (1992)

22. A. Quinten, C. Sklarczyk and W. Arnold, Observation of stable crack

growth in Al_2O_3-ceramics by acoustic microscopy and acoustic emission, *Proc. Int. Symp. Acoustical Imaging*, **18**, 221, Plenum Press, NY, Ed. G. Wade (1990)

23. D. Dornfeld and C. Handy, Slip detection using acoustic emissin signal analysis, Proc. IEEE Int. Conf. Robotics and Automation, Raleigh, NC, p. 1868 (1987)

24. M. K. Jouaneh, R. Lemaster and F. C. Beall, Study of acoustic emission in sliding motion, *J. Acoustic Emission*, **10**, 83 (1992)

25. K. Dransfeld, Kilomegacycle Ultrasonics, *Scientific American*, **208**, June (1963)

Chapter 8

Friction force microscopy experiments

8.1 Material-specific contrast of friction force microscopy

The ability of FFM to be material specific is of general interest for the whole field of scanning probe microscopy (SPM), where contrast mechanisms are searched that give information in addition to topography. However, lateral forces are not independent of topography, but are influenced by local gradients (for more details see chapter IV). In order to exclude these topography effects, the sample should be flat. A surface that consists of atomically smooth terraces is ideal. The topography effect is restricted to the step regions, whereas a pure friction contrast originates from the flat terraces. On rough surfaces the non-dissipative part of the lateral forces due to the topography effect I (see chapter IV) can be separated by subtracting back and forward scan. However, the topography effect II due to changes of long-range forces and contact area is more difficult to be separated.

8.1.1 Langmuir-Blodgett films

Langmuir-Blodgett (LB) films are found to be ideal samples for FFM. Any number of layers can be transferred. Using smooth substrates, such as silicon wafers or mica, films of high quality can be prepared. An example is given in Fig. 8.2 where two bilayers of Cd-arachidate were transferred onto a silicon wafer[2]. From macroscopic experiments it is known, that Cd-arachidate is a good model system for boundary lubrication[3]. Novotny et al.[4] performed pin-on-disc measurements on coated Si-wafers and found that wear rates could be lowered by a factor of 10^5. On the microscopic scale, it is found by force microscopy that the films are well ordered, forming 2-d crystals of close-packed molecules with a periodicity of about 5Å
citemeyer-nature. Molecular lattice imaging was achieved on multilayers but not on monolayers, indicating that the monolayers might be less ordered ("liquid state" of bond orientations)[6]. The corrugation heights was found to be dependent on the chain length: Cd-stearate (C18: 0.1-0.15nm); Cd-arachidate (C20: 0.18-0.22); Cd-behenate (C22: 0.35-0.38);Cd-lignocerate (C24: 0.4-

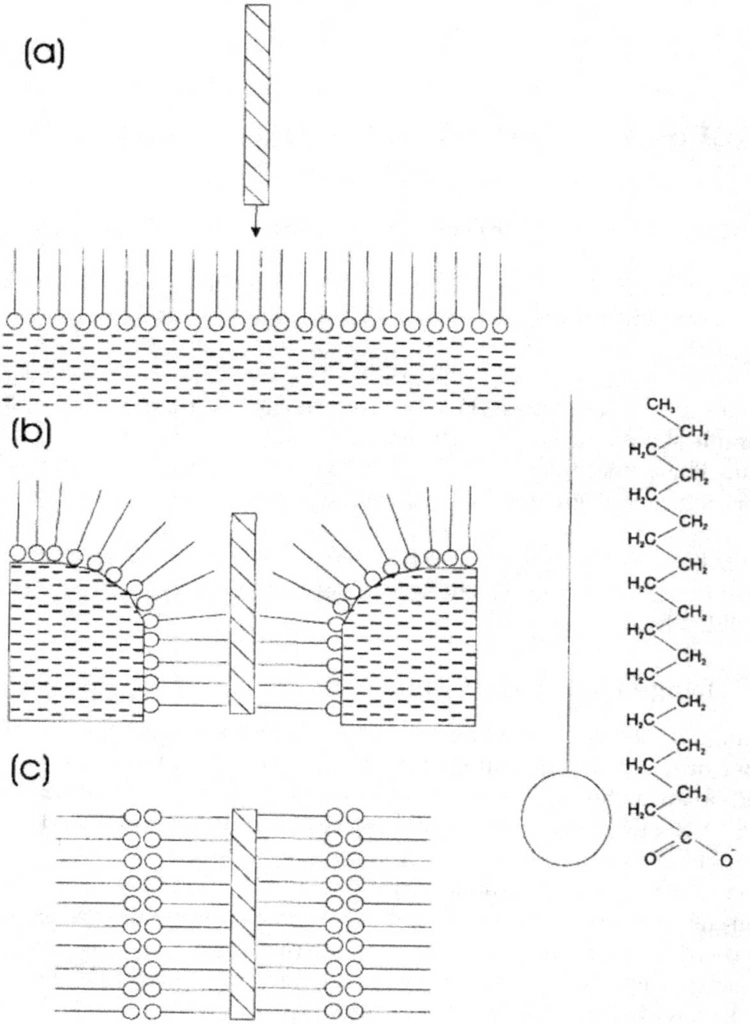

Figure 8.1: The Langmuir-Blodgett technique: (a) The hydrophobic sample is immersed into the liquid. (b) Transfer of the film under constant pressure (c) A bilayer is formed, when the sample is withdrawn from the liquid. Right side: Arachidic acid as an example for an amphiphilic molecule.

0.45nm), which was related to the influence of London dispersion forces, which is larger for longer chains. A systematic study of Shaper et al. of LB-films of saturated fatty acid multilayers with different lengths, has shown that step heights measured with AFM are different from x-ray data, which is related to elastic deformation of the sample by the probing tip [a]. Larger elastic deformations are observed for shorter chain lengths due to reduced intermolecular interactions of the shorter aliphatic chains. In summary, the AFM-measurements have shown that at least bilayers are well ordered and elastic deformations lead to reduced step heights (especially on films with short chain lengths).

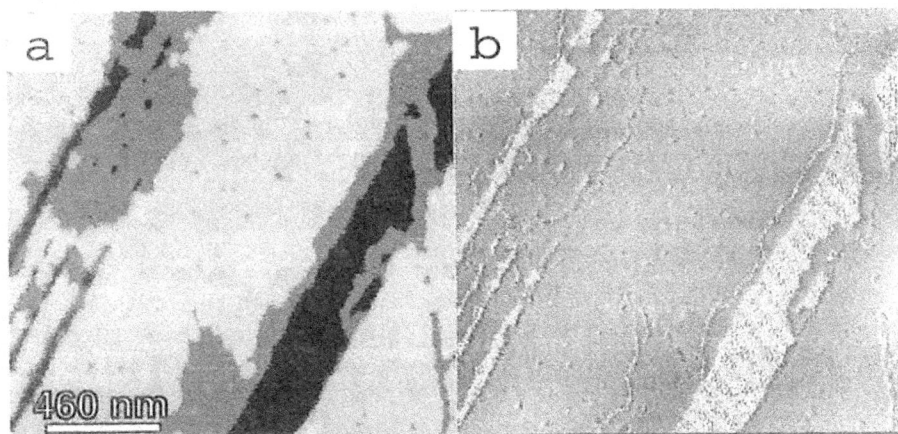

Figure 8.2: (a) Topography image of a Cd-arachidate film. The dark level corresponds to the silicate substrate. The grey level corresponds to the first bilayer (54Å high) and the bright level is related to the second bilayer. (b) Lateral force image.

The first FFM-measurements on LB-films were performed on Cd-arachidate films[1,2]. A reduction of friction on the film covered areas of a double bilayer of Cd-arachidate compared to the substrate was observed. Friction is found to be independent of the film thickness; one 5.4nm high bilayer appears to be sufficient to lubricate the surface. However, on single-bilayer films of Cd-arachidate local variations of friction were observed, which were related to differences in coupling to the substrate. The loading dependence of friction on the terrace was found to be very weak, whereas at the step edges an increase of lateral force was observed. Above 10nN loading, the initial stages of wear were observed. Small islands could be moved in their entirety, which allowed to determine the shear strength between the bilayer films. A value of

---

[a] AFM yields step heights as function of the number of methylene groups x: z [nm] $= -1.63$ nm $+ 0.2 \times$ nm , whereas x-ray yields z [nm] $= 0.39$ nm $+ 0.12 \times$ nm.

$\tau{=}1$MPa was found, which was found to be in agreement with surface force apparatus measurements at low loads.

More complex systems of phase separated LB films of mixtures of hydrocarbons and fluorocarbons on silicon were investigated by Overney et al.[7,8]. The relative friction of the hydrocarbon, fluorocarbon, and silicon surfaces is found to be 1:4:10. Surprisingly, the fluorocarbons are less effective in reducing friction than the hydrocarbons which was also observed by surface force apparatus (SFA) measurements[9]. The particular advantage of the fluorocarbons, which makes them so valuable in technological applications, is their resistance to rupture, as observed by both SFA[9] and force microscopy.

8.1.2   Anorganic thin films

Some examples of UHV-FFM on anorganic thin films were discussed in chapter II. Other examples of material-specific imaging of FFM are exfoliated MoS_2-platelets on mica[11], metallic islands on semiconductors and insulators[12].

8.1.3   Carbon surfaces

Mate[13] has studied various forms of carbon: C_{60}, amorphous carbon, hydrogen terminated diamond and graphite, where highest friction coefficient is found for C_{60} (0.8) and lowest for graphite (0.01). Intermediate values were found for amorphous carbon (0.33) and hydrogen terminated diamond (0.05-0.3). The value for diamond was found to depend on loading. At low loads ($<$400nN) a friction coefficient of 0.3 was observed. At higher loads, the friction coefficient drops into the range of 0.05-0.15. Similar values are also observed on the macroscopic scale. A study of amorphous carbon films by Perry et al.[14] has shown the strong influence of the hydrogen contents on the tribological properties. Namely, both adhesion and friction coefficients were found to increase with increasing hydrogen contents. (e.g., $\mu \approx 0.3$ for about 10at% and $\mu \approx 0.8$ for 35at% hydrogen contents). An oxidized tungsten tip was used as probing tip. These studies are of special relevance of disk drives, which are usually coated with carbon protective coatings and covered with perfluorinated lubricants. (See Mate and Homola[15]).

8.1.4   Silicon and silicon oxides

Silicon and its oxides play a major role in semiconductor industry. Apart from its electronic properties, the mechanical properties become of interest because of the development of micromachinery, such as micron-sized or sub-millimeter-sized motors, moveable mirrors for computer displays, pumps for medical applications or microfabricated scanning probe microscopes. In order

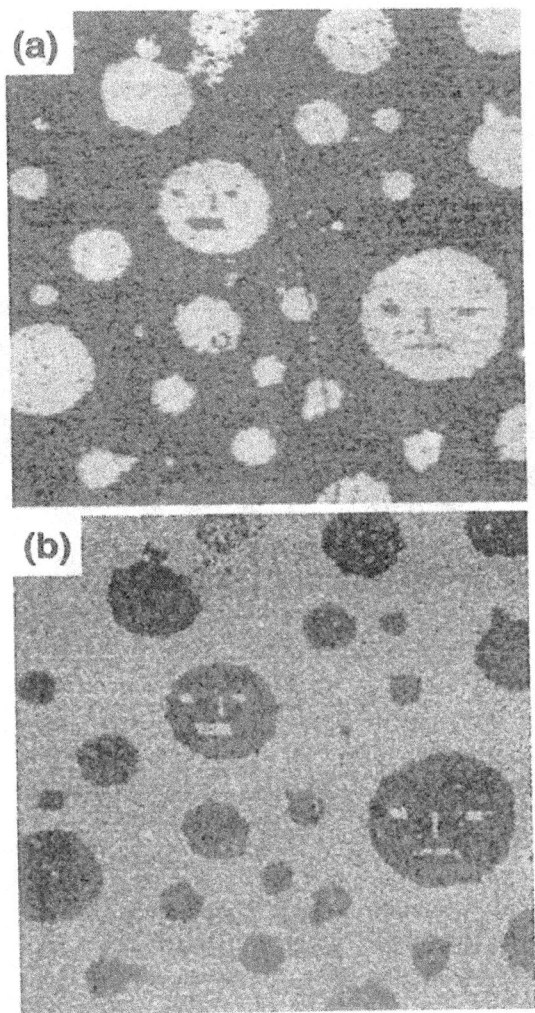

Figure 8.3: (a) Topography image of mixed Langmuir-Blodgett film. Areas of hydrocarbons (hydrocarbon carboxylate $C_{21}H_{43}COO^-$) are surrounded by a "sea" of fluorocarbons (flurocarbon carboxylate $C_9F_{19}C_2H_4OCC_2H_4COO^-$). (b) Lateral force image. Within the hydrocarbon islands, areas have been removed by the action of the probing tip (increased loading, slow speed). The fluorocarbon areas, being more resistant to rupture, were not destroyed at the same operation conditions. From[10].

to minimize the power consumption of these devices, it is necessary to minimize friction. A model system has been studied, where standard photo mask lithography is used to structurize the surface of a Si(110) surface[16]. 2 μm-wide stripes of 150nm thick silicon oxide are adjacent to 4μm wide stripes of hydrogen passivated silicon (cf. Fig. 8.4) The surfaces were etched in 1:100 molar HF/H$_2$O solution for 1-3min. After rinsing in water for some minutes, the samples were immediately transferred into the FFM-chamber and immersed into dried nitrogen. Incomplete or inhomogeneous etch was revealed in the friction force map. Thus, the etch conditions could be optimized by characterization of the surfaces by FFM.

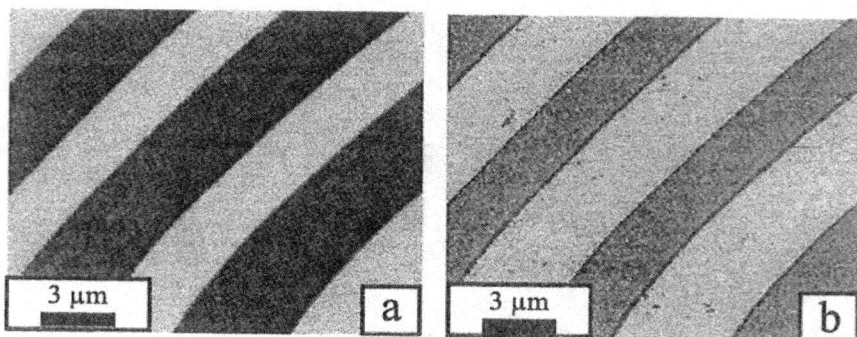

Figure 8.4: (a) Topography image of Si/SiO$_2$ grid etched in 1:100 HF/H$_2$O. The higher part (bright) corresponds to the SiO$_2$-covered regions which are 15nm above the lower hydrogen passivated silicon regions (dark). (b) Corresponding friction force map: Higher friction is observed on the Si:H whereas the friction on SiO$_2$ is lowered by a factor of two.

In addition, XPS-studies give further evidence about the rather high degree of passivation of the surfaces (10% fluorine bonds, 5% hydroxide bonds, 85% hydrogen bonds)[17]. Therefore, the patterns in Fig. 8.4 were related to hydrogen passivated silicon (lower level in topography) and silicon oxide (higher level in topography). The friction force map immediately shows that the friction is higher on the passivated silicon compared with the silicon oxide areas. The application of 2d-histogram technique gives the quantitative loading dependence of friction as shown in Fig. 8.5 . The details of the calibration procedure are given in the appendix. Apparently, the correlation between friction and normal load is very high. Two regions can be distinguished corresponding to the higher friction on Si:H and the lower friction on silicon oxide. The functional dependence is rather linear: $F_F^i = \mu_i \cdot F_n$ corresponding to a friction coefficient of $\mu_{Si:H} = 0.6$ and $\mu_{SiO_2} = 0.3$. In the context of the extended adhesion model, this behaviour can be interpreted as the case of big $\alpha \approx \mu$ where τ_0 is negligable. Alternatively, nm-roughness of the probing tip might lead to multi-

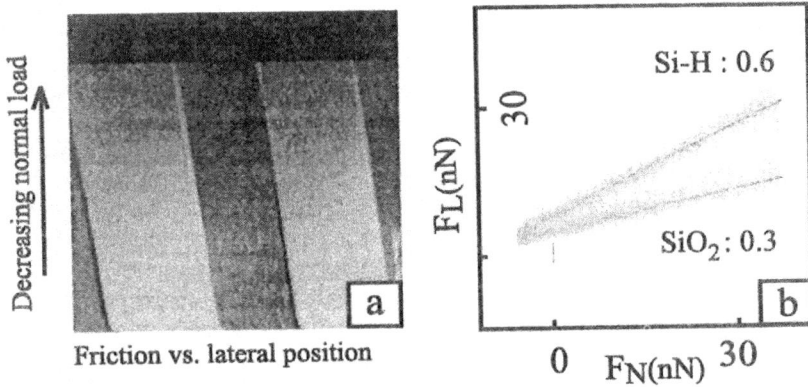

Figure 8.5: (a) Friction force map measured with decreasing load. (b) Corresponding 2d-histogram. The (F_L, F_N)-pairs pile up in distinct regions. The functional dependence is given by simple linear dependences: $F_F^i = \mu_i \cdot F_n$ corresponding to friction coefficients of $\mu_{Si:H} = 0.6$ and $\mu_{SiO_2} = 0.3$.

asperity contacts, which give a linear loading dependence in close analogy to macroscopic experiments.

Teuschler et al.[18] have investigated hydrogen-passivated Si(100) surfaces, which were patterned with SFM-techniques (Writing process: Voltage applied to the conductive probing tip). Then, these modified areas were characterized with friction force microscopy. An increase of friction was observed on the structurized areas, which were presumably related to silicon oxide formed by field enhanced oxidation. The discrepancy between the data from Scandella et al. (large friction on passivated silicon and smaller friction on silicon oxide) and the data from Teuschler et al. (small friction on passivated silicon and larger friction on oxidized silicon) may be related to the different structure of the silicon oxides, which were prepared in different ways. Thus, the friction contrast does not only depend on the elemental composition, but also depends on other factors, such as cristallinity.

8.1.5 III-V Semiconductors

R. Garcia and coworkers could demonstrate that FFM is able to map InP/InGaAs heterostructures with 3nm spatial resolution[19]. The samples have been grown by molecular beam epitaxy and then cleaved along the (110) face for examination. Sharpened Si_3N_4 (nominal tip radius of 10nm) were used. The spatial resolution to detect different semiconductors was tested with a sample made of a stack of layers of 2,3,4,5 and 10nm thicknesses of InP alternated with $In_{0.53}Ga_{0.47}As$. The composition of the InGaAs was chosen to be lattice matched to InP (0.587nm). Alternating regions of higher and lower friction could be detected with a resolution of about 3nm, which was found to be consistent with the estimated contact diameter.

The chemical sensitivity of FFM was tested on a step graded $In_x Ga_{1-x} As$ sample grown on GaAs(001). The indium composition was changed from 0 to 60% in 10% steps. As shown in Fig. 8.6 , 10% changes in indium composition can be clearly distinguished with the frictional forces. Using higher loads, the chemical sensitivity could be improved. However, the lateral resolution is better at lower loads. Thus, a compromise between lateral resolution and chemical sensitivity has to be made. These measurements could be performed in ambient conditions, which is ideal for technological applications.

The authors have carefully checked the mechanisms of FFM[20]. First, they measured an energy which is dissipated per atom of about 0.14eV, which is smaller than cohesive energies of III-V semiconductors. They did not observe changes of the surface due to repeated scanning. Thus, they conclude that wearless friction is observed. Comparison with local compliances have shown that there is no direct correlation between elasticity and friction. E.g., they find higher friction on InP compared to InSb but the elastic deformation was smaller. However, large elastic deformation is observed on InSb compared to InGaAs, where friction is found to be large as well. The authors report that measurements of adhesion hysteresis correlate well with frictional forces as predicted by Yoshizawa et al.[69].

Recently, Tamayo et al. have applied FFM to quantum dot structures (InSb on InP and InAs on InP) and found submonolayer sensitivity[21]. On InAs dots, no frictional contrast relative to InP(001) could be observed, which was related to the formation of a wetting layer of InAs on InP. On InSb dots, frictional contrast contrast could be observed. Comparison with photoluminescence data, led the authors to the conclusion, that FFM is sensitive to the presence of a submonolayer (0.7ML). Within the flat parts, no variations of friction were observed, which indicates the formation of an $In_x Sb_{1-x} P$ alloy.

8.2   Anisotropy of friction

The question of anisotropy of friction is of fundamental interest. From the study of electrical resistance of sliding charge density waves, it is established that friction of extended, periodic surfaces depends on the relative orientation[22]. At angles with high degree of commensurability, local maxima of friction are observed. On the other hand, one expects zero friction for the incommensurate case. There are theoretical investigations that relate commensurability with friction[23] (See also chapter IV).

In the field of nanotribology, Hirano et al.[24] showed that friction between

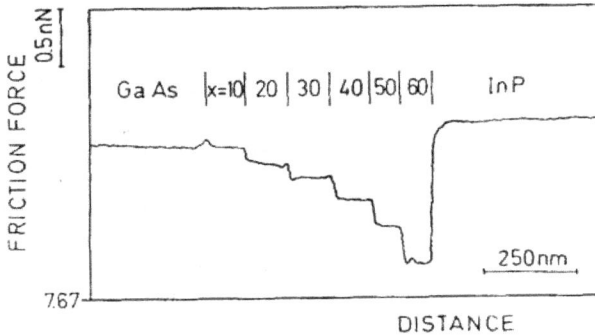

Figure 8.6: Frictional force cross-section across a step graded $In_xGa_{1-x}As$ sample. The indium composition (x) has been changed in 10% steps from GaS to $In_{0.6}Ga_{0.4}As$. The structure is terminated with an InP capping layer. The cross-section is an average of 300 FFM scan lines. From[20].

two mica sheets in contact depends on the crystallographic orientation by a factor of four. Based on theoretical arguments and on their experiments, Hirano et al. predicted the existence of "superlubricity", where sliding with zero friction might exist. Recently, they observed "superlubricity" with a monocrystalline tungsten tip ((011)-facet in close proximity to the sample) on Si(001) in UHV. By rotating the tungsten tip relative the Si-sample, they could find a minimum force of $3 \cdot 10^{-9}$N, which corresponds to the incommensurate case. For the commensurate case an increased frictional force of $8 \cdot 10^{-8}$N was found.

Morita et al.[26] investigated the 2-dimensional stick-slip as a function of orientation of the sample relative to the tip. They could show that the probing tip moves preferentially in certain crystallographic directions. However, this orientation dependence is simply reflecting the symmetry of the sample. The 2d-Tomlinson model is found to be sufficient to explain these atomic stick-slip data.

An interesting case of anisotropy in friction has been studied by Overney et al.[27]. A silicon nitride tip was scanned across a domain boundary of a lipid bilayer film. Due to the different orientation of the molecules, a change of friction by a factor of 1.4 was observed. Remarkably, the contact diameter was of molecular dimension, which was determined from the width of the transition region between the domains. Thus, these measurements represent the first anisotropy measurements of friction of a molecular contact on an anisotropic surface. Overney et al. could relate the change of friction between the domains to the dependence of atomic-scale stick-slip on the orientation of the molecules. The orientation dependence of atomic-scale stick-slip has been further addressed by Takano and Fujihira on Langmuir-Blodgett films with complex unit cells[28]. The inorganic surface of triglycine sulfate (TGS), a ferroelectric crystal, has been studied by Bluhm et al.[29]. They observed an orientation dependence of friction between terraces that are separated by half unit cell steps. By rotating the sample physically (not just the scan direction) from 0 ° to 180 ° a contrast reversal was observed. The structural model of the surface shows that the glycinium molecules (NH_3CH_2COOH) are tilted relative to the surface normal and change their orientation by 180° between terraces, which are separated by half unit cells. This sawtooth-like orientation of the molecules, which changes orientation between the terraces, makes plausible that the friction of a FFM tip changes, when scanning parallel or antiparallel to the molecules. In analogy to the sliding over a cat fur, one expects a dependence of friction on the direction. Unfortunately, this interesting, experimental observation has not been addressed from the theoretical side. A recent study of Gourdon et al.[30] has shown a dependence of friction as a function of direction on thiolipid Langmuir-Blodgett-films. A star-shaped island was observed, where domains with different molecule orientation found. From their friction study, Gourdon et al. conclude that the molecules have radial tilt, which is directed towards the center of the "star".

An important technical aspect of anisotropy measurements is the anisotropy of the cantilever itself. The spring constant is varying as a function of angle (see Kerssemakers et al.[31]). Therefore, it is essential to perform the experiments only in one direction relative to the cantilever axis (preferentially perpendicular to the long axis of the cantilever) and to rotate the sample. If the scanning direction is along the long axis of the cantilever, buckling effects have to be taken into account[32].

In the context of nanosled experiments, only few experiments have been performed to test the orientation dependence of the nanosled motion (Frictional force to move adsorbate islands as a function of angle). Lüthi et al. observed, in first approximation, no dependence of friction as a function of orientation for the case of C_{60}-islands on NaCl(001)[33]. Even rotations of the C_{60}-islands on NaCl(001) were observed. In contrast, Sheehan and Lieber[34] observed that MoO_3 islands on MoS_2 slide only along low index MoS_2 directions ($\langle$ 1000 $\rangle$-directions). The islands could not be slided in other directions. The difference between the experiments of Lüthi et al. and Sheehan et al. might be explained by the different mismatches. In the case of large mismatch (C_{60} on NaCl), a weak dependence of orientation is expected, whereas in the case of small mismatch (MoO_3 on MoS_2) a stronger orientation dependence is plausible.

8.3 Role of environment

The environment plays an essential role in all tribological situations. Increased humidity can cause the formation of water layers on hydrophilic surfaces, which can affect the tribological properties of these surfaces. The formation of liquid necks around asperities changes the interaction drastically, e.g., capillary forces become important. The smoothness of probing tips by the coverage with water may also cause a different loading dependence (single vs. multi-asperity contact).

The immersion into liquids removes capillary forces, but solvent exclusion forces may become important. Different liquid environments can give different contrasts.

8.3.1 Humidity dependence: Mica

Putman et al. investigated the loading dependence of a FFM-tip on mica[35]. At low humidities, a rather linear behaviour of friction vs. normal force was observed. At higher humidities, a non-linear dependence was observed with the same probing tip. Putman et al. interpreted their results in terms of the nanometer-scale-roughness of the probing tip which is smoothened by a condensed water film, leading to a single asperity contact at high humidities in contrast to a multi-asperity contact at low humidities.

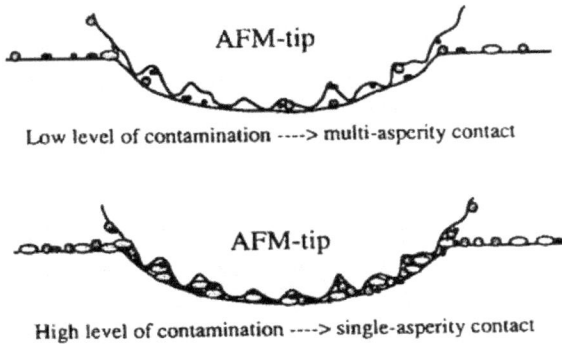

Low level of contamination ----> multi-asperity contact

High level of contamination ----> single-asperity contact

Figure 8.7: Schematical diagram of the influence of humidity in FFM. At higher humidity, a liquid film condenses on the tip, leading to a single asperity contact. From[35].

8.3.2 Humidity dependence: MoS$_2$-platelets on Mica and Al$_2$O$_3$

MoS$_2$ belongs to the solid film lubricants, being frequently used in space applications, surface science instrumentation and ceramic machinery. It has been recognized that shear takes place between or on the basal planes[36].

Here, we discuss experiments on platelets of MoS$_2$, prepared by the exfoliation technique and deposited on the atomically flat surfaces of mica and sapphire[37]. Experiments were performed in controlled humidities from 0% up to 90%. Si-cantilevers[38] and Si$_3$N$_4$-cantilevers[39] were used. Both probing tips were covered by silicon oxide, leading to consistent results for both types of tips. In the case of the MoS$_2$/mica sample, a dramatic dependence of friction as a function of humidity is found: In dried nitrogen higher friction is found on MoS$_2$ compared to mica (cf. Fig. 8.8). Between 40% to 60% a contrast reversal is observed, where friction on mica is higher than on MoS$_2$. Finally, the contrast is reversed again at high humidities. This qualitative behaviour is confirmed with a series of 2d-histograms at different humidities. As shown in Fig. 8.9 the dependence of friction vs. normal load of MoS$_2$ is well separated from mica at low humidities. Around 20% to 50% the slope of the $F_L^{mica}(F_N)$-curve increases, crossing the $F_L^{MoS_2}(F_N)$-curve. It also becomes evident that the curves show strong deviations from the linear dependence but rather agree with a $F_N^{2/3}$-dependence. This observation is in agreement with the observations of Putman et al. on mica, where a non-linear dependence was observed at higher humidities [35]. Thus, we might interpret these changes of the loading dependence as the transformation of a multi-asperity contact to a single-asperity contact due to the formation of a water film on the tip. Alternatively, it has been proposed that the extended adhesion model explains this observation by a decrease of the parameter α with increasing humidity[41]. This decrease of α becomes justified by the presence of water molecules on the surfaces leading to different interfacial shear strengths. The strong dependence of friction on mica is probably related to the hydrophilic nature of mica, corresponding to stronger interactions with the water molecules. At very high humidities, the $F_L^{mica}(F_N)$-curve and the $F_L^{MoS_2}(F_N)$-curve are nearly indistinguishable, but rather are related to a water on water sliding $F_L^{H_2O}(F_N)$ which is only slightly modified by the substrate. In the case of the MoS$_2$/Al$_2$O$_3$-system, the situation is rather different. Friction on both materials is only weakly dependent on humidity, corresponding to their hydrophobic nature. At all humidities, Al$_2$O$_3$ is found to exhibit lower friction than MoS$_2$. Again, we observe rather strong deviations from a linear relationship, corresponding to the "small α"-case. The results are illustrated in Fig. 8.10. Both systems, MoS$_2$/mica and MoS$_2$/Al$_2$O$_3$ are summarized in Fig. 8.11.

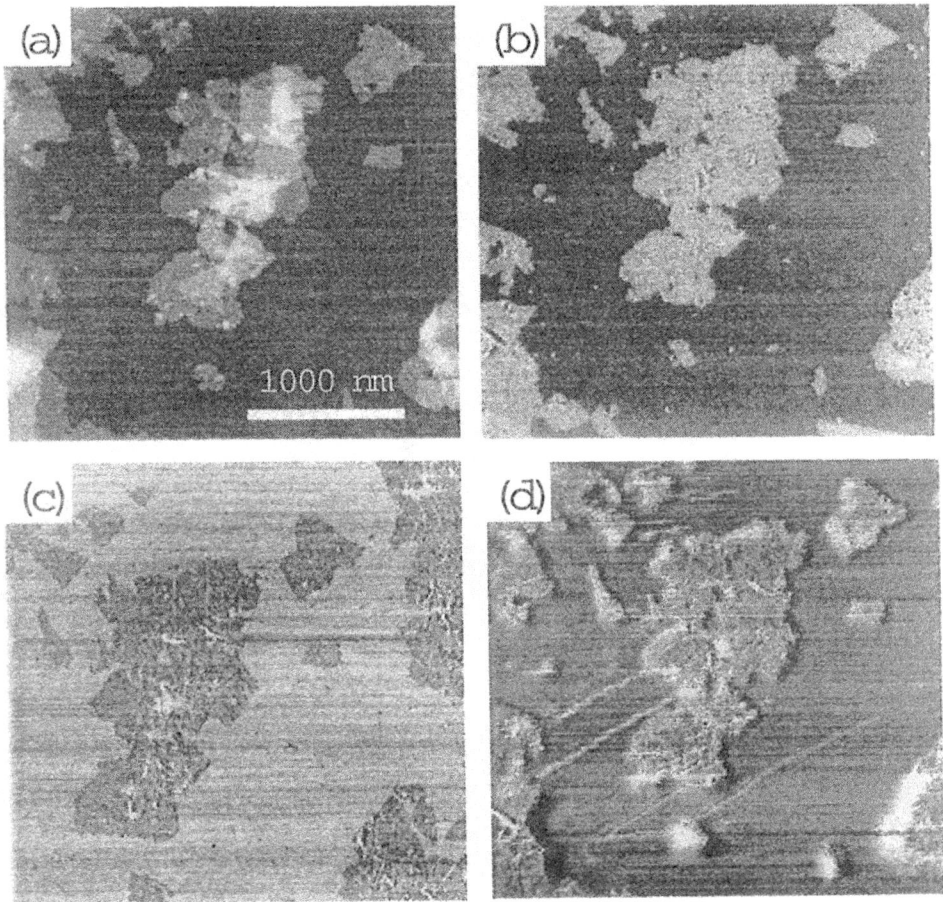

Figure 8.8: (a) Topography image.(b)-(d) Friction force maps on MoS$_2$-platelets deposited on mica in different humidities. Around 40% a contrast reversal is observed. (b) dry nitrogen (c) 40% (d) 80%

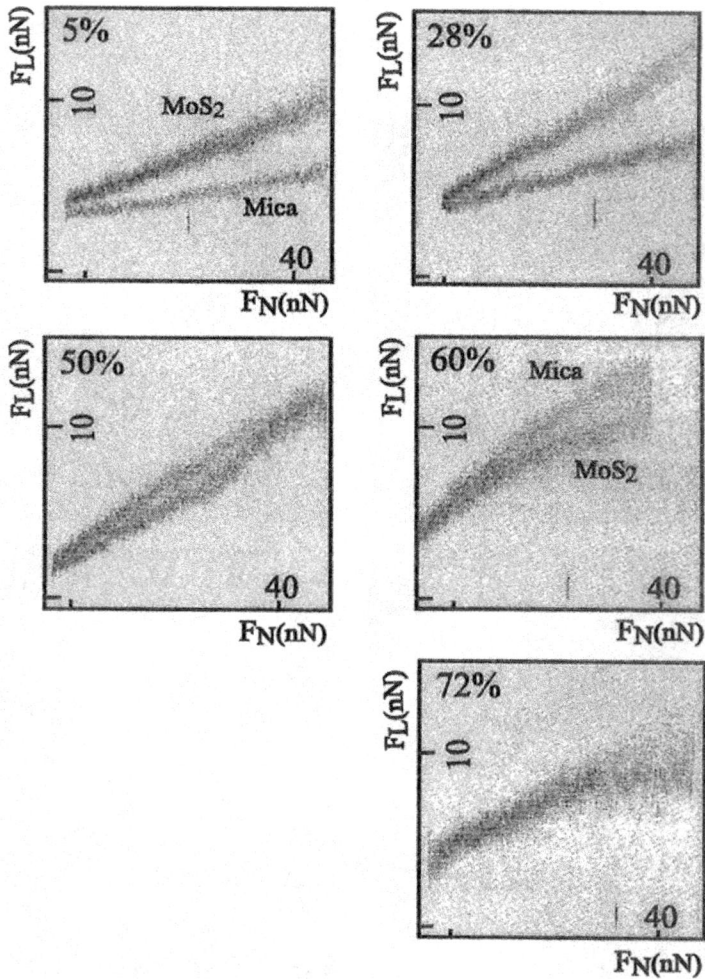

Figure 8.9: 2d-histograms of the MoS_2/mica measurements at different humidities. The strong dependence of friction on the hydrophilic mica surface is observed, whereas the contrast on MoS_2 is not strongly affected. At very high humidities, the curves are nearly indistinguishable, corresponding to the case of sliding of water on water.

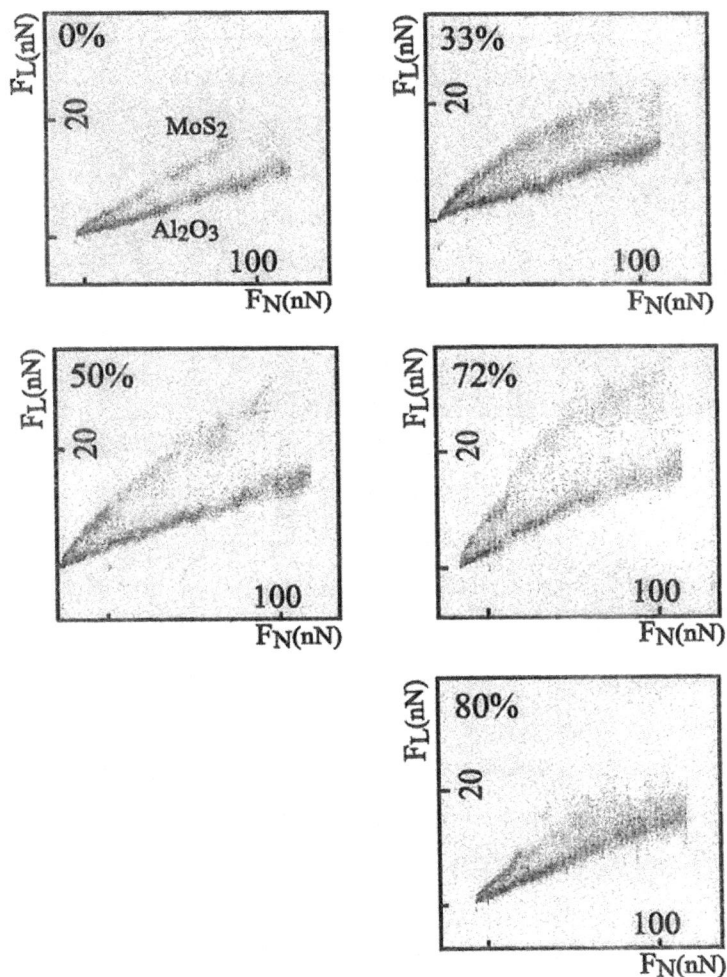

Figure 8.10: 2d-histograms of the MoS_2/Al_2O_3 measurements at different humidities. Both materials are only weakly affected by the different humidities. At very high humidities, the interactions are probably dominated by the water-water interaction.

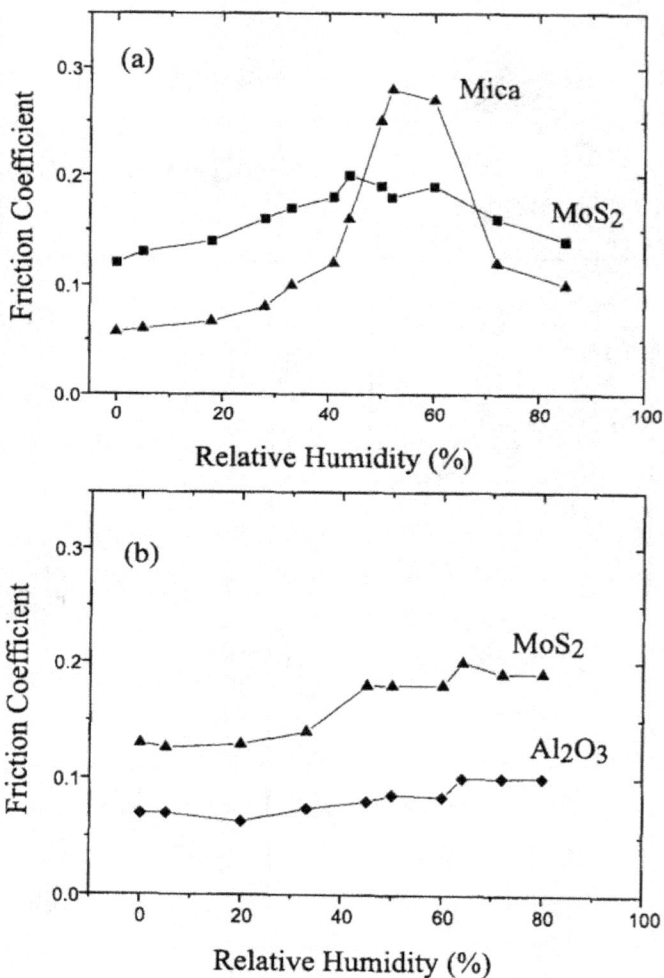

Figure 8.11: Collection of the results on the systems of MoS_2/mica and MoS_2/Al_2O_3. The strong dependence of friction on mica becomes obvious.

8.4 Chemical nature of probing tip

Friction not only depends on the sample surface but it also depends on the probing tip. Under ambient conditions, the dependence on tip material is not predominant. Most surfaces (semiconductors or metals) are covered by thin oxide films. In addition, contaminants, such as hydrocarbons or water, play a significant role. Therefore, the change of the tip material in air only gives rise to minor changes of the tribological properties. On clean surfaces in UHV the influence is much more distinct. Therefore, a careful preparation of the probing tip is essential for fundamental experiments in nanotribology.

8.4.1 Diamond on diamond

Germann et al.[42] prepared a diamond tip which was slided over the surfaces of diamond (001) and (111) in UHV. Atomic stick-slip was observed on both faces. On the diamond (001), periodicities were found that are consistent with the 2x1 reconstruction. With the help of force vs. distance curves, the tip radius (R=30nm) and the contact diameter $a = (3F_N R\frac{1-\nu^2}{2E})^{1/3}$=1.6nm were determined, where ν is the Poisson ratio and E is the Youngs modulus. The authors found a vanishing load dependence of the lateral force, which means that the lateral force remained constant in first approximation. This observation is in contradiction to the Hertz theory which predicts that the contact area of the single asperity and thus the lateral force increases with $F_N^{2/3}$. Even a multi-asperity contact would yield a linear dependence. Molecular dynamics simulations by Harrison et al.[43] of diamond on diamond predict zero friction along a particular direction and also find a weak dependence on load. Sokoloff[44] developed a model that predicts zero friction between smooth surfaces, unless the surfaces are commensurate.

8.4.2 PTFE on silicon

A second key experiment, which shows the crucial importance of the chemical nature of the probing tip, was performed on the surface Si(111)7x7[45]. Being a standard in scanning tunneling microscopy (STM), this surface is an ideal test sample for AFM. First experiments with uncoated Si-tips, covered by native oxide, showed drastic wear behaviour. No reproducible imaging was possible and large adhesive forces were built up during the contact, causing plastic deformation of the sample. Then, the tips were coated with different metals, such as Pt, Au, Ag, Cr, Pt/C. However, no improvement could be made. Apparently, the dangling bonds of the Si(111)7x7 reacted with the atoms of the probing tips, forming more stable configurations. A breakthrough could be achieved by coating the tip with PTFE (poly-tetrafluorethylene). A simple mechanical deposition technique of PTFE was applied. Before application on the Si-surface, the tip is scanned over a PTFE-surface, which leads to the transfer of about one monolayer of PTFE.

This kind of behaviour is known from the field of polymer wear[47,46]. Using this *in situ* tip preparation, the Si(111)7x7 surface can be imaged without damage. The resolution is even sufficient to image the corner holes. The adatoms in the unit cell yield some periodic contrast, but are not resolved individually. Thus, the contact diameter is estimated to be about 1nm. With typical forces of 10^{-9}-10^{-8}N, the 7x7 reconstruction is found to remain stable with pressures of 1-10GPa. PTFE has two functions in this experiment: 1) PTFE is a lubricant, reducing adhesion and friction. 2) PTFE does not react with the dangling bonds of Si(111)7x7. One reason for the inertness of the PTFE-coating are the strong bonds between fluorine and carbon (F-C: 552kJ/mol) which is comparable with the bond strengths of silicon and fluorine (Si-F: 552±2kJ/mol). The bonds between Si-C (451.5kJ/mol) and Si-Si (327±10kJ/mol) are weaker. Thus, the F-C-bonds are so strong that they are not broken by the presence of the reactive silicon surface. Generally, the chemical modification and passivation of the probing tip is important for the investigations in areas, such as catalysis, tribochemistry and corrosion, where reactive surfaces have to be imaged by force microscopy.

8.4.3   SAM on SAM

The group of Lieber at Harvard have used chemically modified probing tips to measure friction and adhesion on patterned self-assembled films (SAM)[48,49,50]. The probing tip (Si$_3$N$_4$) was coated with 30Å of Cr (adhesion layer) and 1000Å of Au. Then, the Au-coated tips were covalently modified with SAMs terminating in specific groups, such as methyl terminated groups (CH$_3$) or acid terminated groups (COOH). Second, optical lithography was used to define patterns of SAMs with different functionality. The adhesive interactions between the tip and sample, determined from force vs. distance curves, showed the following trend: COOH/COOH (2.3±0.8nN) > CH$_3$/CH$_3$ (1.0±0.4nN) > COOH/CH$_3$ (0.3±0.2nN). This is consistent with their expectation that the hydrophilic COOH groups, which can form hydrogen bonds, interact stronger than the hydrophobic CH$_3$ groups. Taking into account the tip radius (determined from SEM-measurements) of 55nm[b] in combination with the JKR-model, the pull-off force can be used to estimate the work of adhesion W_{SMT} to sep-

---

[b] Remarkably, the authors have calculated the effective tip radius, including the roughness of the sample surface with an effective curvature of the gold islands of R_s=500nm and the tip radius, determined with SEM of R_t=60nm. Then, the effective tip radius is given by R_{eff}= $R_t R_s / (R_t + R_s)$=55nm.

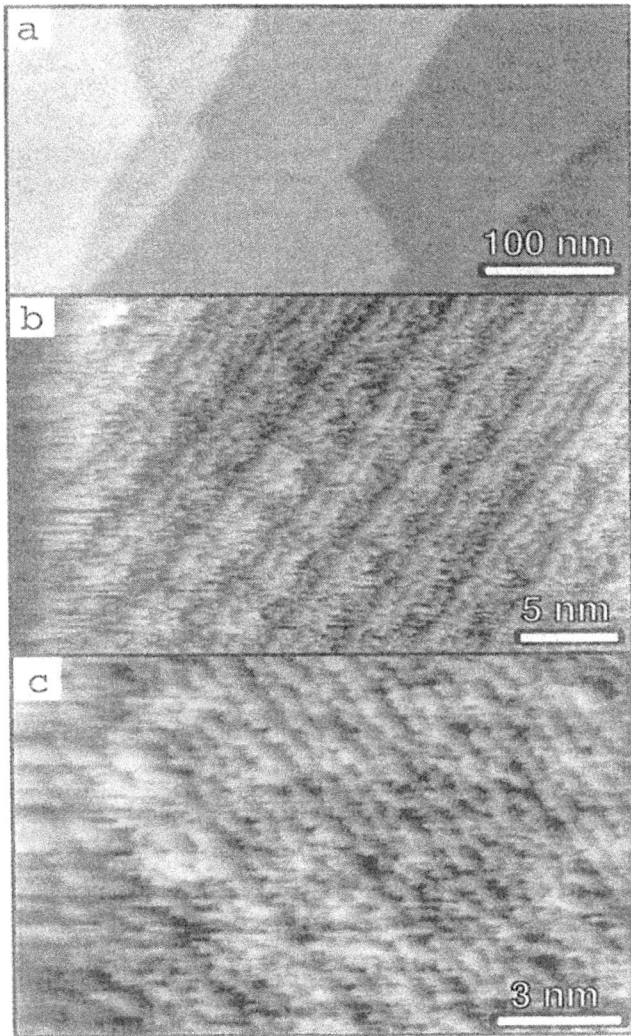

Figure 8.12: (a) Topography image of the Si(111)7x7 surface measured in contact mode with a chemically modified tip (PTFE coated Si-tip). The steps are 3Å high. (b) Lateral force image of the Si(111)7x7 showing the unit cells of the 7x7 reconstruction and some internal structure. Lateral forces vary typically 5-20nN. (c) Corner holes of the 7x7 reconstruction are resolved in the lateral force image.

arate sample (S) and tip (T) in medium (M) (liquid, vapour or vacuum):

$$F_{adh} = -\frac{3}{2}\pi \cdot R \cdot W_{SMT} \tag{8.1}$$

The work of adhesion is given by the surface free energies of the sample and the tip γ_{SM}, γ_{TM} in contact with the medium, and the interfacial free energy γ_{ST}:

$$W_{SMT} = \gamma_{SM} + \gamma_{TM} - \gamma_{ST} \tag{8.2}$$

If sample and tip have identical surfaces (e.g., CH_3/CH_3), then $\gamma_{ST}=0$ and $\gamma_{SM}=\gamma_{TM}$ and the work of adhesion is equal to twice the surface free energy in the medium $W_{SMT} = 2\gamma$. Experimentally, a surface free energy of $\gamma(CH_3/\text{ethanol})=1.9\text{mJ/m}^2$ is found in ethanol, which is consistent with contact angle measurement of ethanol on CH_3-terminated SAM's: The contact angle of ethanol on CH_3-terminated SAM's is about $\theta=40°$, the surface tension of ethanol is $\gamma_{lv}(\text{ethanol}) =22.5\text{mJ/m}^2$, the surface free energy of CH_3-SAM's in vacuum is approximately 19.5mJ/m^2[51,52]. According to the Young's equation

$$\gamma_{sl} = \gamma_{sv} - \gamma_{lv}cos\theta, \tag{8.3}$$

the surface free energy of CH_3 in ethanol is given by: $\gamma_{sl}=2.3\text{mJ/m}^2$, which is in reasonable agreement with the adhesive force measurements.

In addition, the surface free energy of the COOH-terminated surface could be determined[49]: $\gamma_{sl}(\text{COOH/ethanol})=4.5\text{mJ/m}^2$ with force measurements. Remarkably, contact angle measurements cannot be used in the case of COOH, since this high free energy surface is readily wet by ethanol. In addition, the interfacial free energy $\gamma_{CH_3/COOH}=5.8\text{mJ/m}^2$ could be calculated, which explains the strong reduction of adhesive forces in the mixed case.

The contact area at the pull-off point was estimated to be about 3.1nm^2, which corresponds to about 15 functional groups on the sample and the tip. For tip radii of about 10nm, even single molecular contacts are predicted.

Adhesion measurements in water showed that the strong influence of electrostatic interactions. NH_3 and COOH-terminated tips and/or sample surfaces were found to be charged, which gave a large contribution to the adhesive force[49].

A similar trend as in the adhesion measurements was observed for the friction measurements: Large friction for COOH-terminated tips on COOH-terminated regions than on CH_3 terminated regions. Whereas, CH_3-terminated tips gave large friction on CH_3-terminated sample regions and lower friction on COOH-terminated regions. Frisbie et al. mentioned that the friction contrast appeared only above a threshold of 3nN. Otherwise, a rather linear loading dependence was observed, where friction coefficients μ were determined from the slopes of the frictional force vs. normal force curves: $\mu=$

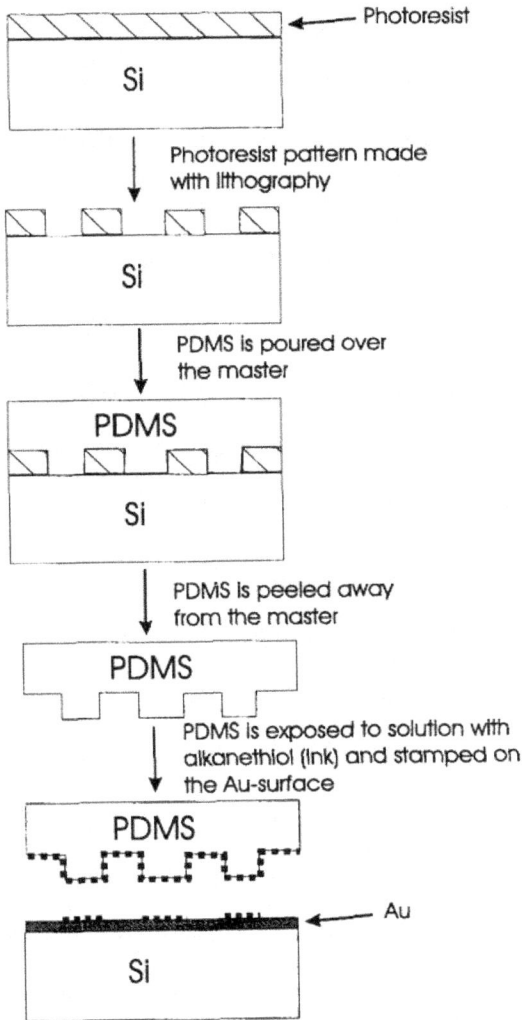

Figure 8.13: Polymeric stamping technique from the Whitesides group[60], also called microcontact printing (μCP) of patterned self-assembled monolayers (SAM) of alkanethiolates on gold surfaces. The stamp is fabricated by pouring polydimethylsiloxane (PDMS) on the master. The stamp is "inked" with alkanethiol. A patterned SAM is deposited with the stamp.

2.5, 0.8 and 0.4 for COOH/COOH, CH_3/CH_3 and COOH/CH_3. In one case (CH_3 on CH_3), a non-linear analysis has been used by Noy et al.[50].

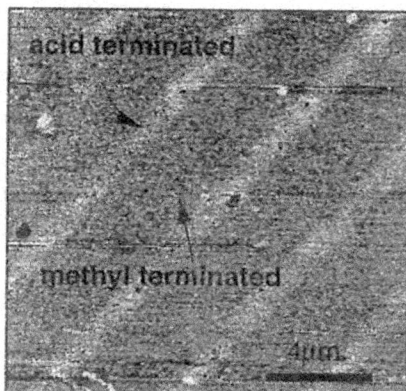

Figure 8.14: FFM-measurement with a Methyl-terminated SAM-covered tip on areas covered with methyl-terminated SAMs (dark) and acid-terminated SAMs (bright) measured in dried nitrogen. Large friction is found on the acid-terminated areas, which may be related to preferred adsorption of contaminants (acid imaging). The films were deposited with the polymeric stamping technique from the Whitesides group [60]. See also Delamarche et al.[61]. Courtesy of M. Kubon, Basel.

Based on these measurements, Frisbie et al. suggest that adhesion is directly correlated with friction, at least when systems in the same aggregate state are compared with each other (e.g., solid films with different functional groups). Previously, SFA-measurements were discussed (see chapter II), where friction is related to adhesion hysteresis. However, in this case different phases (solid-like and liquid-like) were compared with each other.

8.4.4 Chemical force microscopy

The term "chemical force microscopy" was introduced by Frisbie et al. in order to emphasize that the functionalized FFM-tips give more chemical specificity: Depending on the functional group of the probing tip, different contrasts were observed. In analogy to the molecular recognition experiments from Florin et al.[53], Lee et al.[54] or Dammer et al.[55], they envision attaching a specific oligonucleotide or receptor to the probe tip and then mapping friction forces on a surface that contains an array of different nucleotide sequences or ligands to find those having the strongest interaction, corresponding to the complimentary ligand-receptor pairs. Examples of applications of FFM to biological materials have been given by Marti et al.[56] and Eng et al.[57]. A nice example of a chemical modification of the probing tip has been given by Dai et al.[58]

who attached a nanotube to the end of silicon probing tip. Due to the high aspect ratio imaging of rough surfaces becomes possible. Also, the long range forces are reduced, which may lead to higher resolution in contact mode.

There are some limitations to the chemical specificity of force microscopy with functionalized tips:

- *Capillary forces:*

  Measurements in air are dominated by capillary forces[59], which are 1-2 orders of magnitude larger than more specific chemical interactions. These capillary forces will thus obscure small differences in molecular forces. Mainly, hydrophilic surfaces will give increased contrast due to the condensation of water on these areas[60]. Although, capillary forces are to be avoided in most cases, moderately hydrophobic tips are suitable to distinguish hydrophilic from hydrophobic areas on heterogeneous surfaces in air. This mode is also called acid-imaging, because the acid-terminated surfaces are covered by water and cause strongest capillary forces, respectively adhesion or friction forces. Measurements, performed in dry inert gases[62,63] may reduce the thickness of adsorbate films, but it is difficult to exclude them completely. Measurements in ultrahigh vacuum may be ideal, but are often not compatible with the chosen organic systems (high vapor pressures). The capillary effect can be eliminated in liquids, however the influence of solvent exclusion has to be taken into account.

- *Solvent exclusion:*

  Adhesive forces are not only given by the bondings between tip and sample, but are also influenced by the presence of the fluid. A systematic study of adhesive forces with different functionalized tip/sample combinations in different liquids (water, ethanol, n-hexadecane) by Sinniah et al. has shown that adhesive forces are strongly influenced by solvent exclusion. Adhesive forces in water with hydrophobic surfaces are larger than with hydrophilic surfaces. In ethanol and n-hexadecane adhesive forces are reduced. In water, these adhesive forces are dominated by the work required to exclude the solvent from the tip-sample interface. In ethanol these macroscopic solvent exclusion is not sufficient to explain the data. Microscopic concepts, like the increased fluidity at the chain endings, leading to less interdigitations between these monolayers, are proposed by Sinniah. Although, the influence of solvent exclusion may appear as an additional complication for interpretation of the data, it has been shown by Sinniah et al. that optimum contrast on heterogeneous copolymers can be achieved by an appropriate choice of the functional group of the tip to distinguish between hydrophobic and hydrophilic blocks[51].

- *pH-Dependence:*

  Adhesive force and friction measurements in aqueous solutions depend on the pH[65,66]. Depending on the degree of ionization, electrostatic forces arise between the charged surfaces, which is measurable with force vs. distance curves. These experiments are in close analogy to the contact angle measurements vs. pH[64]. If the experimentalist is aware of this effect, it can be a useful tool: At the iso-electric point charge compensation is observable, corresponding to minimum pull-off forces. Thus, adhesive force vs. pH measurement give the opportunity to measure local iso-electric points or pK-values. This type of experiments has been called force titration. In order not to influence the Debye length by the change of pH (inlet of acid or base), it is favourable to measure in electrolytes (buffered solution) where the Debye length is approximately constant for an appropriate range of pHs.

- *Elasticity:*

  Contrasts in friction force microscopy are influenced by the surface compliances of sample and tip. It has been shown by Koleske et al.[67], that Langmuir-Blodgett films with identical end group, but with different chain lengths ($CH_3(CH_2)_{22}COOH$ vs. $CH_3(CH_2)_{14}COOH$). have different adhesion values. Therefore, in addition to the short-range chemical force a more long-range phenomenon, such as elastic deformation due to the repulsive forces, must be effective. Comparing the compressional modulus derived from film pressure vs. area isotherms of the pure components, the adhesion difference can be qualitatively understood. Also, Koleske et al. observed reduced step heights in topography, which is in agreement with the elastic deformation of the films due to the presence of the probing tip. They find that the shorter chain length films deform more than the longer chain length. A quantitative analysis was difficult due to unknown plastic or viscoelastic deformations. The influence of local variations of the Youngs modulus on SAMs has also been suggested by Bar et al.[68].

8.5 Traditional and new concepts to understand the material-specific contrasts of FFM

A fundamental understanding of contrast mechanisms depends not only on experimental work, but also requires theoretical models. Most of the present ideas are based on empirical models:

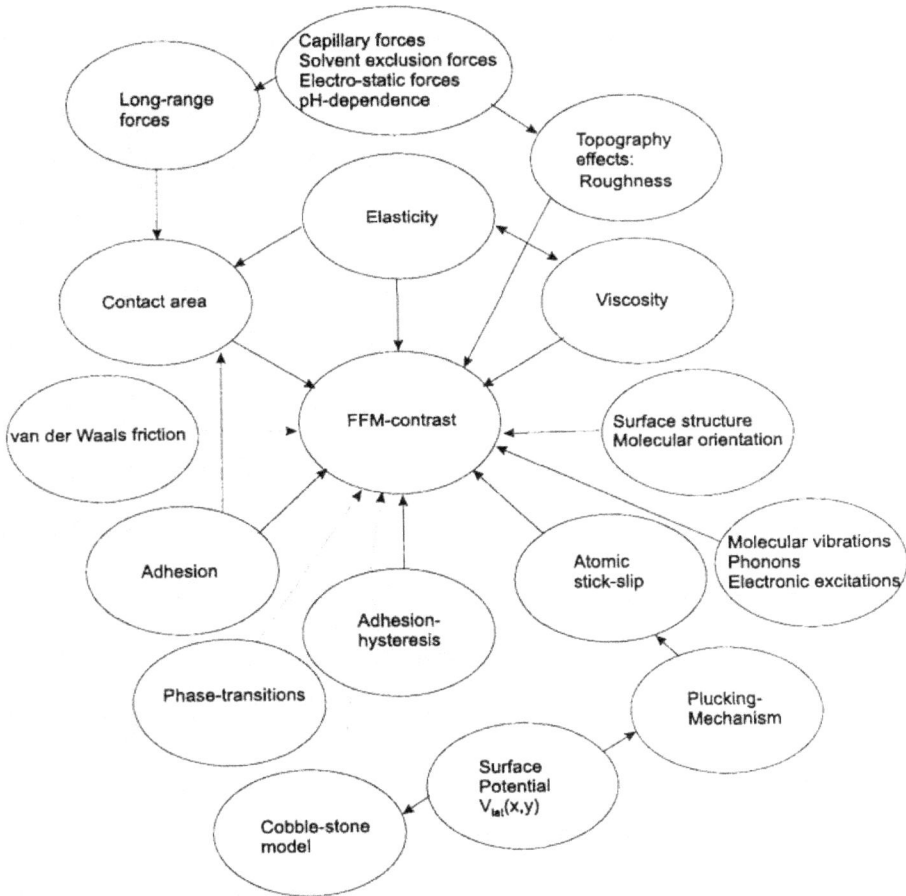

Figure 8.15: Contrast Mechanisms of FFM

1. Adhesion plays traditionally an important role in the understanding of friction and wear. Adhesive forces include van der Waals forces, capillary forces, electrostatic forces and short-range chemical forces (including metallic adhesion and polarization forces). An increase of adhesive forces leads to an increase in contact area, which also increases friction. Apart from this rather trivial effect, there is some hope that adhesion due to short-range chemical forces might be more intimately related to friction. In some cases[50] rather good correlation between adhesion and friction were found. However, examples were found, where high adhesion is accompanied by low frication[69]. Alternatively, adhesion hysteresis was proposed from SFA-experiments to be the relevant parameter to be compared with friction. There is some experimental evidence[65,20] that adhesion hysteresis might be a relevant process in FFM.

2. Elasticity plays a similar role than adhesion. Local variations of sample elasticity cause changes of the contact area and thus changes friction, too. Local elasticity is measurable by force microscopy. A systematic study on mixtures of fluorocarbons and hydrocarbons has shown that there exist similarities between the friction force map and the elasticity map. However, the correlation is not always fulfilled[8]. Similar discrepancies were observed by Garcia et al.[20] on semiconductors.

3. The plucking mechanism, proposed by Tomlinson[70] and Prandtl[71] predicts that friction depends on the potential $V_{lat}(x,y)$ ($F_{lat} = \partial V_{lat}/\partial x$) and on the weakest lateral spring constant[72]. Already, Mate et al.[73,74] observed that a slip occurs at $\approx 10^{-6}$N for a 155N/m spring, which is substantially lower than the $\approx 5 \cdot 10^{-5}$N onset observed for a 2500N/m spring. If the lateral spring constant of the cantilever is larger than the effective sample spring, the instabilities occur in the sample and can be characterized by measuring the slope during the stick period[75]. Lateral stiffness measurements give the opportunity to calculate the contact diameter. A more detailed description of the plucking mechanism is given in Chapter IV.

4. Adsorbed molecules (e.g. lubricants) or surface atoms are presumably first excited by the action of the tip. The amount of energy that can be transferred to such a molecule depends on the degrees of freedom, such as bond stretching, rotation, bond bending. In a second stage, the vibrational motion is transferred to the substrate, e.g. in the form of phonons or electronic excitations. Local inhomogeneities of the substrate can lead to a different coupling of the adsorbed molecules, which is also observed in the friction force map[2]. Some aspects of dissipation mechanisms are discussed in chapter V.

5. The structure of the surface influences the measurements as well. Overney et al.[27] observed that the tilt angle of molecules changes the friction forces significantly. Different faces of a crystal can have different surface phonons or plasmons, which can affect the dissipation process.

6. As discussed above, the chemical nature of the tip can play an important role and has always to be taken into account.

In summary, there are many contrast mechanisms, which can influence friction. There is a strong need for fundamental models from which predictions are made that can be confirmed or denied by FFM. MD-simulations play a central role and are discussed in this book. Other models relate the concepts of commensurability to friction[44,24,76]. Ultimately, a theory would be desirable that makes similar predictions for friction than the BCS-theory for superconductivity, incorporating parameters, such as the coupling of phonons or the density of states at the Fermi edge.

1. E. Meyer, R.M. Overney, L. Howald, D. Brodbeck, R. Lüthi and H.-J. Güntherodt, in *Fundamentals of Friction*, edited by I.L. Singer and H.M. Pollock, Series E: Applied Sciences, Vol. 220, Kluwer Academic Publishers, pp. 427-436 (1992).

2. E. Meyer, R.M. Overney, L. Howald, R. Lüthi, J. Frommer, H.-J. Güntherodt, Friction and wear of Langmuir-Blodgett films observed by friction force microscopy, *Phys. Rev. Lett.* **69**, 1777-1780 (1992).

3. F.P. Bowden and D. Tabor, *The Friction and Lubrication of Solids*, Clarendon Press, Oxford, (1950).

4. V. Novotny, J.D. Swalen, J.P. and Rabe, Tribology of Langmuir-Blodgett films, *Langmuir* **5**, 485-489 (1989).

5. E. Meyer, L. Howald, R.M. Overney, H. Heinzelmann, J. Frommer, H.-J. Güntherodt, T. Wagner, H. Schier and S. Roth, Molecular-resolution images of Langmuir-Blodgett films using atomic force microscopy, *Nature* **349**, 398-399 (1991).

6. A. Shaper, L. Wolthaus, D. Möbius and T.M. Jovin, Surface Morphology and Stability of Langmuir-Blodgett Mono- and Multilayers of Saturated Fatty Acids by Scanning Force Microscopy, *Langmuir* **9**, 2178-2184 (1993).

7. R.M. Overney, E. Meyer, J. Frommer, D. Brodbeck, R. Lüthi, L. Howald, H.-J. Güntherodt, M. Fujihira, H. Takano and Y. Gotoh, Friction measurements of phase-separated thin films with a modified atomic force microscope, *Nature* **359**, 133-135 (1992).

8. R.M. Overney, E. Meyer, J. Frommer, H.-J. Güntherodt, M. Fujihira, H. Takano and Y. Gotoh, Force microscopy study of friction and elastic

compliance of phase-separated organic thin films, *Langmuir* **10**, 1281-1286 (1994).

9. B.J. Briscoe and D.C.B. Evans, The shear properties of Langmuir-Blodgett layers, *Proc. R. Soc. Lond. A* **380**, 389-407 (1982).

10. E. Meyer, R.M. Overney, R. Lüthi, D. Brodbeck, L. Howald, J. Frommer, H.-J. Güntherodt, O. Wolter, M. Fujihira, H. Takano and Y. Gotoh, Friction measurements of phase-separated thin films with a modified atomic force microscope, *Thin Solid Films* **220**, 132 (1992).

11. L. Scandella, A. Schumacher, N. Kruse, R. Prins, E. Meyer, R. Lüthi, L. Howald, H.-J. Güntherodt, Tribology of ultra-thin MoS_2 platelets on mica: studied by scanning force microscopy, *Thin Solid Films* **240**, 101-104 (1994).

12. O. Marti, J. Colchero and J. Mlynek, Friction and forces on an atomic scale, Proceedings NATO ARW Lyon, July 6-10, Kluwer Academic Publishers (1992).
J. Colchero, O. Marti, J. Mlynek, A. Humbert, C.R. Henry and C. Chapon, C. (1991), Palladium clusters on mica: A study by scanning force microscopy, *J. Vac. Sci. Technol. B* **9**, 794-797 (1991).

13. M. Mate, Nanotribology studies of carbon surfaces by force microscopy, *Wear* **168**, 17-20 (1993).

14. Scott S. Perry, C.M. Mate, R.L. White and G.A. Somorjai, Bonding and tribological properties of perfluorinated lubricants and hydrogenated amorphous carbon films, *IEEE Transactions on Magnetics* **32**, 115-121 (1996).

15. C.M. Mate and A.M. Homola, Molecular tribology of disk drives, in B.Bhushan (ed.), *Micro/Nanotribology and its Applications*, pp. 647-661, Kluwer Academic Publisher, (1997).

16. L. Scandella, E. Meyer, L. Howald, R. Lüthi, M. Guggisberg, J. Gobrecht and H.-J. Güntherodt, Friction forces on hydrogen passivated (110) silicon and silicon dioxide studied by scanning force microscopy, *J. Vac. Sci. Technol.B* **14**, 1255-1258 (1989).

17. D. Gräf et al., *J. Vac. Sci. Technol.* **A7**, 808 (1989).

18. T. Teuschler, K. Mahr, M. Hundhausen and L. Ley, Nanometer-scale modificatiion of the tribological properties of Si(100) by scanning force microscope, *Appl. Phys. Lett.* **66**, 2499-2501 (1995).

19. J. Tamayo, L. González, Y. González and R. García, *Appl. Phys. Lett.* **68**, 2297 (1996).

20. R. García, J. Tamayo, L. González and Y. González, Compositional characterization of II-V semiconductor heterostructures by friction force microscopy, in *Micro/Nanotribology and its Applications*, B. Bhushan (ed.), Kluwer Academic Publishers, p. 275-282, (1997).

21. J. Tamayo, R. García, T. Utzmeier and F. Briones, Submonolayer

sensitivity of InSb on InP determined by friction force microscopy, *Phys. Rev. B* **55**, R13436-R13439 (1997).

22. G. Grüner, The dynamics of charge-density waves, *Rev. Mod. Phys.* **60**, 1129-1181 (1988).
    G. Grüner, Nonlinear and frequency-dependent transport phenomena in low-dimensional conductors, *Physica 8D* **119**, 1-34 (1983).
    L. Sneddon, Sliding charge-density waves. I. dc properties, *Phys. Rev. B.* **29**, 719-727 (1984).

23. P. Bak, Commensurate and incommensurate phases, *Rep. Prog. Phys.* **45**, 587-629 (1982).
    J.E. Sacco, J.B. Sokoloff and Widom, Dynamical friction in sliding condensed-matter systems, *Phys. Rev. B.* **20**, 5071-5083 (1979).

24. M. Hirano and K. Shinjo, Atomistic locking and friction, *Phys. Rev. B* **41** 11837-11851 (1990).
    M. Hirano, K. Shinjo, R. Kaneko, Y. Murata, Anisotropy of frictional forces in muscovite mica, *Phys. Rev. Lett.* **67**, 2642-2645 (1991).
    K. Shinjo and M. Hirano, Dynamics of friction: superlubricant state, *Surf. Sci.* **283**, 473-478 (1993).

25. M. Hirano, K. Shinjo, R. Kaneko and Y. Murata, Observation of superlubricity by scanning tunneling microscopy, *Phys. Rev. Lett.* **78** 1448-1451 (1997).

26. S. Morita, S. Fujisawa and Y. Sugawara, *Surf. Sci. Rep.* **23**, 3 (1996).

27. R.M. Overney, H. Takano, M. Fujihira, W. Paulus and H. Ringsdorf, *Phys. Rev. Lett.* **72**, 3546 (1994).
    R. Overney in in *Forces in Scanning Probe Methods*, Eds. H.-J. Güntherodt, D. Anselmetti and E. Meyer, NATO ASI Series E: Applied Sciences Vol. 286, Kluwer Academic publishers, p. 307-312, (1995).

28. H. Takano and M. Fujihira, Study of molecular scale friction on stearic acid crystals by friction force microscopy, *J. Vac. Sci. Technol. B* **14**, 1272-1275 (1996).

29. H. Bluhm, U.D. Schwarz, K.P. Meyer, R. Wiesendanger, Anisotropy of sliding friction on the triglycine sulfate (010) surface, *Appl. Phys. A* **61**, 525-533 (1995).

30. D. Gourdon, N.A. Burnham, A. Kulik, E. Dupas, F. Oulevey, G. Gremaud, D. Stamou, M. Liley, Z. Dienes, H. Vogel and C. Duschl, The dependence of friction anisotropies on the molecular organisation of LB films as observed by AFM, *Tribology Letters* **3**, 317-324 (1997).

31. J. Kerssemakers and J.Th.M. De Hosson, Influence of spring stiffness and anisotropy on stick-slip atomic force microscopy imaging, *J. Appl. Phys.* **80**, 623-632 (1996).

32. R. J. Warmack, X.-Y. Zheng, T. Thundat and D.P. Allison, Friction

effects in the deflection of atomic force microscope cantilevers, *Rev. Sci. Instrum.* **65** 394-399 (1994).

33. R. Lüthi, E. Meyer, H. Haefke, L. Howald, W. Gutmannsbauer and H.-J. Güntherodt, *Science* **266** 1979 (1994).

34. P.E. Sheehan, C.M. Lieber, Nanotribology and Nanofabrication of MoO$_3$ Structures by Atomic Force Microscopy, *Science* **272**, 1158 (1996).

35. C.A.J. Putmann, M. Igarshi and R. Kaneko, Experimental observation of single-asperity friction at the atomic scale, *Appl. Phys. Lett.* **66**, 3221 (1995).

36. I.L. Singer, in I.L. Singer and H.M. Pollock (eds.), *Fundamentals in Friction: Macroscopic and Microscopic Processes*, Kluwer, Dordrecht, p. 237 (1993).

37. A. Schumacher, N. Kruse, R. Prins, E. Meyer, R. Lüthi, L. Howald, H.-J. Güntherodt and L. Scandella, Influence of humidity on friction measurements of supported MoS$_2$ single layers, *J. Vac. Sci. Technol. B* **14**, 1264-1267 (1996).

38. NANOSENSORS, Dr. Olaf Wolter GmbH, IMO-Building, Im Amtmann 6, 35578 Wetzlar-Blankenfeld, Germany, Tel. (+49) 6441 97 88 40 Fax: (+49) 6441 97 8841
e-mail: nanosensors@compuserve.com

39. Park Scientific Instruments, 1171 Borregas Avenue, Sunnyvale, CA 94089, USA
Tel: +1 408 747-1600 Fax: +1 408 747-1601
e-mail: Info@park.com
website: http://www.park.com

40. M. Hu, X.-d. Xiao, D.F. Ogletree, and M. Salmeron, "Atomic scale friction and wear of mica", *Surf. Sci.* **327**, 358-370 (1995).

41. E. Meyer et al. in *Physics of Sliding Friction*, edited by B.N.J. Persson and E. Tosatti, Series E: Applied Sciences, Vol. 311, Kluwer Academic Publishers (1996).

42. G.J. Germann, S.R. Cohen, G. Neubauer, G.M. McClelland and H. Seki, Atomic scale friction of a diamond on diamond(100) and (111) surfaces, *J. Appl. Phys.* **73**, 163-167 (1993).

43. J.A. Harrison, C.T. White, R.J. Colton and W. Brenner, Nanoscale investigation of indentation, adhesion and fracture of diamond (111) surfaces, *Surf. Sci.* **271**, 57-67 (1992).

44. J.B. Sokoloff, Theory of energy dissipation in sliding crystal surfaces, *Phys. Rev. B* **42**, 760-765 (1990).

45. L. Howald, R. Lüthi, E. Meyer, P. Güthner and H.-J. Güntherodt, (1994). Scanning force microscopy on the Si(111)7x7 surface reconstruction, *Z. Phys. B*, **93**, 267-268 (1994).

L. Howald, R. Lüthi, E. Meyer, H. Rudin and H.-J. Güntherodt, Atomic force microscopy on the Si(111) surface, *Phys Rev.* *B* **51**, 5484 (1995).

46. B.J. Briscoe, *J. Am. Chem. Soc. Symp. Ser.*, (ed. L.-H. Lee), **287**, 151-170 (1985).

47. J.C. Wittmann and P. Smith, Highly oriented thin films of poly(tetrafluoroethylene) as a substrate for oriented growth of materials, *Nature* **352**, 414-417 (1991).

48. C.D. Frisbie, L.F. Rozsnyai, A. Noy, M.S. Wrighton and C.M. Lieber, Functional Group Imaging by Chemical Force Microscopy, *Science* **265**, 2071-2074 (1994).

49. A. Noy, D.C. Frisbie, L.F. Rozsnyai, M.S. Wrighton and C.M. Lieber, Chemical Force Microscopy: Exploiting Chemically-Modified Tips to Quantify Adhesion, Friction, and Functional Group Distributions in Molecular Assemblies, *J. Am. Chem. Soc.* **117**, 7943-7951 (1995).

50. A. Noy, D.V. Vezenov and C.M. Lieber, Chemical Force Microscopy, *Annu. Rev. Mater. Sci.* **27**, 381-421 (1997).

51. S.K. Sinniah, A.B. Steel, C.J. Miller and J.E. Reutt-Robey, Solvent Exclusion and Chemical Contrast in Scanning Force Microscopy, *J. Am. Chem. Soc.* **118**, 8925-8931 (1996).

52. M.K. Chaudhury and G.M. Whitesides, Correlation Between Surface Free Energy and Surface Constitution, *Science* **255**, 1230-1232 (1992).

53. E.-L. Florin, V.T. Moy and H.E. Gaub, *Science* **264**, 415 (1994).

54. G.U. Lee, D.A. Kidwell and R.J. Colton, *Langmuir* **10**, 354 (1994).

55. U. Dammer,O. Popescu, P. Wagner, D. Anselmetti, H.-J. Güntherodt and G. Misevic Binding Strength Between Cell Adhesion Proteoglycans Measured by Atomic Force Microscopy, *Science* **267**, 1173 (1995).

56. O. Marti, J. Colchero, H. Bielefeldt, M. Hipp and A. Linder, Scanning probe microscopy: applications in biology and physics, *Microsc. Microanal. Microstruct.* **4**, 429-440 (1993).

57. L.M. Eng, K.D. Jandt and D. Descouts, A combined scanning tunneling, scanning force, frictional force, and attractive force microsope, *Rev. Sci. Instrum.* **65**, 390-393 (1994).
    F. Zenhausern, M. Adrian, B.Ten Heggeler-Bordier, L.M. Eng and P. Descouts, DNA and RNA Polymerase/DNA Complex Imaged by Scanning Force Microscopy: Influence of Molecular-Scale Friction,*Scanning* **14**, 212-217 (1992).

58. H. Dai, J.H. Hafner, A.G. Rinzler, D.T. Colbert and R.E. Smalley, Nanotubes as nanoprobes in scanning probe microscopy, *Nature* **384**, 147-150 (1996).

59. A.L. Weisenhorn, P. Maivald, H.-J. Butt and P.K. Hansma, *Phys. Rev.* *B* **45**, 11226-11232 (1992).

60. J. Wilbur, H.A. Biebuyck, J.C. MacDonald and J.M. Whitesides, Scanning Force Microspies Can Image Patterned Self-Assembled Monolayers, *Langmuir* **11**, 825-831 (1995).
    J.L. Wilbur, A. Kumar, E. Kim and G.M. Whitesides, Microfabrication by Microcontact Printing of Self-Assembled Monolayers, *Advanced Materials* **6**, 600-604 (1994).
61. E. Delamarche, B. Michel, H.A. Biebuyck and C. Gerber, Golden Interfaces: The surface of self-assembled monolayers, *Advanced Materials* **8**, 719-729 (1996).
62. J.-B. Green, M.T. McDermott, M.D. Porter and L.M. Siperko, Nanometer-Scale Mapping of Chemically Distinct Domains at Well-Defined Organic Interfacs Using Friction Force Microscopy, *J. Phys. Chem.* **99**, 10960-10965 (1995).
63. R.C. Thomas, P. Tangyunyong, J.E. Houston, T.A. Michalske and P.M. Crooks, *J. Am. Chem. Soc.* **98**, 4493 (1995).
64. S.R. Holmes-Farley, R.H. Reamey, T.J. Darthy, J. Deutch and G.M. Whitesides, *Langmuir* **4**, 921 (1988).
65. A. Marti, G. Hähner, N. D. Spencer, Sensitivity of Frictional Forces to pH on a Nanometer Scale: A Lateral Force Microscopy Study , *Langmuir* **11**, 4632 (1995).
    G. Hähner, A. Marti and N.D. Spencer, The influence of pH on friction between oxide surfaces in electrolytes, studied with lateral force microscopy: application as a nanochemical imaging technique, *Tribology Letters* **3**, 359-365 (1997).
66. D.V. Vezenov, A. Noy, L.F. Rosznyai, C.M. Lieber, *J. Am. Chem. Soc.*, 2006 (1997).
67. D.D. Koleske, W.R. Barger, G.U. Lee and R.J. Colton, Scanning probe microscope study of mixed chain-length phase-segregated Langmuir-Blodgett monolayers, *Mat. Res. Soc. Symp. Proc.* **464**, 377 (1997).
68. G. Bar, S. Rubin, a.N. Parikh, B.I. Swanson, T.A. Zawodzinski and M.-H. Whangbo, Scanning Force Microscopy Study of Patterned Monolayers of Alkanethiols on Gold. Importance of Tip-Sample Contact Area in Interpreting Force Modulation and Friction Force Microscopy Images, *Langmuir* **13**, 373-377 (1997).
69. H. Yoshizawa, Y.-L. Chen and J. Israelachivili, Mechanisms of Interfacial Friction 1, *J. of Phys. Chem.* **97**, 4128 (1993).
70. G.A. Tomlinson, A molecular theory of friction, *Phyl. Mag. and J. of Science* **7**, 905-939 (1929).
71. L. Prandtl, (1913). See e.g. Ein Gedankenmodell zur kinetischen Theorie der festen Koerper, *Z. angew. Math. Mechanik* **8**, 85-106 (1928).
72. D. Tománek, W. Zhong and H. Thomas, Calculation of an atomically

modulated friction force in atomic force microscopy, *Europhys. Lett.* **15**, 887-892 (1991).

73. C.M. Mate, G.M. McClelland, R. Erlandsson, and S. Chiang, Atomic-scale Friction of a Tungsten Tip on a Graphite Surface, *Phys. Rev. Lett.* **59**, 1942 (1987).

74. McClelland, G.M., Mate, C.M., Erlandsson, R. and Chiang, S. (1988), Direct observation of friction at the atomic scale, *Mat. Res. Soc. Symp. Proc.* **119**, 81-87.

75. J. Colchero et al. in *Forces in Scanning Probe Methods*, Eds. H.-J. Güntherodt, D. Anselmetti and E. Meyer, NATO ASI Series E: Applied Sciences Vol. 286, Kluwer Academic publishers, p. 345-352, (1995).

76. B.N.J. Perrson, Theory of friction and boundary lubrication, *Phys. Rev. B* **48**, 18140-18158 (1993).

Chapter 9

Appendix: Instrumental aspects of force microscopy

The aim of this chapter is to give some useful information for users of force microscopes and those, who intend to purchase or build such an instrument. Some of the manufacturers of scanning probe microscopes[2] and cantilevers[3] are listed. The sections of this chapter include some basic knowledge about AFM in general, such as design criteria, deflection sensors and calibration procedures for piezoelectric scanners. Finally, calibrations of normal and lateral forces are described.

9.1 Cantilevers

Scanning force microscopy is based on the measurement of the force between the probing tip and the sample, where the probing tip is attached to a cantilever-type spring. Thus, the force acting on the probing tip will elastically deform the cantilever. For shortness, the combination of cantilever and probing tip is referred to as the "cantilever" or "lever". If the spring constant c_B is known, the net force F can be derived directly from the deflection Δz according to the equation

$$F = c_B \cdot \Delta z \qquad (9.1)$$

With a weak spring of 0.1N/m and a deflection of 0.01nm, forces of 10^{-12}N can be detected. Alternatively, the force gradient between probing tip and sample is measured by detecting the frequency shift of the cantilever:

$$\omega_1' \approx \sqrt{\frac{c_{eff}}{m}} = \sqrt{\frac{c_B - F'}{m}} \qquad (9.2)$$

where ω_1' is the shifted resonance frequency, m is the mass of the cantilever, c_{eff} is the effective spring constant, which is increased/decreased by a repulsive/attractive force gradient $F' = \frac{\partial F}{\partial z}$. For small force gradients ($|F'| << c_B$), the shifted resonance frequency is approximately given by

$$\omega_1' = \omega_0(1 - \frac{F'}{2 \cdot c_B})$$ (9.3)

where ω_1 is the first resonance frequency of the free cantilever. The frequency shift $\Delta\omega_1 = F'/2c_B$ is then proportional to the force gradient F'. A repulsive force gradient causes an increase of the resonance frequency whereas an attractive force gradient $F' > 0$ causes a decrease of the resonance frequency. Scanning with constant resonance frequency corresponds to lines of constant force gradient. Note, that this approximation is only valid for small oscillation amplitudes and relatively large distances between probing tip and sample. For large amplitude oscillations it has been shown that the frequency shift also depends on amplitude and may not be directly related to force gradients (see also chapter II).

9.1.1   Design principles of cantilevers

Thermal vibrations

In order to measure small forces the spring constant has to be chosen as small as possible. Thermal excitations set a lower limit to the spring constant. If we apply the equipartition theorem to the rectangular cantilever with a spring constant c_B, we get the amplitude $\langle\psi_1\rangle$ of the thermal vibrations of the first eigenmode to be approximatively

$$\langle\psi_1\rangle^2 = 0.9707 k_B T/c_B \approx k_B T/c_B$$ (9.4)

which is similar to the simple harmonic oscillator. For the higher eigenmodes of order n, we obtain approximately

$$\langle\psi_n\rangle^2 = \frac{k_B T}{c_B} \frac{192}{((2n-1)\pi)^4}$$ (9.5)

$\langle\psi_n\rangle$ is inversely proportional to n^2. Consequently, the first eigenmode is dominating. A spring constant of 0.01 N/m results in an amplitude of 6.4 Å. If we apply a repulsive / attractive force to the probing tip the cantilever is stabilized/ destabilized and the amplitude $\langle\psi_1\rangle$ is reduced / increased significantly. In first approximation the amplitude is

$$\langle\psi_1\rangle^2 = \frac{k_B T}{1.0302(c_B - 0.9707 F')} \frac{k_B T}{c_B - F'} = \frac{k_B T}{c_{eff}}$$ (9.6)

where F' is the force derivative and $c_{eff} = c_B - F'$ the effective spring constant. If we assume a repulsive force gradient of 50N/m, the thermal vibration is reduced from 6.4Å to about 0.1Å .

Resonance frequency

The first eigenfrequency of the cantilever should be maximized to reduce the influence of ambient vibrations and acoustical perturbations. For a rectangular cantilever, the first eigenfrequency ω_0 is given by

$$\omega_1 = (1.8751)^2 \frac{t}{l^2}(\frac{E}{12\rho})^{1/2} = \frac{(1.8751)^2}{\sqrt{3}}\sqrt{c_B/m} \tag{9.7}$$

where t, l are the thickness and length, E the Youngs modulus and ρ the mass density. In the second part of the equation the resonance frequency is expressed in terms of the spring constant c_B and the mass m. In order to sustain a high resonance frequency, while reducing the spring constant, it is necessary to reduce the mass of the cantilever. Therefore, the dimensions have to be chosen as small as possible.

9.1.2 Minimum forces and Q-factor

The minimum force F_{min}, that can be measured by a force microscope is given by

$$F_{min} = \sqrt{\frac{4 \cdot c_B \cdot k_B T \cdot B}{Q \cdot \omega_0}} \tag{9.8}$$

and the minimum force gradient F'_{min} is given by

$$F'_{min} = \frac{1}{A}\sqrt{\frac{4 \cdot c_B \cdot k_B T \cdot B}{Q \cdot \omega_0}} \tag{9.9}$$

where c_B is the spring constant of the cantilever, k_B the Boltzmann constant, T the temperature, B the bandwidth, Q the Q-factor of the cantilever, A the amplitude of the cantilever oscillation and ω_0 the first resonance frequency of the cantilever. Small forces can be measured with small spring constant, high Q-factor and high resonance frequency. Spring constant and resonance frequency are not independent of each other ($\omega_0 \approx \sqrt{c_B/m}$). Approximately, we get

$$F_{min} = \sqrt{\frac{4 \cdot \omega_0 \cdot m \cdot k_B T \cdot B}{Q}} \tag{9.10}$$

Therefore, the smallest forces can be measured with a small resonance frequency, small mass and high Q. The lower limit for the frequency is given by the ambient vibrations and typically should not be below some kHz. Dependent on the state of miniaturization, the mass is minimized. At present, the smallest mass of cantilevers is in the range of some nano grams. Ultimately, one could reach some tens of pico grams. The optimization of the Q-factor is not at all trivial and needs some basic understanding of the dissipation mechanisms. The highest Q-factors, measured in high vacuum, are of the order of 10^6. For typical parameters of c_B=5N/m, ω_0=70kHz, Q=10^5, T=300K and B=1kHz, F_{min}=1.09·10^{-13}N. In practice, not the force but the force gradient is determined according to equation 9.9. For the above mentioned example with an excitation amplitude A=10nm, the force gradient is 10^{-5}N/m. For simple force laws of the form $F = \frac{const}{z^n}$, the force is related to the force gradient by $F = -\frac{n \cdot F'}{z}$. Ultimately, forces of the order of 10^{-19}N can be measured with high Q-factors at low temperature[20].

9.1.3 Preparation of cantilevers

Up to now there are two main techniques that are used for the production of levers. First, a thin wire or piece of metallic foil is bent and etched electrochemically. As known from the preparation of STM tips a radius of curvature less than 1000Å can be prepared by this method. The miniaturization of these levers is limited by the manual skill of the experimentalist. Nevertheless, this method is common in the field of magnetic force microscopy where ferromagnetic probing tips are required.

The second method for cantilever preparation involves microfabrication techniques . Several generations of cantilevers already exist. The first generation are simple cantilevers of rectangular and triangular shape. SiO$_2$ cantilever are etched out of an oxidized Si wafer. Standard photo masks are used to define the shape of the levers. With the exception of of the cantilever thickness, the geometrical dimensions of length and width are well known and highly reproducible, which facilitates the calibration of the cantilevers. Later, it is shown that the additional measurement of the resonance frequency of the cantilever also gives a reasonable determination of the thickness of the cantilever.

Small fragments of various materials, such as diamond, are glued to the end of these cantilevers, providing a reasonable probing tip as can be shown with scanning electron microscope (SEM). E.g, the end of a diamond piece may be formed by a tetrahedron confined by crystallographic faces[15]. Some progress has been made in the use of Si$_3$N$_4$ instead of SiO$_2$. Si$_3$N$_4$-levers are less fragile and the thickness can be reduced from 1.5μm to 0.3μm. The second generation of cantilevers had integrated tips with small radius of curvature. In 1989 two groups have succeeded in producing such cantilevers for the first time. Tom Albrecht at Stanford University has developed a procedure, where pyramidal pits are etched into the silicon wafer. Afterwards a Si$_3$N$_4$-film is deposited which follows the contours of the silicon. The Si$_3$N$_4$ is patterned into the shape of the cantilevers. When the silicon is etched away in the region of the lever, the free standing levers have pyramidal tips which are a replica of

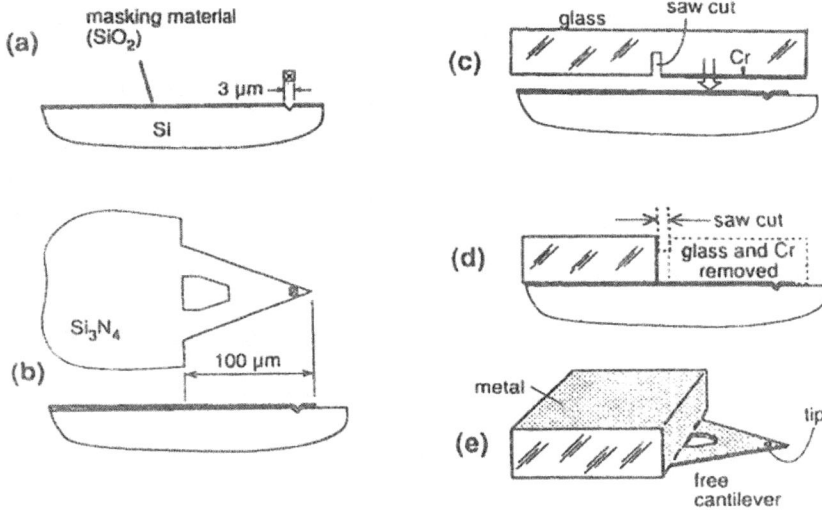

Figure 9.1: Microfabrication of silicon nitride cantilevers with integrated tips from[1]. (a) A pyramidal pit is etched in the silicon waver. (b) A silicon nitride film is deposited. (c) A glass plate anodically bonded to the nitride surface. (d) A saw cut releases removes the Cr-covered areas of the glass plate. (e) A metallic coating is deposited on the rear side of the cantilever.

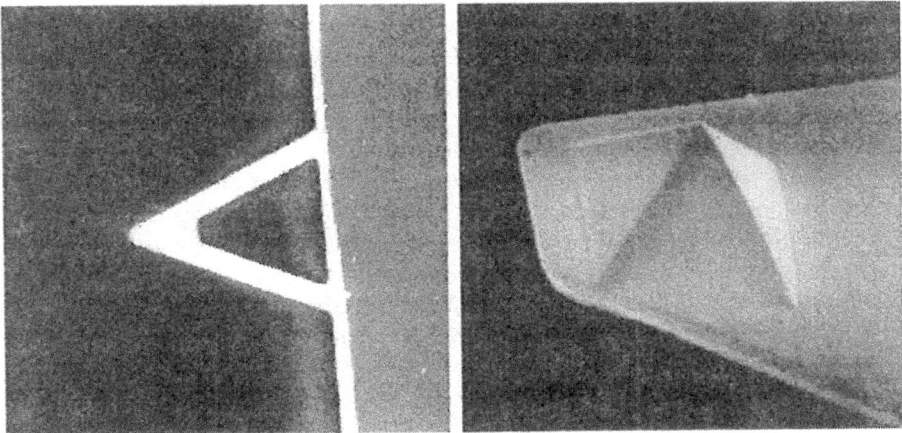

Figure 9.2: Scanning electron microscopy images of a triangular silicon nitride cantilever with integrated probing tip, manufactured by Park Scientific. Tip height is about 2 μm. See[50]. SEM-images by J.-P. Ramseyer.

the previous pyramidal pits in the silicon (see Fig. 9.1). This method yields tips with a sharpness of less than 300Å. The group of Olaf Wolter at IBM Sindelfingen has applied a different process to a (100)-silicon wafer fabricating a silicon cantilever with integrated tip. The sharpness is below 100Å. Due to the crystalline state of the material, this method has the intrinsic advantage that the very end of the tip is confined by crystallographic faces. It appears promising to apply field ion microscopy techniques to these Si-tips. In analogy to the single crystalline tungsten tips prepared by Fink et al. monatomic tips one might be able to prepare monatomic tips. At present, there are a variety of microfabricated cantilevers with integrated probing tips that are commercially available[3]. Roughly, the cantilevers can be divided up into two groups:

1. The simpler, less expensive Si_3N_4 levers that have radii of curvature of 10-100nm. They are very robust and of great value for contact force microscopy. Within different series and even within the same wafer, the spring constant and also the sharpness of the probing tips can vary up to a factor of four. Some manufacturers provide also oxide sharpened versions that have smaller probing tip radii (<30nm) but are more expensive. For some experiments on soft materials, such as polymers or biological materials, the small radii of curvature are not always the best choice, but a slightly blunter tip can provide less destructive imaging and thus also better, reproducible imaging conditions.

2. The silicon cantilevers are of great value for force microscopy. Especially, the high Q-factors make them ideal for non-contact force microscopy, where small force gradients are to be detected. Also, the geometrical dimensions and the radii of curvature are quite reproducible and facilitate calibration procedures. The thickness variations of the cantilevers are rather easily corrected by measuring the resonance frequency. For contact force microscopy, the rather small radii of curvature cause high pressures at the contact zone. In combination with the reactive nature of the Si-bonds, the Si-cantilevers are not always ideal for contact imaging of delicate, soft surfaces. Tapping mode turned out to be a real improvement where the interaction force is reduced and the imaging is less influenced by destructive tip-sample interactions.

Modifications of probing tips are possible. E.g., the probing tip can be oxide sharpened, as shown in Fig. 9.4 (left). Alternatively, contamination tips can be deposited with a high resolution SEM (see Fig. 9.4 (right)). Deposition of carbon nanotubes or diamond coatings are other possibilities. Further modifications are described in chapter VIII (chemical modifications of probing tips).

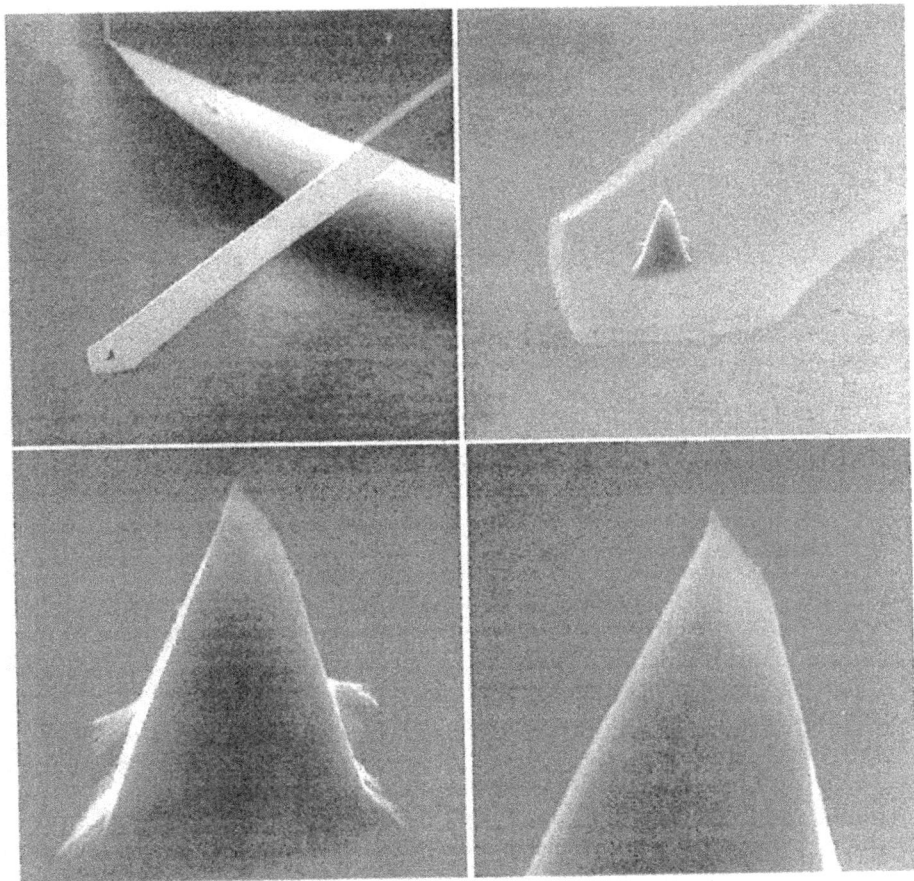

Figure 9.3: Scanning electron microscopy images of a rectangular silicon cantilever with integrated probing tip, manufactured by Nanosensors. Height of the probing tip is 12.5μm.[49]. SEM-images by J.-P. Ramseyer.

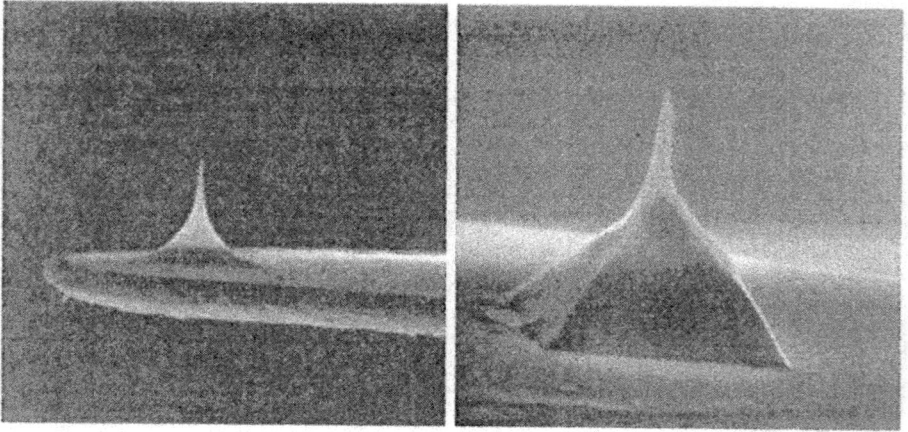

Figure 9.4: Scanning electron microscopy images of modified triangular cantilevers with integrated probing tip, manufactured by Park Scientific[50]. Left: oxidize sharpened. Right: Contamination tip has been deposited with field emission electron microscope. SEM-images by J.-P. Ramseyer.

9.2 Microscopes

9.2.1 Deflection sensors: Techniques to measure small cantilever deflections

There are different techniques to detect the small displacement of the lever (cf. Fig. 9.5). Originally, Binnig et al. proposed electron tunneling as the deflection sensor[4]. Due to the exponential decay of tunneling current with the distance between the electrodes it is possible to measure distances of less than 10^{-2}Å with the STM[5]. Other sensors incorporate optical interferometry[6,7] or the reflection of a laser beam from a mirror mounted on the rear side of the lever[8,9,10]. Both setups achieve a vertical resolution of the same order. Actually, the force sensitivity is already limited by the thermal noise of the cantilever. Even capacitance methods seem to have the potential to achieve such high resolutions[11,12]. Recently, methods, such as piezoresistance[13] and piezoelectricity[14], have been used to microfabricate the whole force sensor, including the deflection sensor.

Electron tunneling

Measuring the tunneling current between the tip and the conductive rear side of the lever is a very sensitive technique. A decrease in the gap of 1Å will increase the current by an order of magnitude. Experimentally, a z-resolution

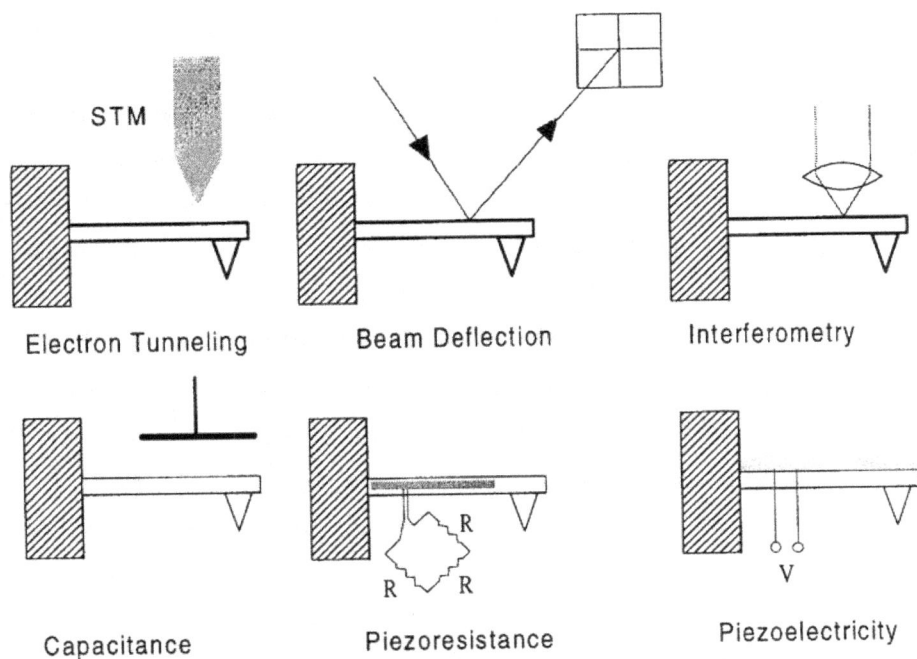

Figure 9.5: Deflection sensors for scanning force microscopy.

of 0.01Å can be achieved with a bandwidth of few thousand Hz. Especially, high speed scan rates of about 100 - 1000 Hz per scan line can be performed by digitizing the tunneling current instead of the feed-back output signal. Due to the high mechanical stability and low thermal drift rates the tunneling SFM can be operated in the dc mode reliable and efficient[15,16]. However, some aspects of the tunneling SFM have to be examined carefully . First the interaction between tunneling tip and lever is not simply negligible. As has been shown by Dürig et al. an STM causes forces of the order of 10^{-9}N[17]. Secondly, the roughness of the rear side of the lever can influence the SFM-image[16].

Optical interference

In 1986 McClelland et al.[7,18] and Martin et al.[6,19] have presented the first optical interferometers based on homodyne and heterodyne interferometry. The main advantages compared to the tunneling SFM are

1. The interaction between the laser beam and the lever can be neglected.

2. Due to the large beam diameter the optical SFM are insensitive to the roughness of the rear side of the lever.

Especially, for dynamic force microscopy, also called non-contact force microscopy, optical SFMs have shown to be favourable achieving a sensitivity of 10^{-4}Å$/\sqrt{Hz}$. Forces as small as 10^{-16}N could be measured[20]. Optical SFM's developed towards more stable and compact designs using fiber optical devices[21,22,23] or differential interferometers based on the Normarski principle[24,25,26]. An intriguing simple version of an interferometer has been presented by Sarid et al. where the reflected light from the cantilever is fed back into the laser diode cavity[27]. The above mentioned instruments combine the excellent sensitivity in the high frequency regime with a good dc-stability. These detectors already achieve a force sensitivity limited only by the thermal noise of the cantilever. A lateral modulation technique has been suggested by Göddenhenrich et al. to measure both normal and lateral forces with one interferometer[29].

Laser beam deflection

Gerhard Meyer and Nabel M. Amer[8] have developed an optical method which appears suitable for measurements in the dc-mode achieving atomic resolution . At the same time, Marti et al. have developed independently a similar instrument[9,10]. A laser beam is reflected off the rear side of the cantilever. The deflection is monitored with a position sensitive detector (PSD). The signal-to-noise ratio S/N is given by

$$S/N = \frac{\sqrt{a^3 b}}{l} \frac{\sqrt{IRR_s}}{\sqrt{B}} \Delta z \qquad (9.11)$$

where a,b are the dimensions of the mirror, l the cantilever lengths, R the mirror reflectivity, R_s the spectral responsitivity of the PSD, I the laser intensity and B the bandwidth. For a signal-to-noise ratio of 1 a spectral sensitivity of $10^{-4} \text{Å}/\sqrt{Hz}$ can be calculated. In the dc-mode a z-resolution of about 0.1Å has been found. Actually, this design has been already implemented in UHV[30,31]. A lock-in technique for measuring friction at high scan velocities has been introduced by Colchero et al.[32].

Capacitance

The group of McClelland at IBM Almaden has demonstrated that capacitance detection is suitable for the detection of both lateral and normal forces and can be implemented into ultrahigh vacuum[11]. Göddenhenrich et al.[12] optimized this method for magnetic force microscopy. The main problem of the technique is the electrostatic attraction between the cantilever and the electrodes, which leads to instabilities of weak cantilevers. An advantage is the suitability of the detector for microfabrication, which has been demonstrated by the group of de Rooij in Neuchatel[33]. The measurement of small capacitances is not easy, but can be achieved with high frequency modulation techniques (typically >10MHz).

Piezoresistivity and piezoelectricity

At present, two different sensors were reported that could be microfabricated with piezoresistive detection. Tortonese, Barrett and Quate presented cantilevers with a piezoresistive layer (highly doped silicon region)[13]. Giessibl et al. were the first to implement the piezoresistive cantilevers into UHV and could demonstrate good performance in contact mode where atomic-scale features are observed on MoS$_2$ were observed. In non-contact mode even better performance is found because of reduced 1/f-noise. Examples are monatomic steps on KCl[37] and atomic resolution imaging of Si(111)7x7[38]. Brugger et al.[34] and Kassing et al.[35] could demonstrate that both lateral and normal forces can be detected with piezoresistive detection, using cantilevers with two similar Wheatstone bridges. Itoh et al. could show the performance of an SFM based on piezoelectricity[14].

9.3   Calibration procedures

9.3.1   Calibration of scanner

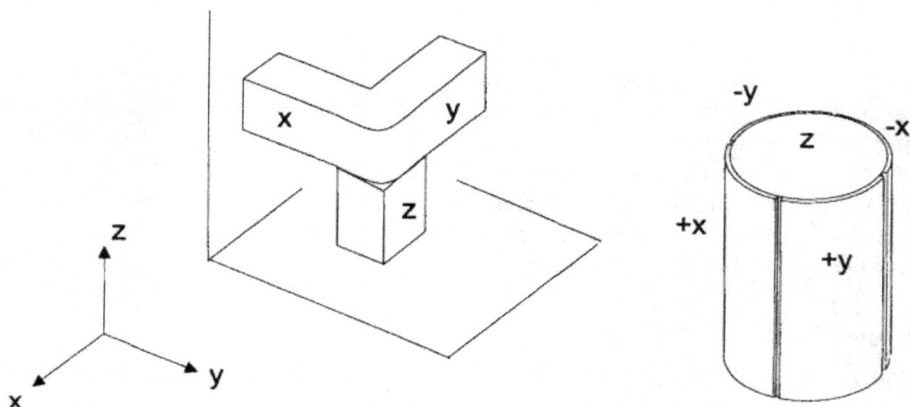

Figure 9.6: Piezoelectric scanners: Left side: Tripod scanner. Three bars are used to control the x,y and z-displacements. Rather orthogonal movements are achieved. Right side: Piezoelectric tube scanner. The tube is poled in radial direction. The outer electrode is segmented in four electrodes. Similar to a bimorph, the tube is bent laterally (x,y-directions), when the voltage is applied to one of these quadrant electrodes. The bending is increased when the voltage is applied with negative sign to the opposite electrode as shown in the figure. In addition, the inner electrode is used for the z-motion.

For both scanning in x- and y-directions and adjusting the z-position with the help of the feed-back loop, piezoelectric elements are used. A voltage is applied across the piezoelectric material. In most cases, the transverse piezoelectric effect is used, which means that the material will extend or contract perpendicular to the applied field $\vec{E}$[39]. The change in length for a simple rod of length L is given by $\Delta L = L \cdot |\vec{E}| \cdot d_{31}$, where d_{31} is the piezoelectric coefficient. Scanners have to be designed according to the desired scan range and resonance frequency of the microscope. For low temperature experiments, the reduced sensitivity has to be taken into account. In UHV, the bake-out temperature gives a lower limit for the Curie-temperature. Taking these preconditions into account, the piezoelectric elements can be selected. A list of manufacturers of piezoceramics for SFM is listed below[48]. The most common material is lead zirconate titanate (PZT). Different Zr/Ti ratio yield a variety of materials that differ in Curie temperature, depolarization voltage and piezoelectric coefficients. Different geometries, such as tripod scanner[5], tube scanner[40,41], bimorph scanner[42,43] or piezostack-scanner[44] have been implemented. The most common geometry of a scanner is the tube scanner. The sensitivity in z-direction is given by:

$$\Delta z = d_{31} \cdot V \cdot \frac{L}{H} \tag{9.12}$$

where d_{31} is the piezoelectric coefficient (for PZT-5H d_{31}=-2.62Å/V). The deflection in x-direction (y-direction is analogous) is given by

$$\Delta x = \frac{L^2}{2 \cdot R} = 2 \cdot \sqrt{2} d_{31} \cdot V \cdot \frac{L^2}{\pi \cdot D \cdot H} \tag{9.13}$$

where $R = \pi D H / (4\sqrt{2} d_{31} V)$ is the curvature of bending. Usually, the x-sensitivity is defined by

$$s_x = \Delta x / V = 2 \cdot \sqrt{2} d_{31} \cdot \frac{L^2}{\pi \cdot D \cdot H} \tag{9.14}$$

Thus, a PZT-5H scanner with a diameter of D=12.7mm, a wall thickness

Material Properties	PZT-5A	PZT-5H	PZT-8
d_{31} (Å/V)	-1.73	-2.62	-0.95
d_{33} (Å/V)	3.80	5.83	2.20
Dielectric constant	1725	3450	1050
Curie temperature (°C)	350	190	300
Mechanical Q	100	65	960

Table 9.1: Properties of PZT-materials

of H=1mm and a length of L=12.7mm gives a sensitivity of 30Å/V. The sensitivity is reduced by a factor of two when the voltage is applied only to one quadrant. For more details about calculations of piezo sensitivities see Ref.[45,46,47].

9.3.2 Calibration of lateral forces

The calibration procedure is one of the essential parts of FFM-experiments. Each cantilever should be characterized accurately. Manufacturer's data are

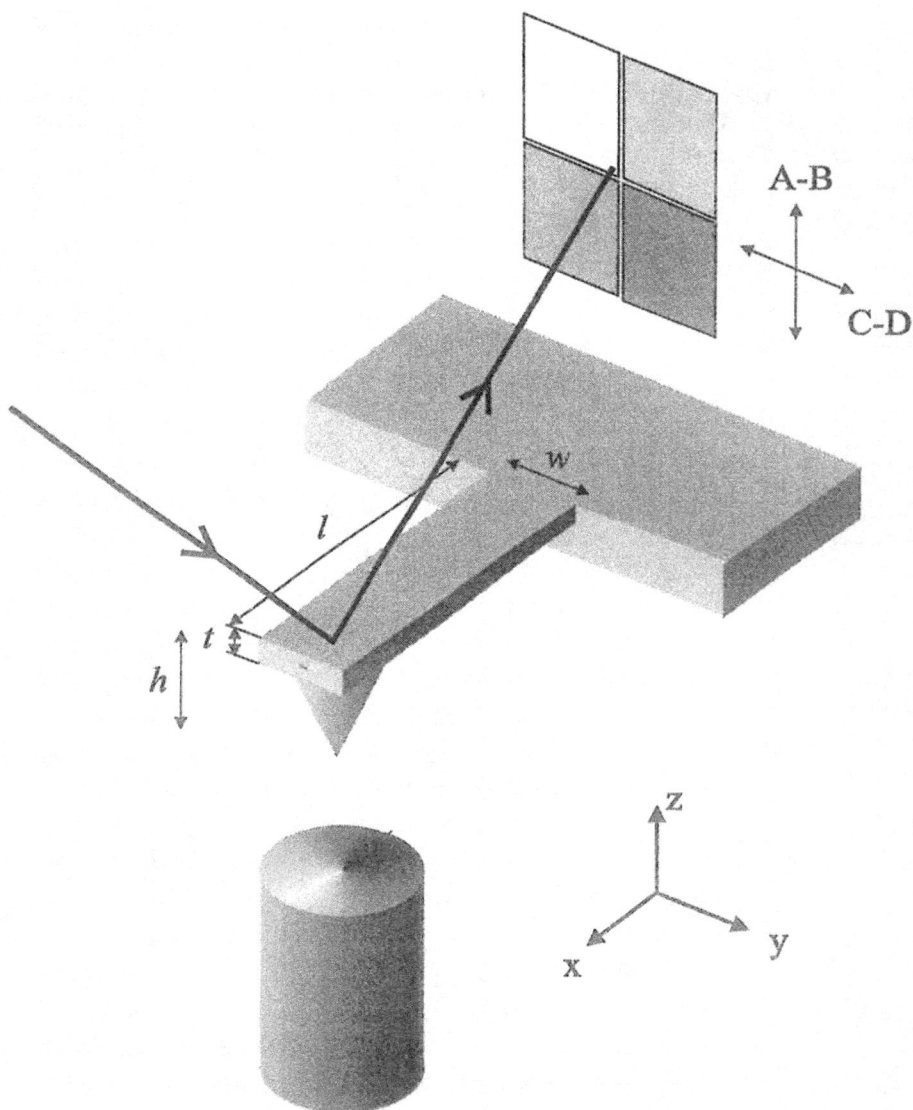

Figure 9.7: Schematic diagram of the beam-deflection AFM. The relevant dimensions of the rectangular cantilever are indicated: length l, width w, thickness t and height of the tip h (note: $h = h_{eff} + t/2$. Normal and lateral forces are measured with normal and torsional motions of the cantilever. A laser beam is reflected off the rear side of the cantilever. Angular deflections of the laser beam are measured with a position sensitive detector (4-quadrant photo diode). The A-B-signal is proportional to the normal force and the C-D-signal is proportional to the torsional force.

usually not sufficient and can lead to errors of up to a factor 10. Thus, each cantilever has to be characterized. One way is to use an electron microscope and to determine all the relevant parameters, such as: Tip radius R; height of tip h ; width, thickness and length of cantilever (w, t, l) and position of tip on the cantilever. In addition, elastic constants are needed: Youngs modulus E, shear modulus G. Having determined all these parameters, the normal spring constant c_B and the torsion spring constant c_t for a rectangular cantilever are given by:

$$c_B = \frac{E \cdot w \cdot t^3}{4 \cdot l^3} \qquad (9.15)$$

$$c_t = \frac{G \cdot w \cdot t^3}{3 \cdot h^2 \cdot l} \qquad (9.16)$$

Where E is the Youngs modulus and $G = E/(2(1+\nu))$ For commercially avail-

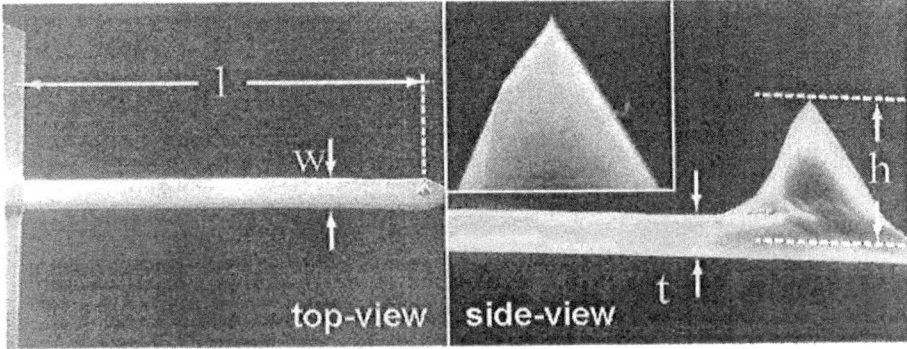

Figure 9.8: SEM-images that show the relevant dimensions of a rectangular cantilever from[49]: Length l, width w, thickness t, height of the tip h. Typical dimensions for contact levers are: $l=445\mu$m, $w=43\mu$m, $t=4.5\mu$m, $h_{eff}=12.5\mu$m and $h = h_{eff} + t/2 = 14.75\mu$m. Courtesy of Roland Lüthi.[51]

able silicon cantilevers[49,50], the elastic properties are well-defined and the first resonance frequency in normal direction $f_1 = \omega_1/2\pi$ can be used to determine the thickness of the cantilever more accurately[52]:

$$t = \frac{2 \cdot \sqrt{12}\pi}{1.875104^2} \sqrt{\frac{\rho}{E}} f_1 \cdot l^2 = 7.23 \times 10^{-4} s/m \cdot f_1 \cdot l^2 \qquad (9.17)$$

where ρ is the density of the cantilever (For silicon cantilevers: $\rho=2.33 \cdot 10^3$kg/m^3, $E=1.69 \cdot 10^{11}$N/m^2).

Thus, the procedure is more simplified: The lateral dimensions of width and length (w, l) can be determined with an optical microscope. Usually, these dimensions are quite reproducible by current microfabrication procedures. The height of the tip h can vary a few microns and should be checked with an optical or electron microscope. Alternatively, Cleveland et al.[53] have suggested that small particles of defined mass are dropped on the cantilever. The corresponding decrease of the resonance frequency due to the mass change is used for calibration of normal spring constant c_B. Similar approaches may be possible for lateral spring constants, but were not reported so far. Bhushan et al. have suggested to use a combination of two springs, where one is well-defined (large stainless steel spring) and to measure the common deflections[54]. Hutter et al. suggested to measure the thermal spectrum of the cantilever and to fit the Lorentzian curve, where the spring constant is the free fit parameter[55].

For the beam-deflection-type AFM the sensitivity of the photodetector S_z [nm/V] has to be determined by measuring force vs. distance curves on reasonably hard surfaces (e.g., Al_2O_3) where elastic deformations can be neglected and the movement of the z-piezo z_s equals the deflection of the cantilever z_t. It is found that the laser focus should be well defined in order to position the laser beam above the probing tip and to achieve accurate calibrations.

Force-distance curves should be performed before and after spectroscopy measurements in order to exclude changes of the sensitivity S_z, e.g. caused by variations of laser intensity or laser beam position. Following this procedure, the forces in normal and lateral directions are given by the difference signals of the 4-quadrant detector in normal direction U_{A-B} and lateral direction U_{C-D}:

$$F_N = c_B \cdot S_z \cdot U_{A-B} \tag{9.18}$$

$$F_L = \frac{3}{2} \cdot c_t \frac{h}{l} \cdot S_z \cdot U_{C-D} = S_x \cdot U_{C-D} \tag{9.19}$$

If the laser beam is not positioned above the probing tip, which is close to the end of the cantilever, then the distance from the end of tip $x - l$ has to be taken into account (see[61]):

$$F_N = c_B \cdot z(l) = c_B \cdot S_z \cdot U_{A-B} \tag{9.20}$$

$$F_L = c_t \frac{hl}{x} \cdot \frac{S_z \frac{\partial \bar{z}}{\partial x}(x)}{\bar{z}(l)} \cdot U_{C-D} = S_x \cdot U_{C-D} \tag{9.21}$$

where $\bar{z}(x) = \frac{6 \cdot l x^2 - 2 x^2}{E \cdot w \cdot t^3}$ is the normalized bending curve of the rectangular cantilever.

Triangular cantilevers are more difficult to be accurately calibrated, both the thickness of the cantilever, the length of the tip and the position of the tip on the cantilever can lead to large deviations of the spring constants from manufacturer data. Thus, it becomes necessary to characterize each cantilever with electron microscopy. Using this procedure and the formulae given by

Neumeister et al. a reasonable accuracy of 10% could be achieved[56]. See also the reference list about lateral spring constant calculations[57].

Alternative, experimental methods to calibrate normal and lateral forces are given in the literature. Marti et al.[60] have shown that a beam-deflection AFM can be calibrated with a thin glass plate of thickness T, which is positioned into the optical path. Then, the inclination angle α_{tilt} is changed, which gives a calibration of the detector. A change of the tilt angle α_{tilt} causes a movement of the laser beam by a distance

$$Y = T(1 - \frac{1}{n})\alpha_{tilt} \tag{9.22}$$

where n is the index of refraction of the glass plate. The same movement of the laser beam is observed, when the cantilever is bent by the distance Δz_t:

$$Y = 2L(\frac{3\Delta z_t}{2l}) \tag{9.23}$$

where L is the distance between cantilever and photo diode and l the length of the cantilever. Δz_t is proportional to the measured signal $\Delta z_t = S_z \Delta U_{A-B}$. Thus, we can determine S_z independently:

$$S_z = \frac{T}{3}\frac{l}{L}(1 - \frac{1}{n})(\frac{\alpha_{tilt}}{\Delta U_{A-B}}) \tag{9.24}$$

Similarly, the calibration of the lateral forces is achieved by tilting the glass plate in the $C - D$-direction by an angle α^x_{tilt}:

$$X = T(1 - \frac{1}{n})\alpha^x_{tilt} \tag{9.25}$$

The same movement is caused by lateral force at the tip apex:

$$X = \frac{2F_L L}{c_t h} \tag{9.26}$$

where h is the effective length of the tip. Thus, we can determine

$$S_x = c_t h \frac{T}{2L}(1 - \frac{1}{n})(\frac{\alpha^x_{tilt}}{\Delta U_{C-D}}) \tag{9.27}$$

by the thin glass plate tilting method in an independent way. For this method, some space has to be available for the glass plate, which is possible for most commercial instruments.

One has to be aware, that small misalignments will cause cross-talk between lateral and normal forces:

$$X' = X\cos\gamma - Y\sin\gamma \tag{9.28}$$
$$Y' = X\sin\gamma + Y\cos\gamma \tag{9.29}$$

For well oriented photo diodes the misalignment angle γ is below $1°$. A simple check, whether the normal force is influenced by friction, is to scan back and forward and to observe whether the topography signal depends on direction. In some cases (heterogeneous samples), one may observe that the islands turn into hills, which is a bad sign. The lateral force signal may also be influenced by topography (the torsional spring constant is larger than the normal spring constant!) and has to be examined carefully. If one is only interested in the dissipative part of the lateral forces (friction), one can take the difference between forward and backward scan divided by two, which is not influenced by the normal forces.

Alternatively, Ogletree et al.[61] have suggested an in-situ calibration of lateral forces on samples with well-defined morphology. The method is based on comparing lateral force signals on surfaces with different slopes. Then, the topography effect I, discussed in chapter IV, is used to calibrated the lateral forces. In first approximation the non-dissipative part of the lateral force is given by the local slope

$$F_L = \frac{\partial s}{\partial x} F_N \qquad (9.30)$$

More precise approximations are given in the paper of Ogletree et al.[61]. Suitable samples are either microfabricated structures with well-defined slopes or samples with facets, such as (103) and (101) facets of $SrTiO_3$ (See also Sheiko et al.[62]). A good introduction of error analysis of lateral force calibration is given by Schwarz et al.[63]. An important aspect is the influence of feed-back oscillations on friction. Schwarz et al. have shown that small oscillations due to the increased gain of the feed-back loop cause a reduction of friction. Therefore, the gain of the feed-back loop should be selected below a critical value.

The radius of curvature of the tip can be determined with a scanning electron microscopy. Otherwise, well-defined structures, such as step sites[58,62] or whiskers[59] can be imaged. The image of those high-aspect ratio structures is a convolution with the tip structure. A simple deconvolution algorithm allows to extract the radius of curvature of the probing tip.

9.4   Modes of operation

9.4.1   Imaging modes

In analogy to the modes of STM, such as constant current and constant height mode, the new property, namely the force, can be kept constant or varied during scanning. In addition, one may also use dynamic modes, where the cantilever is oscillated. Hence, the following most common modes can be distinguished.

- *Equiforce mode:* In principle the equiforce mode is the most important mode being the easiest to interpret. During the scan process the deflection of the lever and therefore the force is kept constant by the feed-back loop. This mode is the most important mode for contact mode AFM and also for FFM.

- *Variable deflection mode:* The feed-back is disabled or slowed down. During the scan process the deflections are measured by the detector. If the relative variations of the force are not too large this mode can be interpreted in a similar manner as with the the equiforce mode. In contrast to the equiforce mode, higher scan rates can be achieved, because the feed-back has not to be used. At large separations (non-contact mode), the variable deflection mode yields simply a force image which is of interest to fields, such as magnetic or electrostatic force microscopy.

- *Constant frequency mode:* The cantilever is excited close or at the resonance frequency. During lateral scanning, the frequency of the cantilever is kept constant. Historically, this mode is also called constant force gradient mode, which is only a valid expression for small amplitudes. Typically, the cantilever is operated in a non-contact regime or near-contact regime. Thus, frictional forces as discussed in this book are not present. However, long-range dissipation can be observed, which is currently investigated.

- *Tapping mode:* The cantilever is excited close or at the resonance frequency. During lateral scanning, the amplitude is kept constant. The oscillation amplitudes are rather large (>10nm) and the tip gets into contact during part of the oscillation cycle. Thus, dissipative processes may be present, but are usually not observable in the lateral force signal. However, phase contrast has been observed. Essentially, the phase difference between excitation signal and lever oscillation is measured. This phase difference is related to local variations of the viscosity.

- *Spectroscopic modes*, being common in STM, are not so far developed in SFM. One example has been given by Mate et al. The authors performed spatially resolved force-distance curves on a liquid polymer determining not only capillary forces but also the thickness variations of the liquid film[64]. 2d-histogram techniques are discussed in chapter II. Properties, such as adhesion, local compliance, viscosity or friction can be investigated as a function of properties, such as normal force, velocity or scan direction.

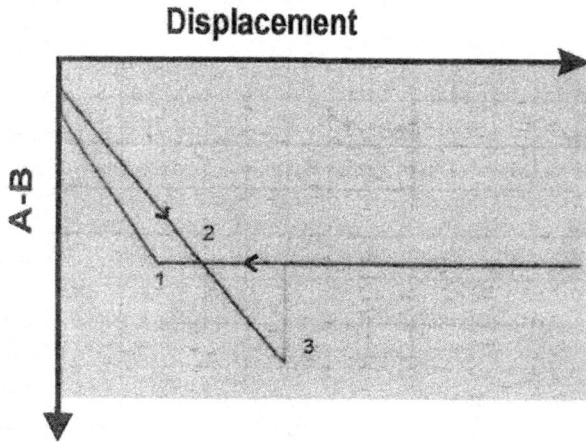

Figure 9.9: The normal deflection $A - B$ is plotted vs. the sample movement z_s. Due to the low spring constant, an instability occurs. The sample was an oxidized silicon wafer, measured with a silicon tip in dried nitrogen.

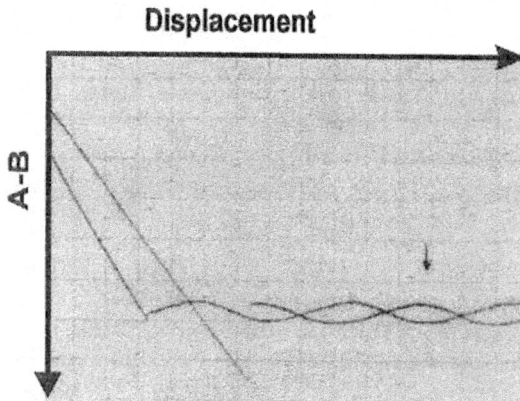

Figure 9.10: Same as Fig. 9.9, but with bad focus, which leads to interference effects between the beam reflected from the sample and the beam reflected from the cantilever.

9.4.2 Force vs. distance curves

SFM can measure the force as a function of distance between the probing tip and the sample. The $z_t(z_s)$-plots contains information about the interaction between sample and tip. z_s is the movement of the sample and z_t is the movement of the lever and tip. Both movements are approximately perpendicular to the sample surface. Multiplicating z_t with the spring constant c_B the force $F = c_B \cdot z_t$ can determined. Neglecting elastic deformations of the sample and tip the interaction distance between tip and sample d is given by $d = z_t - z_s$. Thus we can determine the force-distance dependence $F(d)$ from the $z_t(z_s)$-plot.

Firstly, the simplest mechanisms of such $z_t(z_s)$-plots will be explained (cf. Fig. 9.9). When the sample is approached towards the probing tip the lever bends due to the attractive force. At point 1 the gradient of the attractive force surpasses the spring constant c_B which leads to a first point of instability. Afterwards, the level of zero net force is passed, which means that the attractive and repulsive force cancel each other. In the repulsive regime the sample just pushes the lever. When the sample is retracted again point 2 is passed. Then, the maximum attractive force, called adhesive force or pull-off force, is reached (point 3), where a second instability occurs and the tip jumps out of contact. Finally, we reach again the free lever (no measurable interaction between tip and sample). Fig. 9.9 shows an experimental $z_t(z_s)$-plot on silicon measured in dried nitrogen. From this plot we can determine parameters such as the pull-off force, which is also called adhesion. Several phenomena such as capillary forces, tip shape and piezo creep effects impede a more quantitative determination of the interaction under ambient conditions. Weisenhorn et al. could actually demonstrate the influence of capillary forces by comparing $z_t(z_s)$-plots in air and water[65]. If the lever was fully immersed into water capillary forces can be excluded. A significantly decreased pull-off force of 10^{-9}N in water compared to 10^{-8} to 10^{-7}N in air has been observed. Furthermore, piezo creep effects can be minimized with faster acquisition cycles or actively corrected z-piezos (e.g., with capacitance detectors). See also chapter VIII for adhesion measurements and molecular recognition experiments. More details on the relationship between adhesion and friction are given in chapters II and VIII.

Fig. 9.10 shows a $z_t(z_s)$-plot, where the focus was not optimized. Here, interference between the laser beam reflected from the sample and the beam reflected from the cantilever occurs. The distance between the interference maxima, $d_{max} = \lambda / \sin\theta$ is related to the wavelength of the laser source, λ (typically about 620nm), where the angle of incidence of the laser beam relative to the sample surface, θ, is taken into account. This interference effect can also disturb the lateral force imaging. Fig. 9.11 shows a FFM-image of MoS_2-platelets on mica, as discussed in chapter VIII. However, the focus was rather bad ($>30\mu m$), which caused interference patterns, which are pronounced in the FFM-image. The use of optimized optics with small-area focus and cantilevers with gold coating ($>20nm$) can minimize these effects. Alternatively, light emitting diodes can be used, which have a broader spectrum and small coherence length, which eliminates this problem.

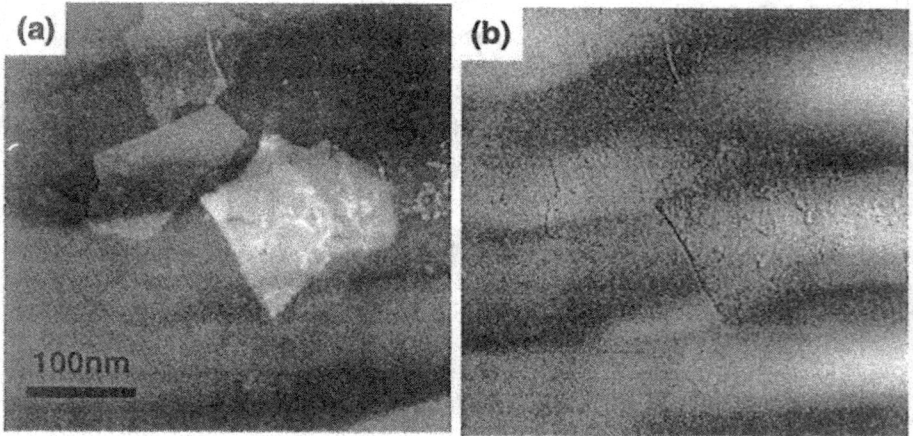

Figure 9.11: FFM-image of MoS_2-platelets on mica, imaged with a bad focus, which leads to interference patterns (stripes across the whole image). (a) topography (b) lateral force map. For more details about the sample see chapter VIII, where images without interference are shown.

1. T. Albrecht, S. Akamine, T.E. Carver, and C.F. Quate, *J. Vac. Sci. Techn. A* **8**, 3386 (1990).
2. Some of the SPM-manufacturers are listed below (November 97):

Omicron Vakuumphysik GmbH, Idsteinerstr. 78, D-65232 Taunusstein, Germany
Tel: +49 (0)61 28 987 0 Fax: +49 (0)61 28 987 185
e-mail: sales@omicron-instruments.com
website: http://www.omicron-instruments.com

Park Scientific Instruments, 1171 Borregas Avenue, Sunnyvale, CA 94089, USA
Tel: +1 408 747-1600 Fax: +1 408 747-1601
e-mail: Info@park.com
website: http://www.park.com

Digital Instruments, 112 Robin Hill road, Santa Barbara, California 93103, USA
Tel: +1 800 873 9750 Fax: +1 805 967 7717
e-mail: info@di.com
website: http://www.di.com

TopoMetrix Corporation, 5403 Betsy Ross Drive, Santa Clara, CA 95054-1162 Tel: 408.982.9700 Fax: 408.982.9751
e-mail: info@topometrix.com
website: http://www.topometrix.com

Nanosurf AG, Austrasse 4, CH-4410 Liestal, Switzerland
Tel:+41 61 9030611 Fax:+41 61 9030613
e-mail: info@nanosurf.ch
website: http://www.nanosurf.ch

CSEM Instruments, Rue Jaquet Droz 1, CH-2007 Neuchatel, Switzerland
Tel: +41 32 720 5111 Fax: +41 32 720 5730
e-mail: instruments@csemne.ch
website: http://www.csem.ch/instrum

Burleigh Instruments, Inc., P.O. Box E, Burleigh Park Fishers, NY 14453-0755 Tel. (716) 924-9355 Fax. (716) 924-9072
e-mail:info@burleigh.com
website: http://www.burleigh.com

Seiko Instruments Inc., Microsystem Dept., Takatsuka Unit, 563
Taktsuka-shinden, Matsudo-shi, Chiba 271, Japan
e-mail: webmaster@sii.co.jp
website: http://www.seiko.com or http://www.sii.co.jp

Quesant Instrument Corporation, 29397 Agoura Road, Suite 104,
Agoura Hills, CA 91031, USA
Tel: +1 (818) 597 0311 Fax: +1 (818) 991 5490 e-mail:
qsales@quesant.com
website: http://www.quesant.com

Surface Imaging Systems GmbH, Kaiserstrae 100, Herzogenrath Tel:
+49 2407 96 147 Fax: +49 2407 96 275
E-Mail: surface.imaging@t-online.de
website: http://www.zeiss.de/mi/rsm_e/ultra_e.html

Besocke Delta Phi GmbH, Auf der Tuchbleiche 8, D-52428 Jlich
Fax: +49 2461 56025

Carl Zeiss,73446 Oberkochen, Germany
Tel.: ++49(0)7364 - 20 - 0 Fax: ++49(0)7364 - 6808
Carl Zeiss Jena GmbH, 07740 Jena, Germany
Tel.: ++49(0)3641 - 64 - 0 Fax: ++49(0)3641 - 64 - 2856
e-mail:info@zeiss.de
website: http://www.zeiss.de/contacts

DME, Danish Micro Engineering A/S, Transformervej 12, DK-2730
Herlev, Denmark
Tel: +45 42 84 92 11 Fax: +45 42 84 91 97

Veeco Instruments Inc., 602 East Montecito Stree, Santa Barbara, CA
93103
Tel: +1 805 963 4431 Fax: +1 805 965 0522
e-mail: info@veeco.com
website: http://www.veeco.com

Oxford Instruments GmbH, Kreuzberger Ring 38, D-65205 Wiesbaden
Tel: +49 611 764-0 Fax: +49 611 764-175
email: Oxford.Instruments.Scientific@t-online.de

JEOL LTD., 1-2 Musashino 3-chome, Akishima Tokyo 196, Japan

Tel: (0425) 42-2187 Fax: (0425) 46-5757
e-mail: info@jeol.com
website:http://www.jeol.com

RHK Technology, Inc., 1750 Hamlin Road, Rochester Hills, MI 48309
Tel: +1 248 656 3116 Fax: +1 248 656 8347
email: info@rhk-tech.com
website: http://www.rhk-tech.com

3. Some of the manufacturers of cantilevers (November 97):
NANOSENSORS, Dr. Olaf Wolter GmbH, IMO-Building, Im Amtmann 6, 35578 Wetzlar-Blankenfeld, Germany, Tel. (+49) 6441 97 88
40 Fax: (+49) 6441 97 8841
e-mail: nanosensors@compuserve.com

Park Scientific Instruments, 1171 Borregas Avenue, Sunnyvale, CA
94089, USA
Tel: +1 408 747-1600 Fax: +1 408 747-1601
e-mail: Info@park.com
website: http://www.park.com

NT-MDT Co., Zelenograd Research Institute of Physical Problems,
1034460 Moscow, Russia, Tel: +7 (095) 535-0305 Fax:+7 (095) 535-
6410
e-mail: spm@ntmdt.zgrad.ru //http://www.ntmdt.ru

Olympus Optical Co., Ltd., Tokyo, Japan.
http://www.olympus.co.jp/LineUp/Technical/Cantilever
/levertopE.html#menu
TopoMetrix Corporation, 5403 Betsy Ross Drive, Santa Clara, CA
95054-1162 Tel: 408.982.9700 Fax: 408.982.9751
e-mail: info@topometrix.com
website: http://www.topometrix.com

Digital Instruments, 112 Robin Hill road, Santa Barbara, California
93103, USA
Tel: +1 800 873 9750 Fax: +1 805 967 7717
e-mail: info@di.com
website: http://www.di.com

Nanonics, (Micropipette Probes), 21 Havaad Haleum St., Jerusalem, Israel
Tel: 972 2 635243 Fax: 972 2 61745

4. G. Binnig, C.F. Quate and Ch. Gerber, Atomic force microscope, *Phys. Rev. Lett.* **56**, 930-933 (1986).
5. G. Binnig, H. Rohrer, Ch. Gerber and E. Weibel, 7x7 reconstruction on Si(111) resolved in real space, *Phys. Rev. Lett.* **50**, 120-123 (1983). G. Binnig and H. Rohrer, Scanning tunneling microscopy, *Helv. Phys. Acta* **55**, 726 (1982).
6. Y. Martin, C.C.Williams and H.K. Wickramasinghe, Atomic force microscope-force mapping and profiling on a sub 100-Å scale, *J. Appl. Phys.* **95**, 4723-4729 (1987).
7. G.M. McClelland, R. Erlandsson and S. Chiang, Atomic force microscopy: general principles an a new implementation, in *Review of Progress in Quantitative Non-Destructrive Evaluation*, edited by D.O. Thompson and D. E. Chimenti (Plenum, New York, 1987), Vol. 6B, pp. 1307-1314.
8. G. Meyer and N.M. Amer, Novel optical approach to atomic force microscopy, *Appl. Phys. Lett.* **53**,1045-1047 (1988).
9. O. Marti, V. Elings, M. Haugan, C.E. Bracker, J. Schneir, B. Drake, S.A.C. Gould, J. Gurley, L. Hellemans, K. Shaw, A.L. Weisenhorn, J. Zasadinski and P.K. Hansma, Scanning probe microscopy of biological samples and other surfaces, *J. of Microscopy* **152**, 803-809 (1986).
10. S. Alexander, L. Hellemans, O. Marti, J. Schneir, V. Elings, P.K. Hansma, M. Longmire and J. Gurley, An atomic-resolution atomic-force microscope implemented using an optical lever, *J. Appl. Phys.* **65**, 164-167 (1989).
11. G.M. McClelland and J.N. Glosli, Friction at the atomic scale, in *Fundamentals of Friction: Macroscopic and Microscopic Processes* edited by I.L. Singer and H.M. Pollock, p. 405-425, NATO ASI Series E: Applied Sciences, Vol. 220, Kluwer Academic Publishers (1992).
12. T. Göddenhenrich, H. Lemke, U. Hartmann and C. Heiden, Force microscope with capacitive displacement detection, *J. Vac. Sci. Techn.* A **8**, 383-387 (1990).
13. M. Tortonese, R.C. Barrett and C.F. Quate, Atomic resolution with an atomic force microscope using piezoresisitive detection, *Appl. Phys. Lett.* **62**, 834-836 (1993).
14. T. Itoh, and T. Suga, Proc. *7th Int. Conf. Sol. Stat. Sens. Act.*, 610, (1993).
15. E. Meyer, H. Heinzelmann, P. Grütter, T. Jung, T. Weisskopf, H.-R. Hidber, R. Lapka, H. Rudin and H.-J. Güntherodt, Comparative study

of lithium fluoride and graphite by atomic force microscopy (AFM), *J. of Microscopy* **152**, 269-280 (1988).

16. H. Hug, T. Jung and H.-J. Güntherodt, A high stability and low drift atomic force microscope, *Rev. Sci. Instr.* **63**, 3900-3904 (1992).

17. U. Dürig, O. Züger and D.W. Pohl, Force sensing in scanning tunneling microscopy: observation of adhesion forces on clean metal surfaces, *J. of Microscopy* **152**, 113-119 (1988).

18. R. Erlandsson, G.M. McClelland, C.M. Mate and S. Chiang, Atomic force microscopy using optical interferometry, *J. Vac. Sci. Technol. A* **6**, 266-270 (1988).

19. Y. Martin and H.K. Wickramasinghe, Magnetic imaging by force microscopy with 1000 Å Resolution, *Appl. Phys. Lett.* **50**,1455-1457 (1988).

20. D. Rugar, O. Züger, S. Hoen, C.S. Yannoni, H.-M. Vieth and R.D. Kendrick, Force detection of nuclear magnetic resonance, *Science* **264**,1560-1563 (1994).

21. D. Rugar, H.J. Mamin, R. Erlandsson, J.E. Stern and B.D. Terris, Force microscope using a fiber-optic displacement sensor, *Rev. Sci. Instrum.* **59**,1045-1047 (1988).

22. D. Rugar, H.J. Mamin, P. Güthner, Improved fiber optic interferometer for atomic force microscopy, *Appl. Phys. Lett.* **55**, 2588-2590 (1989).

23. A. Moser, H.J. Hug, T. Jung, U.D. Schwarz and H.-J. Güntherodt, A miniature fibre optic force microscope scan head, *Meas. Sci. Technol.* **4**, 769-775 (1989).

24. C. Schönenberger and S.F. Alvarado, A differential interferometer for force microscopy, *Rev. Sci. Instrum.* **60**, 3131-3134 (1988).

25. D. Anselmetti, Ch. Gerber, B. Michel, H.-J. Güntherodt and H. Rohrer, Compact, combined scanning tunneling, force microscope, *Rev. Sci. Instrum.* **63**, 3003-3006 (1989).

26. A.J. den Boef, Scanning force microscopy using a simple low-noise interferometer, *Appl. Phys. Lett.* **55**, 439-441 (1989).

27. D. Sarid, D. Iams, V. Weissenberger and L.S. Bell, Compact scanning force microscope using a laser diode, *Opt. Lett.* **13**, 1057-1059 (1988).

28. D. Rugar and P. Grütter, Mechanical parametric amplification and thermomechanical squeezing, *Appl. Phys. Lett.* **67**, 699-702 (1991).

29. T. Göddenhenrich, S. Müller and C. Heiden, A lateral modulation technique for simultaneous topography and friction force measurements with the AFM, *Rev. Sci. Instrum.* **65** 2870-2873 (1994).

30. G. Meyer and N. Amer, Simultaneous measurement of lateral and normal forces with an optical beam deflection atomic force microscope, *Appl. Phys. Lett.* **57**, 2089 (1990).

31. L. Howald, E. Meyer, R. Lüthi, H. Haefke, R. Overney, H. Rudin and H.-J. Güntherodt, Multifunctional probe microscope for facile operation in ultrahigh vacuum, *Appl. Phys. Lett.* **63**, 117-119 (1993).

32. J. Colchero, M. Luna and A.M. Baro, Lock-in technique for measuring friction on a nanometer scale, *Appl. Phys. Lett.*, **68**, 2896-2898 (1996).

33. N. Blanc, J. Brugger, N.F. de Rooij and U. Dürig, Scanning force microscopy in the dynamic mode using microfabricated capacitive sensors, *J. Vac. Sci. Technol. B* **14**, 901-905 (1996).

34. J. Brugger, J. Burger, M. Binggeli, R. Imura and N.F. de Rooij, Lateral force meaurements in a scanning force microscope with piezoresistive sensors, Proceedings ofthe 8th International Conference on Solid-State Sensors and Actuators and Eurosensors IX, Stockholm, Sweden June 25-29, 636-639 (1995).

35. R. Kassing and E. Oesterschulze, Sensors for Scanning Probe Microscopy, in *Micro/Nantotribology and Its Applications* edited by B. Bhushan, p. 35-54, Kluwer Academic Publishers (1997).

36. R. Feynman, There's plenty of room at the bottom, *Engineering and Science* **February**,22-36 (1960).

37. F.J. Giessibl and B.M. Trafas, Piezoresistive cantilevers utilized for scanning tunneling and scanning force microscope in ultrahigh vacuum, *Rev. Sci. Instrum.*, **65** 1923-1929, (1994).

38. F.J. Giessibl, Atomic Resolution of the Silicon(111)7x7 Surface by Atomic Force Microscopy, *Science* **267**, 68-71 (1995).

39. P. Curie, *J. Phys. (Paris)* **8**, 149 (1989).

40. C.P. Germano, *IRE Transactions on Audio* **7**, 96 (1959).

41. G. Binnig and D.P.E. Smith,*Rev. Sci. Instrum.* **57**, 1688 (1986).

42. P. Muralt, D.W. Pohl and W. Denk, *IBM. J. Res. Develop.* **30**, 443 (1986).

43. A. Hammiche, R.P. Webb and I.H. Wilson, *Rev. Sci. Instrum.* **64**, 3333 (1993).

44. D. Rugar, H.-J. Mamin, P. Güthner, *Appl. Phys. Lett.* **55**, 2588 (1986).

45. C.J. Chen, *Appl.Phys. Lett.* **60**, 132 (1992).

46. M.E. Taylor, *Rev. Sci. Instrum.* **64**, 154 (1993).

47. O. Marti and J. Colchero, Scanning probe microscopy instrumentation, in *Forces in Scanning Probe Methods*, Eds. H.-J. Güntherodt, D. Anselmetti and E. Meyer, NATO ASI Series E: Applied Sciences Vol. 286, Kluwer Academic publishers, p. 15-34, (1995).

48. Manufacturers of piezoelectric elements
    Staveley Sensors Inc., EBL Division, 91 Prestige Park Circle, East Hartford, CT 06108

Morgan Matroc, Inc., Vernitron Piezoelectric Division, 232 Forbes Rd.,Bedford, OH 44146

Physik Instrumente (PI) GmbH & Co. Polytec-Platz 5-7, D-76337 Waldbronn, Germany

Valpey-Fisher Corporation, 75 South Street, Hopkinton MA 01748

49. NANOSENSORS, Dr. Olaf Wolter GmbH, IMO-Building, Im Amtmann 6, 35578 Wetzlar-Blankenfeld, Germany
    Tel. (+49) 6441 97 88 40 Fax: (+49) 6441 97 8841
    e-mail: nanosensors@compuserve.com
50. Park Scientific Instruments, 1171 Borregas Avenue, Sunnyvale, CA 94089, USA
    Tel: +1 408 747-1600 Fax: +1 408 747-1601
    e-mail: Info@park.com
    website: http://www.park.com
51. R. Lüthi, PhD-thesis, Untersuchungen zur Nanotribologie und zur Auflösungsgrenze im Ultrahochvakuum mittels Rasterkraftmikroskopie, University of Basel, (1996).
52. M. Nonnenmacher, J. Greschner, O. Wolter, and R. Kassing, *J. Vac. Sci. Technol. B* **9** 1358 (1991).
53. J. Cleveland, S. Manne, D. Bocek and P.K. Hansma, *Rev. Sci. Instrum.* **64**, 403 (1993).
54. B. Bhushan in *Handbook of Micro/Nanotribology*, Edt. B. Bhushan, CRC Press Inc. (1994)
55. J.L. Hutter and J. Bechhoefer, *Rev. Sci. Instrum.* **64**, 1868-1873 (1993).
56. J.M. Neumeister, and W.A. Ducker, *Rev. Sci. Instrum.* **65** 2527 (1994).
57. J.E. Sader,*Rev. Sci. Instrum.* **66** 4583 (1995).
    J.E. Sader and L.R. White, *J. Appl. Phys.* **74** 1 (1993).
    M. Labardi, M. Allegrini, C. Ascoli, C. Frediani and M. Salerno, in *Forces in Scanning Probe Methods*, Eds. H.-J. Güntherodt, D. Anselmetti and E. Meyer, NATO ASI Series E: Applied Sciences Vol. 286, Kluwer Academic publishers, p. 319-324, (1995).
    D.F. Ogletree, R.W. Carpick and M. Salmeron,*Rev. Sci. Instrum.* **67**, 3298-3306 (1996).
    R.J. Warmack, X.-Z. Zhen, T. thundat and D.P. Allison,*Rev. Sci. Instrum.* **65**,394 (1994).
    M.A. Lantz, S.J. O'Shea, A.C.F. Hoole and M.E. Welland, *Appl. Phys. Lett.*, **70**, 970-972 (1997).

58. L. Howald, H. Haefke, R. Lüthi, E. Meyer, G. Gerth, H. Rudin, and H.-J. Güntherodt, *Phys. Rev. B* **49**, 5651-5656 (1993).
59. F. Atamny and A. Baiker, *Surf. Sci.* **323**, L314 (1995).
60. O. Marti J. Colchero and J. Mlynek, Friction and Forces on an atomic scale, in Proceedings of the Nato Advanced Research Workshop on Nanosources and Manipulations of Atoms under High Fields and Temperatures: Applications, July 6-10, 1992, Lyon, Kluwer Academic Publishers, Dordrecht, pp. 253-269 (1993).
61. D.F. Ogletree, R.W. Carpick and M. Salmeron, Calibration of frictional forces in atomic froce microscopy,*Rev. Sci. Instrum.* **67**, 3298-3306 (1996).
62. S.S. Sheiko, M. Möller, E.M.C.M. Reuvekamp and H.W. Zandberger, *Phys. Rev. B* **48**, 5675 (1993).
63. U.D. Schwarz, P. Koster and R. Wiesendanger, Quantitative Analysis of lateral force microscopy, *Rev. Sci. Instrum.* **67**, 2560-2567 (1996).
64. C.M. Mate, M.R. Lorenz and V.J. Novotny, Atomic force microscopy of polymeric liquid films *J. Chem. Phys.*, **90**, 7550-7555 (1989).
    C.M. Mate, M.R. Lorenz and V.J. Novotny, Determination of lubricant film thickness on a particulate disk by atomic force microscopy *IEEE Transactions on Magnetics*, **26**, 1225-1229 (1990).
    C.M. Mate and V.J. Novotny, Molecular conformation and disjoining pressure of polymeric liquid films *J. Chem. Phys.*, **94**, 8420-8427 (1991).
    C.M. Mate, Atomic force microscope study of polymer lubricants on silicon surfaces *Phys. Rev. Lett.*, **68**, 3323-3326, (1992).
65. A. Weisenhorn, P.K. Hansma, T.R. Albrecht and C.F. Quate, Forces in Atomic Force Microscopy in Air and Water, *Appl. Phys. Lett.* **54**, 2651-2653 (1989).

Index